S. Meuer   C. Wittwer   K. Nakagawara
Rapid Cycle Real-Time PCR

Springer
*Berlin
Heidelberg
New York
Barcelona
Hong Kong
London
Milan
Paris
Singapore
Tokyo*

S. Meuer   C. Wittwer   K. Nakagawara (Hrsg.)

Rapid Cycle Real-Time PCR

Methods and Applications

With 133 Figures and 209 Tables

Prof. Dr. med. STEFAN MEUER
Ruprecht Karls-Universität Heidelberg
Institut für Immunologie
Im Neuenheimer Feld 305
69120 Heidelberg, Germany

Prof. Dr. CARL WITTWER
University of Utah Medical School
Department of Pathology
Salt Lake City, UT 84132, USA

Dr. KAN-ICHI NAKAGAWARA
Sendai, Japan
2-3-36 Ohgimachi Miyagino-ku
Sendai 983-0034, Japan

ISBN 3-540-66736-9   Springer Verlag Berlin Heidelberg New York

Library of Congress applied for
Die Deutsche Bibliothek – CIP-Einheitsaufnahme
Rapid cycle real time PCR : methods and applications / Stefan Meuer ... (ed.). – Berlin ; Heidelberg ; New York ; Barcelona ; Budapest ; Hong Kong ; London ; Mailand ; Paris ; Santa Clara ; Singapore ; Tokyo : Springer, 2001
   ISBN 3-540-66736-9

This work is subject to copyright. All rights are reserved, whether the whole or part of the material is concerned, specifically the rights of translation, reprinting, reuse of illustrations, recitation, broadcasting, reproduction on microfilm or in any other way, and storage in data banks. Duplication of this publication or parts thereof is permitted only under the provisions of the German copyright Law of September 9, 1965, in its current version, and permission for use must always be obtained from Springer-Verlag. Violations are liable for prosecution under the German Copyright Law.

Springer-Verlag Berlin Heidelberg New York
a member of BertelsmannSpringer Science+Business Media GmbH

© Springer-Verlag Berlin Heidelberg 2001
Printed in Germany

The use of general descriptive names, registered names, trademarks, etc. in this publication does not imply, even in the absence of a specific statement, that such names are exempt from the relevant protective laws and regulations and therefore free for general use.

Product liability: The publisher cannot guarantee the accuracy of any information about dosage and application thereof contained in this book. In every individual case the user must check such information by consulting the relevant literature.

Cover Design: design & production, 69121 Heidelberg, Germany
Production: ProEdit GmbH, 69126 Heidelberg, Germany
Typesetting: TBS, 69207 Sandhausen, Germany
SPIN 10723838   18/3134 Re – 5 4 3 2 1 0 – Printed on acid free paper

Table of Contents

Rapid Cycle Real-Time PCR: Methods and Applications 1
CARL WITTWER

Part I
Methods

Mutation Detection by Fluorescent Hybridization Probe Melting Curves 11
PHILIP S. BERNARD, ASTRID REISER, GREGORY H. PRITHAM

Quantification on the LightCycler 21
RANDY RASMUSSEN

**Selection of Hybridization Probes for Real-Time Quantification
and Genetic Analysis** 35
OLFERT LANDT

**Using the Nearest Neighbor Model for the Estimation
of Matched and Mismatched Hybridization Probe Melting Points
and Selection of Optimal Probes on the LightCycler** 43
NICOLAS VON AHSEN, EKKEHARD SCHÜTZ

**Quantification of Human Papilloma Virus Type 16
Using Quantitative Competitive PCR on the LightCycler** 57
BRIAN ERICH CAPLIN

**Use of TaqStart Antibody to Increase the Sensitivity
of Herpesvirus Quantitative PCR on the LightCycler** 65
KAREN BRENGEL-PESCE, GÉRARD BARGUES, PATRICE MORAND,
JEAN-MARIE SEIGNEURIN

Part II
Genotyping of Human Germline Variations

High-Speed Detection of α_1-Antitrypsin Deficiency Alleles Pi*S and Pi*Z on the LightCycler 75
CHARALAMPOS ASLANIDIS, GERD SCHMITZ

High-Speed Methylenetetrahydrofolate Reductase C → T 677 Mutation Detection on the LightCycler 83
CHARALAMPOS ASLANIDIS, GERD SCHMITZ

Dual Color Detection of Splice Variants of the c-erbA α (Thyroid Hormone Receptor α) Gene 91
ONNO BAKKER

Detection of Three Major Polymorphisms in the *N*-Acetyltransferase 2 Gene by Melting Peak Analysis Using Fluorogenic Hybridization Probes 97
BRUNHILDE BLÖMEKE

Genotyping of Cytochrome P450 2D6*4 Mutation with Fluorescent Hybridization Probes Using LightCycler 105
JEANETTE BJERKE, CHUNG-CHE CHANG, CHUCK SCHUR, STEVEN WONG, NAZIHA NUWAYHID

Fluorescent Hybridization Probe Detection of the F508del Cystic Fibrosis Allele on the LightCycler 111
CAMERON N. GUNDRY

Genotyping β-globin Mutations (Hb S, Hb C, Hb E) by Multiplexing Probe Color and Melting Temperature 119
MARK G. HERRMANN

Simultaneous Detection of C282Y and H63D Hemochromatosis Mutations Using LCRed 640 and LCRed 705 Labeled Hybridization Probes 127
CINDY A. MEADOWS, BS CLsp (MB), MAREC PHILLIPS, MING Y. HUANG, MS and ELAINE LYON, PH.D

Genotyping of Angiotensin-Converting Enzyme and Angiotensinogen Polymorphisms with the LightCycler System 135
EIICHI SAKAI, MINORI TAJIMA, MITSUKO MORI, REIKO INAGE, MANABU FUKUMOTO, KAN-ICHI NAKAGAWARA

Genotyping of the Most Common Thiopurine Methyltransferase Mutations with the LightCycler 143
EKKEHARD SCHÜTZ, NICOLAS VON AHSEN

**Detection of the Mitochondrial DNA Mutation MELAS3243
Using Hybridization Probes** 153
STEPHANIE KLEINLE, SABINA GALLATI

Part III
Acquired Genetic Alterations in Human Diseases

**Detection of *p53* Allele Deletions in Human Cancer by Quantification
of Genomic Copy Number** 159
JOCHEN WILHELM, ALFRED PINGOUD, MEINHARD HAHN

**Monitoring of Residual Disease in Patients
with Chronic Myelogenous Leukemia Using Specific Fluorescent
Hybridization Probes for Real-Time Quantitative RT-PCR** 171
ANDREAS HOCHHAUS, MICHAEL EMIG, ANDREAS WEISSER,
SUSANNE SAUSSELE, MARTIN MÜLLER, PAUL LA ROSÉE,
CHRISTIAN KUHN, PETER PASCHKA, RÜDIGER HEHLMANN

**Development of Quantitative RT-PCR for the Expression
of Wilms' Tumor WT1 Suppressor Gene in Leukemia on the LightCycler** 187
YOJI ISHIDA, KAZUNORI MURAI

**Real-Time Detection of Minimal Residual Disease by Amplifying
Immunoglobulin Genes in Acute Lymphoblastic Leukemia on the LightCycler** ... 197
MAKOTO NAKAO, BART JANSSEN, CLAUS R. BARTRAM

**HER2/*neu* Gene Amplification Quantified by PCR and Melting Peak Analysis
Using a Single Base Alteration Competitor as an Internal Standard** 207
ELAINE LYON, PH.D, ALISON MILLSON, MT(ASCP), ARMINDA SULI, B.SC.

Quantification of Residual Tumor Cells in Monoclonal B-cell Lymphoma 219
THOMAS PFITZNER, ANDREAS ENGERT, STEFAN BARTH

**Development of PCR-Based Assays for the Detection
of Chromosomal Translocations Using SYBR Green I** 231
SANDRA D. BOHLING, KOJO S. J. ELENITOBA-JOHNSON

Relative Quantification of the HER2/*neu* Oncogene Using SYBR Green I 241
RACHEL WOODS

Part IV
Receptors and Mediators

Development of Quantitative RT-PCR Tests for the Expression of Cytokine Genes on the LightCycler 251
THOMAS GIESE

Quantitative RT-PCR for the Detection of T Cell Receptor Transcripts in T Lymphocytes Populations Using LightCycler Technology 263
E. JOUVIN-MARCHE, I. VIGAN, V. LEROY, P. N. MARCHE

Rapid, Homogeneous Genotyping of Human Platelet Antigen 1 by Fluorescence Resonance Energy Transfer and Probe Melting Curves 273
MARKUS S. NAUCK, HEDI GIERENS, MATTHIAS A. NAUCK, WINFRIED MÄRZ, HEINRICH WIELAND

Development and Validation of an Externally Standardised Quantitative Insulin-like Growth Factor-1 RT-PCR Using LightCycler SYBR Green I Technology .. 281
MICHAEL PFAFFL

An Application of Melting Curve Analysis to Large-Scale Genetic Analysis in Atherosclerotic Disease: Two Linked Polymorphisms of Glycoprotein Ia Gene and Myocardial Infarction in Japanese 293
H. MORITA, H. KURIHARA, Y. YAZAKI and R. NAGAI

Part V
Infectious Organisms

Genotype-Specific Analysis of Hepatitis B Virus DNA on the LightCycler 303
GUNHILD SOMMER, HANS WILL

Rapid and Specific Detection of *Bordetella pertussis* in Clinical Specimens by LightCycler PCR 313
UDO REISCHL, SIEGFRIED BURGGRAF, BIRGIT LEPPMEIER, HANS-JÖRG LINDE, NORBERT LEHN

Rapid and Specific Detection of *Helicobacter pylori* by LightCycler PCR 323
UDO REISCHL, BIRGIT LEPPMEIER, MARKUS HEEP, DANIELA BECK, NORBERT LEHN

Qualitative Detection of Herpes Simplex Virus DNA on the LightCycler 331
HARALD H. KESSLER

**Quantitative Detection of *Cryptosporidium parvum* after In Vitro Excystation
by LightCycler PCR** 341
Petra Krüger, Albrecht Wiedenmann, Despina Tougianidou,
Konrad Botzenhart

Quantitative Analysis of CMV in Infected Mice on the LightCycler System 349
Junichi Honda, Kotaro Oizumi

**Detection and Differentiation of Equine Herpes Virus Type 1
and Type 4 on the LightCycler** 359
Peter Hübert

**Rapid and Quantitative Detection of *Toxoplasma gondii* by PCR
– A LightCycler Application in Prenatal Diagnosis** 365
Jean-Marc Costa, Pauline Ernault, Stéphane Bretagne

**Development of Quantitative PCR Tests for the Detection
of the Orthopox Virus Adsorption Protein Gene (ORF D8L) on the LightCycler** ... 371
Claus-Peter Czerny, Michaela Alex, Jana Pricelius,
Christiane Zeller-Lue

Part VI
Plant Gene Products and Miscellaneous

**Quantification of Genetically Modified Soybeans
in Food with the LightCycler System** 383
Klaus Pietsch, Hans-Ulrich Waiblinger

**Real-Time PCR Monitoring of Estuarine Water Samples for *Pfiesteria piscicida*:
A Dinoflagellate Associated with Fish Kills and Human Illness** 391
Holly Bowers, Torstein Tengs, Mark Herrmann, David Oldach

**Quantification of Retrotransposon XIR-2.5 Copy Number
in Genomes of *Poeciliidae* Species** 399
Meinhard Hahn, Christiane Thömmes, Jochen Wilhelm,
Jamilah Michel

x

Abbreviations

Bp	Base pair
EDTA	Ethylenediamine Tetraacetic Acid
F	Fluorescein
LCRed640	LightCycler Red 640 Fluorescent Dye
LCRed705	LightCycler Red 705 Fluorescent Dye
Nt	Nucleotide(s)
P	Phosphate
PBS	Phosphate Buffered Saline
SDS	Sodium Dodecylsulfate
TE	1xTE Buffer Contains 0.01 M Tris, 0.001 M EDTA, pH 8.0
TE'	1xTE' Buffer Contains 0.01 M Tris, 0.0001 M EDTA, pH 8.0
TEN (=STE)	1xTEN Buffer Contains 0.01 M Tris (pH 8.0), 150 mM NaCl, 0.05% Tween 20
T_m(°C)	Melting Temperature
n-n	nearest neighbor
FRET	Fluorescence Resonance Energy Transfer
ΔH	Enthalpy
ΔS	Entropy
ΔG	Free Energy
TPMT	Thiopurine Methyltransferase
XO	Xanthine Oxidase
AA	Amino Acid
ORF	Open Reading Frame

Rapid Cycle Real-Time PCR: Methods and Applications

Carl Wittwer

Introduction

The heat stable polymerase, *Thermus aquaticus*, was first reported for use in the polymerase chain reaction in 1988 (1). Instead of adding the polymerase each cycle, only one addition of enzyme was needed at the beginning of PCR. Once all reaction components were combined, amplification could proceed automatically merely by temperature cycling the sample. Of course, automated thermal cyclers were not available yet, so most laboratories experienced a new incarnation of monotony, the repetitive manual transfer of PCR samples between 3 different water baths. It was clear that there must be a better way.

During the late 1980's, commercial thermal cyclers were developed that used familiar microfuge tubes placed in metal blocks. Cycle times of 2 to 8 min. were required. Before these thermal cyclers were available, a system based on capillary tubes and air temperature control was devised (2). Because of the high surface area to volume ratio of capillaries and the low heat capacity of air, cycle times of less than 30 sec were possible, a speed improvement of about an order of magnitude (3, 4). Faster cycling with better temperature control also improved PCR specificity (4, 5). "Rapid cycle PCR" became known as PCR with temperature cycles of 20-60 sec (6). Using rapid cycle PCR, 30 cycles of amplification required only 10-30 min.

Rapid cycle PCR suggests a "kinetic" rather than an "equilibrium" paradigm for PCR (7). PCR is usually considered a repetitive process where 3 reactions occur at 3 temperatures for 3 times during each cycle (Figure 1, left). In contrast, the kinetic paradigm emphasizes temperature transitions (Figure 1, right). Denaturation and annealing times are often reduced to "zero" and the temperature may always be changing. Denaturation, annealing and extension occur at different rates depending on the temperature, and multiple reactions may occur simultaneously. The kinetic paradigm is more correct, both theoretically and practically. Sample temperatures do not change instantaneously but occur as smooth transitions (4, 6). Nevertheless, the kinetic paradigm is difficult to accept intuitively. For example, there are many protocols in this book that do not use "zero" sec denaturation times, even though there is no reason to hold denaturation temperatures during cycling. Suggested temperature and time parameters for rapid cycle PCR are given in Table 1 (8).

Conventionally, analysis of PCR products is a separate step performed after PCR is completed. Most commonly, gel electrophoresis is used to assess the size and purity of the products. If analysis could somehow be performed during cycling, no extra analysis steps would be required, the time to results would be faster, contamination concerns could be eliminated with a closed tube system, and

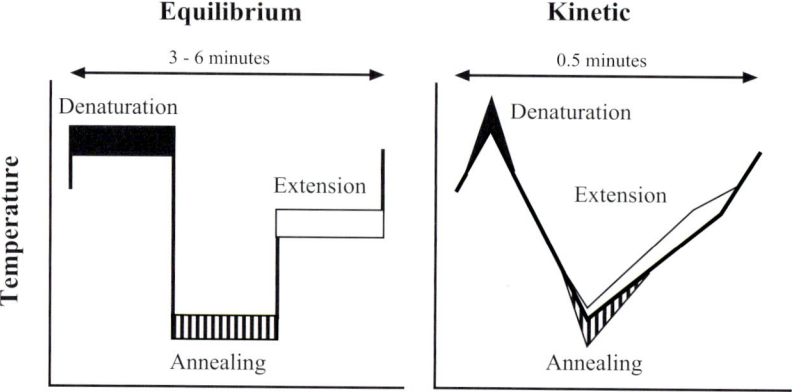

Fig. 1. Equilibrium and Kinetic Paradigms of PCR

Table 1. Suggested Temperature and Time Parameters for Rapid-Cycle PCR Amplification

	Temperature (°C)	Time (sec)
Denaturation	94[a]	0
Annealing	30 + 0.5 * (primer GC%)	0
Extension	74[b]	0.03 * (product length)[c]

NOTE. Reproduced with permission from Brown et al. (1998)
[a] For products with high GC domain, consider adding DMSO of formamide and/or increasing the temperature.
[b] May need to be lower for products with a high AT domain.
[c] For products <100 bp use a 0 sec extension time.

the entire process could be automated once temperature cycling is begun. Analyzing PCR products during amplification has become known as "real-time" PCR. The easiest way to monitor PCR during amplification is with fluorescence. The glass capillary tubes used in rapid cycle PCR are convenient optically clear cuvettes for fluorescent analysis. The combination of rapid temperature cycling with real-time PCR analysis is the subject of this book. Currently, the only commercial instrument that both cycles rapidly and acquires fluorescence in real-time is the LightCycler (Roche Molecular Biochemicals, Mannheim).

Ethidium bromide was the first double strand-specific fluorescent dye shown to be compatible with PCR (6, 9, 10). As PCR progresses, double stranded DNA is synthesized and the fluorescence of ethidium bromide increases. The amount of double stranded PCR product can be monitored each cycle by fluorescence. If fluorescence is plotted against cycle number, the accumulation of PCR products can be visualized on a growth curve, similar to a bacterial growth curve, with an

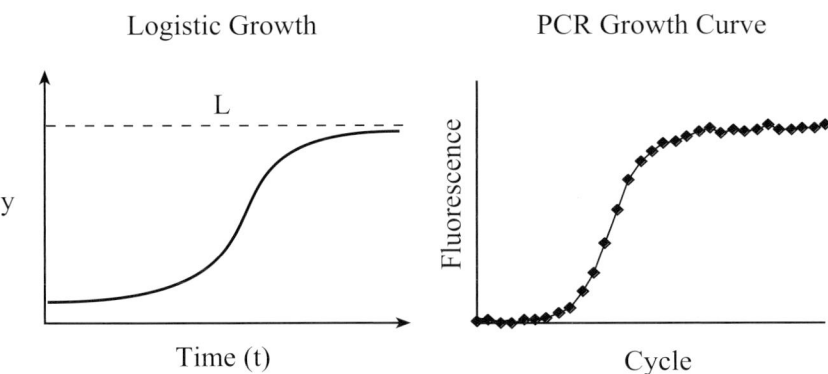

Fig. 2. Comparison of Logistic Growth to a PCR Growth Curve

initial lag phase (below the sensitivity of the fluorimeter), an exponential log phase, and a final plateau phase. The curve is similar to the logistic model of population growth (Figure 2), where the rate of growth is proportional to both the population size y and to the difference L-y, where L is the maximum population that can be supported. For small y, growth is exponential, but as y nears L, the growth rate approaches zero. Today, SYBR Green I is more commonly used than ethidium bromide because of greater sensitivity and a more convenient excitation spectra (11, 12).

Monitoring fluorescence each cycle is a powerful way to quantify the initial number of template copies (10, 13). Higher concentrations of template shift the growth curve to earlier cycles. This shift can be quantified as a fractional cycle number and is inversely proportional to the log of the initial template concentration. Quantification applications form a major section of this book. The chapter by Dr. Rasmussen provides an introduction to quantification methods.

Many applications require only a double strand-specific dye such as SYBR Green I (14, 15). Using a generic dye eliminates the cost and trouble of probe synthesis. However, certain applications require greater sequence specificity, and a variety of fluorescently-labeled oligonucleotide probes can be used to monitor the progress of PCR (16). These include exonuclease ("TaqMan") probes and hybridization probes (11, 12). Double strand-specific DNA dyes, exonuclease probes and hybridization probes are compared in Figure 3. Hybridization probes are particularly interesting because the probe Tm can be easily measured and used for fine sequence analysis, including single base changes (17-20). Mutation detection applications form another major section of this book. The chapter by Dr. Bernard provides an introduction to mutation detection methods.

Measuring fluorescence once each cycle at the same temperature is very useful for quantification. However, much more information is available if fluorescence is monitored continuously during temperature cycling as the temperature changes. Rapid cycle temperature vs time plots for typical SYBR Green I and hybridization probe reactions are shown in Figure 4. Corresponding fluorescence vs time plots

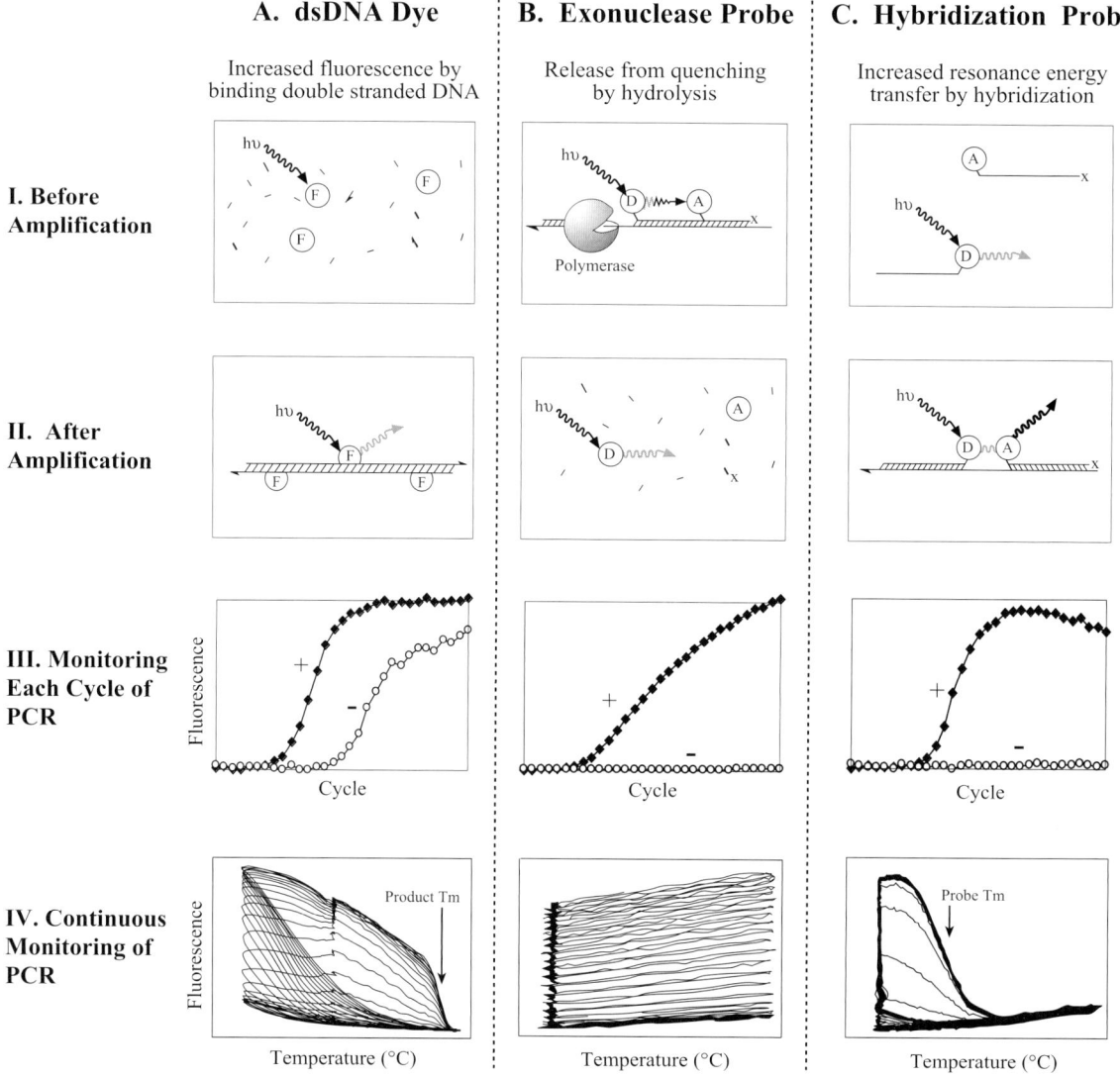

Fig. 3. Three Fluorescence Monitoring Schemes for PCR

and fluorescence vs temperature plots are shown in Figures 5 and 6, respectively. The fluorescence vs temperature plot of SYBR Green I shows that fluorescence is very sensitive to small changes in temperature around the Tm of the PCR product (90 °C). Similarly, the fluorescence vs temperature plot of hybridization probes shows a strong temperature dependence near the Tm of the probe (70 °C). This real-time monitoring of hybridization is extraordinarily powerful. PCR products

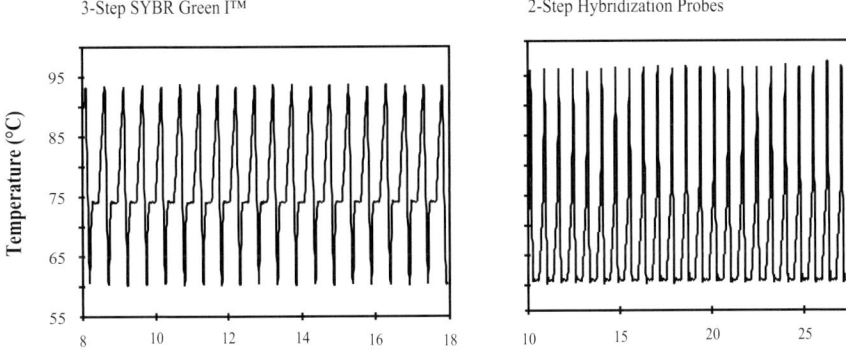

Fig. 4. Temperature vs. Time Tracings

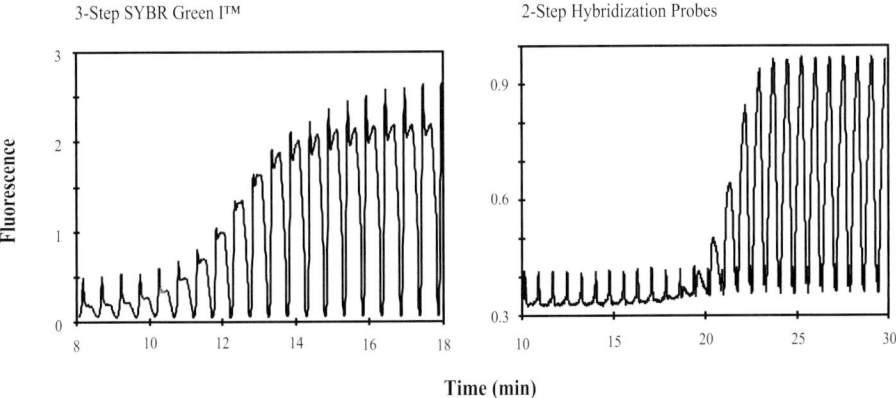

Fig. 5. Fluorescence vs. Time Tracings

can be identified by their Tm (21). Base changes under a hybridization probe can easily be detected and distinguished from most other changes by probe Tm shifts (17–20). The time, temperature and fluorescence history of a sample during rapid cycle, real–time PCR can be summarized as 3-dimensional traces (Figure 7).

There are many times during temperature cycling where fluorescence acquisition is informative (Figure 8). Product melting can measure Tms for product identification (21). Product re-annealing and probe annealing rates can be used to calculate product concentrations (13). Probe melting can assess probe Tm to identify sequence alterations. Primer annealing and extension rates can be monitored to assess and potentially control PCR efficiency. Different dyes and probes can be devised.

New Rapid cycle PCR and its extension into real-time monitoring began in a small research laboratory. Its early development can be followed by the references listed in this Introduction. Today, the LightCycler is distributed worldwide and the international contributors of this book evidence its dissemination. Many

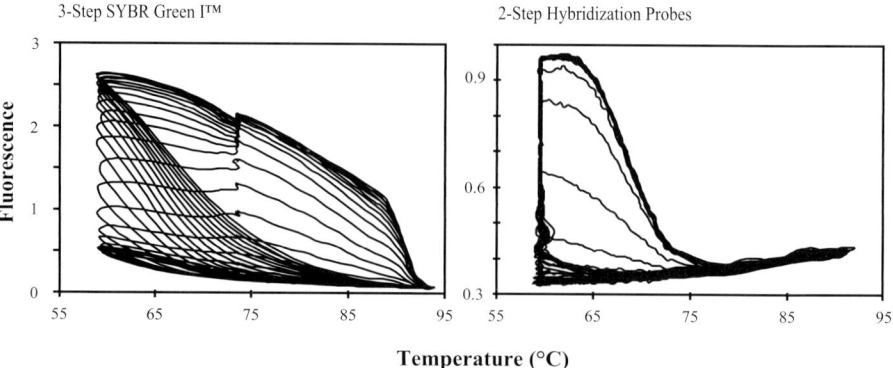

Fig. 6. Fluorescence vs. Temperature Tracings

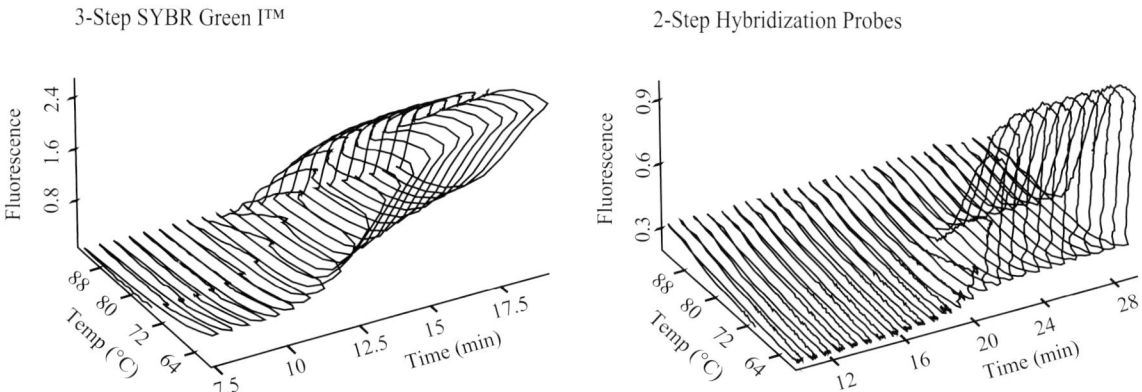

Fig. 7. Time / Temperature / Fluorescence (3D) Tracings

additional references to the current literature will be found in the chapters to follow. We thank the authors, Springer-Verlag, and Roche Molecular Biochemicals for their contributions.

We have attempted to standardize most of the Tables, given author tolerance. The final Mg^{++} concentration listed in the master mix Tables includes any contribution from the buffers or master solutions. In the oligonucleotide Tables, predicted T_ms were calculated as described in the chapter by Drs. von Ahsen and Schütz with the Santa Lucia parameters (22), with the following modifications. If commercial LightCycler buffers were used, the sodium ion concentration was set at 20 mM. The concentration of Mg^{++} was decreased by the dNTP concentration, assuming stoichiometric chelation. C_T was calculated as $C_A - (C_B/2)$ with $C_A > C_B$, the concentrations being [PCR product] and [oligonucleotide]. Best values for

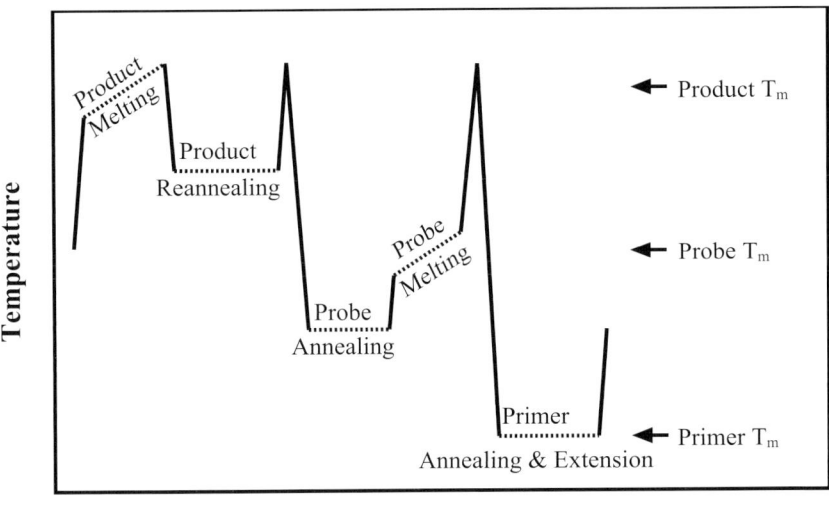

Fig. 8. Interesting Parts of a PCR Cycle for Fluorescence Observation

the [PCR product] and the Na$^+$ equivalency factor for Mg^{++} were determined empirically from a plot of calculated vs measured T_m of 134 different oligonucleotide/template duplexes. The measured Tms for this data set were obtained on several different instruments over 3 years, and gave best fits of 66 for the Mg^{++} factor and 0.2 uM for the [PCR product]. The mean error from the T_m estimation was 1.7 °C. Please note that dangling ends, fluorescent labels, and dUTP incorporation into products have not been specifically considered, although any effects on T_m may be partly compensated for by the empirical values for [PCR product] and the Na$^+$ equivalency factor for Mg^{++}. Since different protocols use different Mg^{++}, primer and probe concentrations, the same oligonucleotide may have a different T_m in different assays.

The book is organized into sections, beginning with an overview of relevant methods. The Methods section includes chapters on mutation detection, quantification, hybridization probe design, probe T_m estimation, competitive quantificative PCR, and "hot start" techniques. Chapters focusing on applications are divided into the following sections: Genotyping of human germline variations, Acquired genetic alterations in human diseases, Receptors and Mediators, Infectious organisms, and Plant gene products. Enjoy!

References:

1. Saiki RK, Gelfand DH, Stoffel S, Scharf SJ, Higuchi R, Horn GT, Mullis KB, Erlich HA (1988). Primer-directed enzymatic amplification of DNA with a thermostable DNA polymerase. Science 239:487-91.

2. Wittwer CT, GC Fillmore, DR Hillyard (1989). Automated polymerase chain reaction in capillary tubes with hot air. Nucl. Acids Res. 17:4353-4357.
3. Wittwer CT, GC Fillmore, DJ Garling. Minimizing the time required for DNA amplification by efficient heat transfer to small samples. Anal. Biochem. 186:328-331, 1990.
4. Wittwer, CT and DJ Garling. Rapid Cycle DNA Amplification. BioTechniques, 10:76-83, 1991.
5. Wittwer, CT, BC Marshall, GB Reed, JL Cherry. Rapid cycle allele-specific amplification: studies with the cystic fibrosis ?F508 locus. Clinical Chemistry, 39:804-809, 1993.
6. Wittwer, CT, GB Reed and KM Ririe. Rapid cycle DNA amplification. In K Mullis, F Ferre and R Gibbs (Eds.), The polymerase chain reaction. Springer-Verlag, Deerfield Beach, FL, pp 174-181, 1994.
7. Wittwer CT and MG Herrmman. Rapid thermal cycling and PCR kinetics, in PCR Methods Manual (Innis M, D Gelfand, and J. Sninsky, eds.), Academic Press, San Diego, 211-229, 1999.
8. Brown RA, MJ Lay, and CT Wittwer. Rapid cycle amplification for construction of competitive templates, in Genetic Engineering with PCR, (Horton, RM and RC Tait, eds.), Horizon Scientific Press, Norfolk, England, 57-70, 1998.
9. Higuchi R, Dollinger G, Walsh PS, Griffith R. Simultaneous amplification and detection of specific DNA sequences. Biotechnology 10:413-7, 1992.
10. Higuchi R, Fockler C, Dollinger G, Watson R. Kinetic PCR analysis: real-time monitoring of DNA amplification reactions. Biotechnology 11:1026-30, 1993.
11. Wittwer CT, MG Herrmann, AA Moss, RP Rasmussen. Continuous fluorescence monitoring of rapid cycle DNA amplification. BioTechniques, 22:130-138, 1997.
12. Wittwer CT, KM Ririe, RV Andrew, DA David, RA Gundry, UJ Balis. The LightCycler™: a microvolume, multisample fluorimeter with rapid temperature control. BioTechniques, 22:176-181, 1997.
13. Wittwer CT, K Ririe, R. Rasmussen. Fluorescence monitoring of rapid cycle PCR for quantification, in Gene Quantification, Ferre, F., ed., Birkhauser, New York, 129-144, 1998.
14. Morrison TB, JJ Weis, and CT Wittwer. Quantification of low-copy transcripts by continuous SYBR Green I monitoring during amplification. BioTechniques, 954-962, 1998.
15. Bay S, CT Wittwer, TC King, KSJ Elelitoba-Johnson. Fluorescence melting curve-based analysis for the detection of the bcl-1/JH translocation in mantle cell lymphoma. Lab. Invest. 79:337-345, 1999.
16. Pritham GH and CT Wittwer. Continuous fluorescent monitoring of PCR. J. Clin. Lig. Assay, 21(4):404-412, 1998.
17. Lay MJ and CT Wittwer. Real-time fluorescence genotyping of factor V Leiden during rapid cycle PCR. Clin. Chem.43:12, 2262-2267, 1997.
18. Bernard PS, MJ Lay, CT Wittwer. Integrated amplification and detection of the C677T point mutation in the methylenetetrahydrofolate reductase gene by fluorescence resonance energy transfer and probe melting curves. Anal. Biochem., 255:101-107, 1998.
19. Bernard PS, RS Ajioka, JP Kushner, and CT Wittwer. Homogeneous multiplex genotyping of hemochromatosis mutations with fluorescent hybridization probes. Am. J. Pathol., 153:1055-1061, 1998.
20. Lyon E, A Millson, T Phan, and CT Wittwer. Detection of base alterations within the region of factor V Leiden by fluorescent melting curves. Mol. Diag., 3:203-210, 1998.
21. Ririe KM, RP Rasmussen, and CT Wittwer. Product differentiation by analysis DNA melting curves during the polymerase chain reaction. Anal. Biochem, 245:154-160, 1997.
22. SantaLucia J. A unified view of polymer, dumbbell, and oligonucleotide DNA nearest-neighbor thermodynamics. Proc. Natl. Acid. Sci. USA, 95:1460-1465, 1998.

Methods   I

Mutation Detection by Fluorescent Hybridization Probe Melting Curves 11
Philip S. Bernard, Astrid Reiser, Gregory H. Pritham

Quantification on the LightCycler 21
Randy Rasmussen

**Selection of Hybridization Probes for Real-Time Quantification
and Genetic Analysis** 35
Olfert Landt

**Using the Nearest Neighbor Model for the Estimation
of Matched and Mismatched Hybridization Probe Melting Points
and Selection of Optimal Probes on the LightCycler** 43
Nicolas von Ahsen, Ekkehard Schütz

**Quantification of Human Papilloma Virus Type 16
Using Quantitative Competitive PCR on the LightCycler** 57
Brian Erich Caplin

**Use of TaqStart Antibody to Increase the Sensitivity
of Herpesvirus Quantitative PCR on the LightCycler** 65
Karen Brengel-Pesce, Gérard Bargues, Patrice Morand,
Jean-Marie Seigneurin

Mutation Detection by Fluorescent Hybridization Probe Melting Curves

Philip S. Bernard*, Astrid Reiser, Gregory H. Pritham

Sequence analysis of nucleic acids is used in many fields of science and for a variety of purposes. Although direct sequencing remains the gold standard for initially characterizing a new region of DNA, it is not a practical solution for the routine analysis of a sequence once it is known. For this reason, molecular techniques that facilitate the analysis of variants in established sequence have widespread application and continue to be developed. Conventionally, the target is first amplified by PCR and then further processed for mutation detection. Recently developed instrumentation couples rapid PCR with fluorescent hybridization probes, allowing target amplification and product analysis to be done consecutively, without sample handling and within 30 min [1].

The LightCycler provides a platform for monitoring fluorescence during temperature transitions [2]. This allows the annealing and denaturation of nucleic acids to be followed in real-time, using dsDNA dyes (e.g., SYBR Green I) or fluorescently-labeled oligonucleotides [3, 4]. During PCR, once per cycle fluorescence acquisition allows an estimate of starting copy number, while after PCR, continuous monitoring during slow heating provides qualitative information about the target. The temperature at which a DNA duplex melts gives information about the particular sequence. The melting of short duplexes, such as with hybridization probes, allows even single base alterations in the amplicon to be identified [5, 6].

Oligonucleotides labeled with donor and acceptor fluorophores have been used for mutation detection on the LightCycler in both a primer/probe and probe/probe format [1]. The first scheme uses a 3′-labeled hybridization probe designed to anneal to a strand extended by an internally labeled primer (Fig. 1A). This method requires internal labeling of the primer and the labeled-primer needs to be positioned near the mutation site, usually within 5 bps for adequate fluorescence resonance energy transfer [5, 6]. The second scheme uses separate 3′ and 5′-labeled probes designed to hybridize adjacently to an unlabeled complementary PCR strand (Fig. 1B). This method allows for genotyping probes to be placed anywhere within an unlabeled primer set [7].

In both schemes, an increase in fluorescence resonance energy transfer is observed as the reaction is cooled and probe/target annealing brings the donor and acceptor fluorophores into close proximity. Reciprocally, as the reaction is

* Philip S. Bernard (✉) (e-mail: phil.bernard@path.med.utah.edu)
  Department of Pathology, University of Utah School of Medicine, Salt Lake City, Utah 84132, USA

A. Primer / Probe Scheme for Resonance Energy Transfer

Baseline Fluorescence

Resonance Energy Transfer

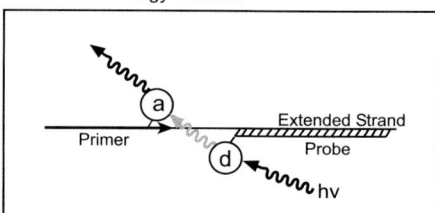

B. Probe / Probe Scheme for Resonance Energy Transfer

Baseline Fluorescence

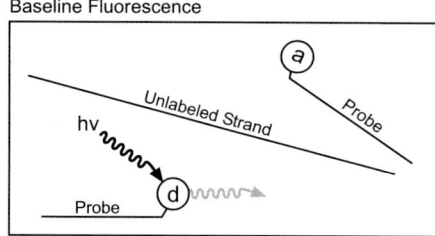

Resonance Energy Transfer

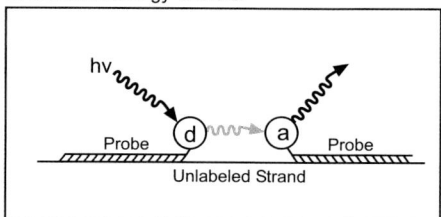

Fig. 1. Two schemes for fluorescence resonance energy transfer with hybridization probes. The primer/probe design (**A**) uses an internally labeled primer to extend and label a strand during PCR. Typically, acceptor (*a*) dyes, such as LCRed640, are used for primer labeling. Fluorescence resonance energy transfer is observed during hybridization of a complementary 3′-fluorescently labeled probe. The LightCycler uses fluorescein as the donor (*d*) dye. In the probe/probe system (**B**), the donor and acceptor fluorophores are conjugated to the 3′ and 5′ ends of different oligonucleotides. Fluorescence resonance energy transfer occurs as the two probes hybridize to adjacent sequences on the same unlabeled complementary strand. Genotyping is performed with this method by designing the probe spanning the variant to have a lower T_m than its adjacent probe. In this way the higher T_m probe, sometimes referred to as the anchor probe, remains annealed through the T_m of the genotyping probe. In addition, any probe used in PCR must have the 3′ end blocked in order to prevent extension by the polymerase. This is accomplished either by the addition of a dye or a phosphate group

heated, the probe/target duplex is denatured, the fluorophores are separated and fluorescence resonance energy transfer drops to background. Although the temperature dependence of hybridization probes can be observed during rapid thermal cycling [5], slower temperature transitions (0.1–0.2°C/s), performed as an addendum to PCR, are optimal for single base discrimination. Hybridization probes have been used for mutation detection by both slow cooling (annealing curves) and slow heating (melting curves) [8], however, the latter is more commonly used and will be the only method discussed.

Each probe/target duplex has a characteristic thermal stability that depends on such factors as length, G:C content, sequence order and Watson-Crick pairing [9,10]. Base pair mismatches shift the stability of a duplex by varying amounts depending on the particular mismatch, the mismatch position, and neighboring base pairs [11–16]. When a probe hybridizes over a sequence variant, a mismatch is formed and the duplex is destabilized. This is reflected by a shift in melting temperature (T_m) from the completely complementary duplex.

Hybridization probe T_m is the temperature at which 50% of the probe has strand-separated from a template and can be estimated from the inflection point of the melting curve or the center of derivative melting curves [17]. Figure 2 shows a heterozygous genotyping for a variant in the apolipoprotein E gene, using a probe/probe scheme. A more thermally stable LCRed640 (acceptor) probe

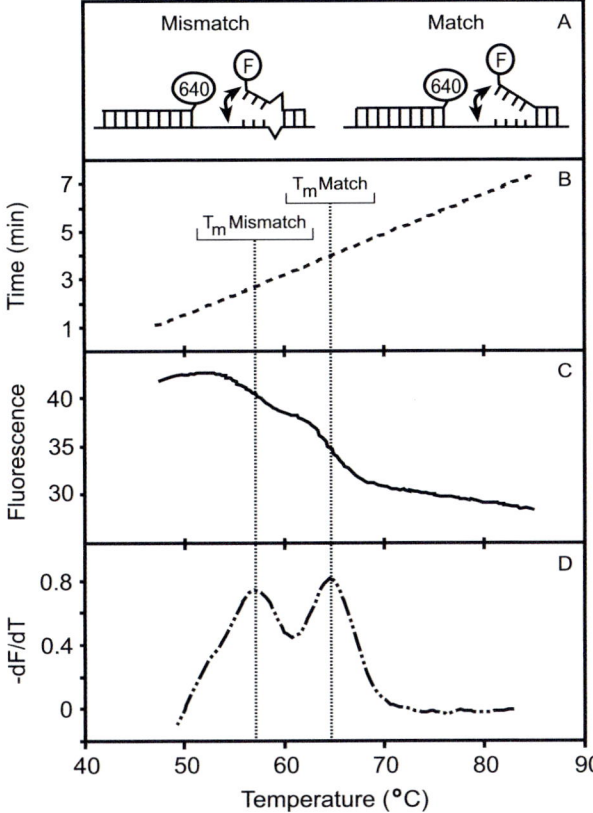

Fig. 2. Schematic display of fluorescent melting curve analysis for genotyping. Data for codon 158 of the apolipoprotein E locus is shown. **A** portrays the equilibrium between annealing and melting at the T_m of matched and mismatched duplexes. **B** plots time vs sample temperature during the melting phase of the reaction. As the solution is heated at 0.1°C/s, the fluorescein probe spanning the alteration site is thermally denatured away from the adjacent LCRed640 acceptor probe. The LCRed640 probe is designed to remain annealed during melting of the fluorescein probe. The point of inflection on a plot of fluorescence vs temperature (**C**) identifies the temperature at which 50% of the fluorescein probe has melted (T_m). Melting temperatures are more easily discerned by plotting a derivative melting curve where the center of the melting peak (**D**) corresponds to the point of inflection. The ordinate of **D** is the negative derivative of fluorescence (F) with respect to temperature (T), or $-dF/dT$. Reprinted with permission from [17]

remains annealed to the template as the genotyping fluorescein (donor) probe is slowly heated through its matched and mismatched template T_ms.

The probe used for genotyping in Fig. 2 was a 17-mer that formed an A:C mismatch positioned 5 bps from the 3′-end of the probe when hybridized to the non-complementary target. When designing a probe, the more destabilizing mismatches should be chosen for the best resolution of a heterozygous sample. This can be done either by changing a base in the probe or switching the strand to which the probe anneals.

Mismatch stability is strongly dependent on the nearest neighbors to the mismatch. A detailed discussion of nearest neighbor thermodynamic parameters can be found in several current papers on the subject [12–16]. A recent article by von Ahsen and Schütz uses nearest neighbor parameters to precisely predict T_ms of matched and mismatched duplexes of fluorescent hybridization probes [18]. The authors also present these findings in this volume. Generally, mismatches flanked by G:C pairs are more stable than those flanked by A:T pairs and mismatches positioned near the center of the probe are more destabilizing than mismatches positioned near the end of the probe. A general ranking of mismatch stability is presented in Table 1.

The estimated T_m from the derivative melting curve is highly reproducible under constant reaction conditions of heating rate, salt concentration, and probe-target concentrations. Multiple replicate samples have a within run variation of only 0.1°C and the between run variation is within 0.5°C [19]. Thus, shifts greater than 1°C from the characteristic and expected derivative melting curve profile are causes for suspicion of new mutations. In fact, aberrant derivative melting curves have effectively identified unexpected variants in factor V (activated protein C resistance), HFE (hereditary hemochromatosis) and CFTR (cystic fibrosis) [6, 7, 19].

In the clinical setting it is important to know the accuracy of a positive screening test. That is, one would like to know the proportion of people who actually have disease from all those who test positive. This value is the positive predictive value of the test. The positive predictive value of a test will increase with the prevalence of the disease and can be affected by the age of disease onset and the age at screening. Testing for the factor V Leiden mutation is used to identify a genetic cause for venous thrombosis in a clinically defined patient population. In this scenario, the disease (venous thrombosis) is present in nearly 100% of the patient population and the predictive value of a positive test is high, especially given the dominant phenotype of the Leiden mutation. However, the predictive value of a positive test decreases if the specificity is decreased. The specificity of hybridization probes decreases if other variants near the mutation of interest cause a similar T_m shift and result in a false positive. For instance, there are two polymorphisms (G1689 A and A1692 C) that might result in false positives using the wild-type probe for genotyping the factor V Leiden (G1691 A) mutation.

Table 1. Empirically determined base pair stabilities from DNA duplexes

G:C > A:T > G:T ≥ G:A > T:T ≥ G:G > A:A ≥ C:C > C:T > C:A

However, these rare polymorphisms can be distinguished from factor V Leiden by T_m [19]. Mutation specific probes are seldom necessary clinically.

Specificity can also be improved by using shorter hybridization probes (i.e., <25 bps). Figure 3 shows how all four possible bases can be differentiated at the center position of a 15-mer probe/target duplex. This is essentially sequencing at a given position under a probe.

The high specificity of fluorescent hybridization probes is complemented by their high stability [20], making them optimal for the clinical laboratory. Hybridization probe genotyping assays that use a single color include factor V Leiden [5,21], methylenetetrahydrofolate reductase [6, 21], prothrombin [21], apoB-3500 [22], human platelet antigen 1 [23], N-acetyltransferase 2 [24], plasminogen activator inhibitor-1 [25], BRCA1 [26], and antiviral resistance-associated mutations in the hepatitis B virus [27].

Further T_m multiplexing can be performed with the use of probes with different T_ms to identify different sites. Figure 4 illustrates the temperature range available for T_m multiplexing by resolving seven discrete melting peaks over 50°C, using probes with a 55% G:C content but varying in length from 11–45 bps. Since the purity of synthesis and the sensitivity to detect mismatches decrease with increasing length, genotyping probes are usually between 15–35 bps. An example of genotyping by T_m multiplexing is provided in Fig. 5 where the two disease-causing mutation sites for hereditary hemochromatosis are simultaneously identified using two different length probes with different T_ms. A higher melting

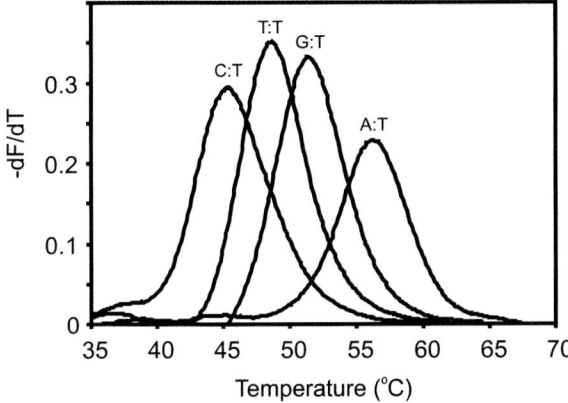

Fig. 3. Derivative melting curves using a model hybridization probe system with different mismatches at the center of a 15-mer probe. In this probe/probe system, four 3'-fluorescein labeled probes that differed only by a single base at the center position were melted in separate tubes from a common synthetic oligonucleotide target. The same acceptor-labeled anchor probe was used for fluorescence resonance energy transfer with the fluorescein-labeled probes. The Watson-Crick pair and the three different mismatches all produced unique T_ms, as shown by their derivative melting curve profiles. All melting curves were generated at a temperature transition rate of 0.1°C/s and final reagent concentrations of 3 mM Mg^{++}, 0.2 μM target, 0.2 μM anchor-probe, and 0.1 μM fluorescene-labeled probe

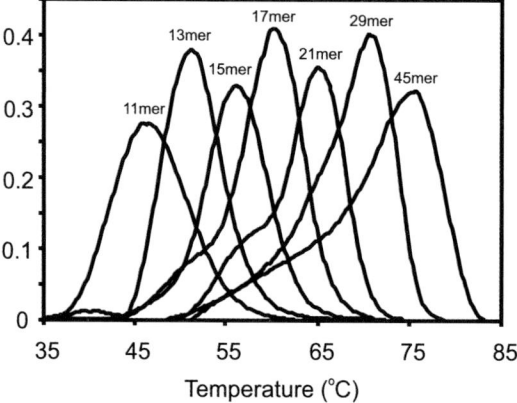

Fig. 4. Derivative melting curves from 3'-fluorescein labeled probes varying in length from 11–45 bps. The samples all contained the same synthesized oligonucleotide target and acceptor-labeled anchor probe. The 3'-fluorescein labeled probes maintained the same 55% G:C content by the addition of an equal number of purines and pyrimidines to the 5' end. Seven discrete peaks are discerned over a temperature range of approximately 50°C using a temperature transition rate of 0.1°C/s. The final reagent concentrations were 3 mM Mg^{++}, 0.2 µM target, 0.2 µM anchor-probe, and 0.1 µM fluorescene-labeled probe

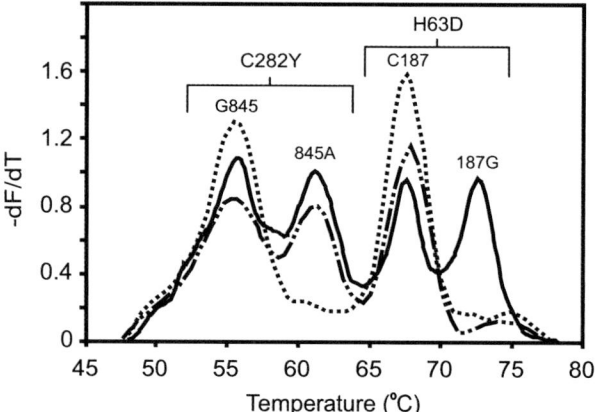

Fig. 5. Derivative melting curves showing homogeneous multiplex genotyping of four alleles in the hereditary hemochromatosis gene. Three samples are shown with different C282Y/H63D genotypes: homozygous G845/homozygous C187 (---), heterozygous G845 A/homozygous C187 (-..-), and heterozygous G845A/heterozygous C187G (—). Genotyping was performed at each site, using a probe/probe scheme with fluorescein and LCRed640 as the donor and acceptor chromophores, respectively. The C282Y (G845A) mutation in exon 4 was spanned by a LCRed640 probe (using a fluorescein-labeled probe as an anchor), whereas a fluorescein probe was used to span the H63D (C187G) mutation in exon 2 (using a LCRed640 probe as an anchor). Multiplex amplification and genotyping of the two sites was performed in the same tube and within 45 min

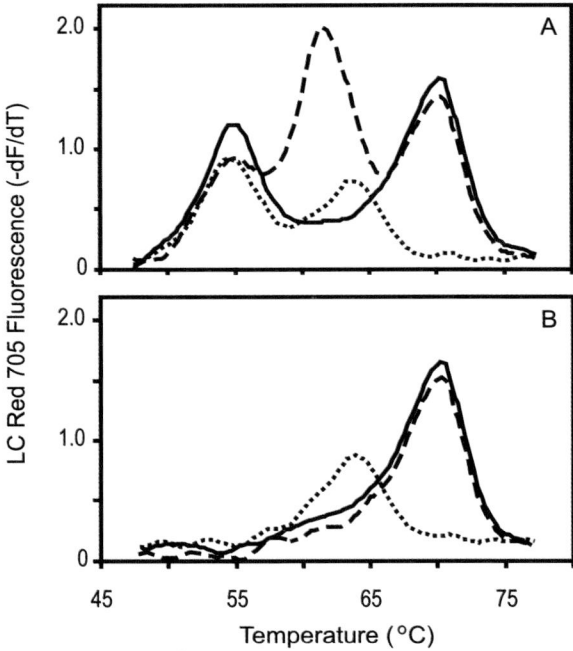

Fig. 6. Multiplex color genotyping of codons 112 and 158 of the ApoE gene. The uncompensated (A) and compensated (B) derivative melting curve data (plotted as $-dF/dT$ vs temperature) is shown for the LCRed705 channel used for analysis of codon 112. The uncompensated data displays melting peaks from codon 158 *(Fig. 2)* due to bleed over from the LCRed640 acceptor dye. The compensated data shows only the genotypes at codon 112. The genotypes shown are: homozygous ε4/homozygous ε3 (---), homozygous ε3/homozygous ε3 (—) and homozygous ε3/heterozygous ε2/ε3 (- -). Reprinted with permission from [17]

33mer probe was used to genotype the site of the C187G variant and a 21mer probe was used for the G845 A variant.

Other probe systems strive to increase the power of variant analysis by color multiplexing rather than T_m multiplexing. Dual-labeled probes that are sequence specific, such as exonuclease probes (TaqMan) and hairpin probes (Molecular Beacons), homogenously genotype during PCR. Each probe is labeled with a different colored reporter dye and is designed to anneal only to its complementary target during PCR for proper scoring. These probes are technically more difficult to design and optimize. For example, some exonuclease probe sequences are poorly cleaved by the polymerase during annealing even when they are completely complementary to the target. This can result in incorrect genotyping, since the allele can only be detected if the probe is digested and the reporter fluorophore is freed from the quencher dye. In addition, the number of variants identified with dual-labeled probes is limited to the number of unique color spectrums resolvable. Currently, as many as six different fluorescent dyes can be combined with a common quencher, but only three dyes can be analyzed in real-time [28]. In contrast, a single hybridization probe can be used to identify multiple variants.

Multiplexing provides internal controls, allows higher throughput, conserves precious sample, and saves on material costs [1]. Hybridization probes can provide higher-order multiplexing than other systems by combining color and T_m. This was first demonstrated in a synthetic system for variants of the apo E gene [17]. Various colored dyes were multiplexed to identify each codon (LCRed705 for codon 112 and LCRed640 for codon 158) and variants at each codon were identified by characteristic T_m shifts (Fig. 6). Algorithms for solution color compensation provide correction of fluorescence crosstalk between reporters [17]. Recently, assays for genotyping on the LightCycler from genomic DNA, using color and T_m have been developed for the α_1-antitrypsin gene [29] and the β-globin gene [30].

Genotyping in solution using both color and T_m can be imagined as a 2-dimensional matrix. The number of different targets identified is the product of the number of colors and the number of T_ms resolvable. Using a full temperature and color range, it is feasible that a multiplex of at least 21 could be reached (i.e., 7 T_ms×3 colors). The homogenous format, ease of synthesis, and density of information provided by hybridization probes allows simple and rapid genotyping that is ideal for the clinical laboratory.

References

1. Bernard PB, CT Wittwer (2000) Homogenous amplification and variant detection by fluorescent hybridization probes. Clin Chem 46: 147–148
2. Wittwer CT, Ririe KM, Andrew RV, David DA, Gundry RA, UJ Balis (1997) The LightCycler: A microvolume multisample fluorimeter with rapid temperature control. Biotechniques 22: 176–181
3. Ririe KM, Rasmussen RP, CT Wittwer (1997) Product differentiation by analysis of DNA melting curves during the Polymerase Chain Reaction. Anal Biochem 245: 154–160
4. Wittwer CT, Herrmann MG, Moss AA, RP Rasmussen (1997) Continuous fluorescence monitoring of rapid cycle DNA amplification. Biotechniques 22: 130–138
5. Lay MJ, CT Wittwer (1997) Real-time fluorescence genotyping of factor V Leiden during rapid-cycle PCR. Clin Chem 43: 2262–2267
6. Bernard PS, Lay MJ, CT Wittwer (1998) Integrated amplification and detection of the C677 T point mutation in the methylenetetrahydrofolate reductase gene by fluorescence resonance energy transfer and probe melting curves. Anal Biochem 255: 101–107
7. Bernard PS, Ajioka RS, Kushner JP, CT Wittwer (1998). Homogenous multiplex genotyping of hemochromatosis mutations with fluorescent hybridization probes. Am J Pathol 153:1055–1061
8. Gundry CN, Bernard PS, Herrmann MG, Reed GH, CT Wittwer (1999). Rapid F508del and F508 C assay using fluorescent hybridization probes. Genetic Testing 3: 365–370
9. Wetmur JG (1995) In: Meyers RA (ed) Molecular Biology and Biotechnology pp 605–608 VCH Publishers Inc., New York, NY
10. SantaLucia J Jr, Allawi HT, Seneviratne PA (1996) Improved nearest-neighbor parameters for predicting DNA duplex stability. Biochemistry 35: 3555–3562
11. Guo Z, Liu Q, Smith LM (1997) Enhanced discrimination of single nucleotide polymorphisms by artificial mismatch hybridization. Nature Biotech 4: 331–335
12. Allawi HT, SantaLucia J Jr (1997). Thermodynamics and NMR of Internal G:T Mismatches in DNA. Biochemistry 36: 10581–10594

13. Allawi HT, SantaLucia J Jr (1998). Thermodynamics of internal C:T mismatches in DNA. Nucleic Acids Res 26: 2694–2701
14. Allawi HT, SantaLucia J Jr (1998). Nearest-neighbor thermodynamics of internal A:C mismatches in DNA: Sequence dependence and pH effects. Biochemistry 37: 9435–9444
15. Allawi HT, SantaLucia J Jr (1998). Nearest neighbor thermodynamic parameters for internal G:A mismatches in DNA. Biochemistry 37: 2170–2179
16. Peyret N, Seneviratne PA, Allawi HT, SantaLucia J Jr (1999). Nearest-neighbor thermodynamics and NMR of DNA sequences with internal A:A, C:C, G:G, and T:T mismatches. Biochemistry 38: 3468–3477
17. Bernard PS, Pritham GH, Wittwer CT (1999) Color multiplexing hybridization probes using the apolipoprotein E locus as a model system for genotyping. Anal Biochem 273: 221–228
18. Schutz E, von Ahsen N (1999) Spreadsheet for thermodynamic melting point prediction of oligonucleotide hybridization with and without mismatches. Biotechniques 27: 1218–1222
19. Lyon E, Millson A, Phan T, Wittwer CT (1998). Detection and identification of base alterations within the region of factor V Leiden by fluorescent melting curves. Mol Diag 3: 203–210
20. Mitchell RS, Stevenson E, Mouritsen CL, Bohling S, Lyon E (1999) A comparison of storage conditions for PCR mixtures with fluorescently labeled probes [Abstract]. J Mol Diag 1: 60
21. von Ahsen N, Schutz E, Armstrong VW Oellerich M (1999) Rapid detection of prothrombotic mutations of prothrombin (G20210), factor V (G1691 A), and methylenetetrahydrofolate reductase (C677 T) by real-time fluorescence PCR with the LightCycler. Clin Chem 45:1875–1878
22. Aslanidis C, Schmitz G (1999) High-speed apolipoprotein E genotyping and apolipoprotein B3500 mutation detection using real-time fluorescence PCR and melting curves. Clin Chem 45: 1094–1097
23. Nauck MS, Gierens H, Nauck MA, Marz W, Wieland H (1999) Rapid genotyping of human platelet antigen 1 (HPA-1) with fluorophore-labeled hybridization probes on the LightCycler. Br J Haematol 105: 803–810
24. Blomeke B, Sieben S, Spotter D, Landt O, Merk HF (1999) Identification of N-acetyltransferase 2 genotypes by continuous monitoring of fluorogenic hybridization probes. Anal Biochem 275: 93–97
25. Nauck MS, Wieland H, Marz W (1999) Rapid, homogenous genotyping of the 4G/5G polymorphism in the promoter region of the PAII gene by fluorescence resonance energy transfer and probe melting curves. Clin Chem 45: 1141–1147
26. Pals G, Pindolia K, Worsham MJ (1999) A rapid and sensitive approach to mutation detection using real-time polymerase chain reaction and melting curve analysis, using BRCA1 as an example. Mol Diag 4: 241–246
27. Can PA, Cook P, Ratcliffe D, Multimer D, Pillay D (1999) Use of real-time PCR and fluorimetry to detect lamivudine resistance-associated mutations in hepatitis B virus. Antimicrob Agents Chemother 43: 1600–1608
28. Lee LG, Livak KJ, Mullah B, Graham RJ, Vinayak RS, Woudenberg TM (1999) Seven-color, homogenous detection of six PCR products. Biotechniques 27: 342–349
29. von Ahsen N, Oellerich M, Schutz E (2000) Use of two reporter dyes without interference in a single-tube rapid cycle PCR: α_1-antitrypsin genotyping by multiplex real-time fluorescence PCR with the LightCycler. Clin Chem 46: 156–161
30. Herrmann MG, Dobrowolski SF, Wittwer CT (2000) Beta-globin genotyping by multiplexing probe Tm and color. Clin Chem 46: 425–428

Quantification on the LightCycler

RANDY RASMUSSEN*

Introduction

Over the last 15 years, PCR has become an essential part of most laboratories involved in biomedical research. PCR amplification turns a few attograms of a specific fragment of nucleic acid (far too little to be analyzed directly or used in biochemical reactions) into as much as a microgram of DNA.

Everyone has heard the "quantitative PCR is an oxymoron" joke, and it is not without some truth. By nature, an exponential amplification is not ideally suited to quantification. Small differences in amplification efficiency between samples can become huge differences in results when they are amplified through 40 doublings. Anyone working with quantitative PCR who forgets this fact is in danger of making mistakes that are measured in orders of magnitude. Nonetheless, the years 1991–1998 saw a 10-fold increase in the number of papers using quantitative PCR methods. Why then the continuing increase in the use of quantitative PCR? It has a sensitivity five orders of magnitude better than the best blotting procedures and a dynamic range of 10 orders of magnitude. This unsurpassable sensitivity and range has made the work of turning PCR into a quantitative tool worthwhile [1].

Quantitative PCR

A PCR reaction profile can be thought of as having three segments: an early background phase, an exponential growth phase (or log phase) and a plateau. The background phase lasts until the signal from the PCR product is greater than the background signal of the system. The exponential growth phase begins when sufficient product has accumulated to be detected above background, and ends when the reaction efficiency falls as the reaction enters the plateau.

During the "log" phase the amplification course is described by the equation:

$$T_n = T_o (E)^n \tag{1}$$

---

* Randy Rasmussen (✉) (e-mail: randy@idahotech.com)
 Idaho Technology, 390 Wakara Way, Salt Lake City, UT 84108 USA

Where T_n is the amount of target sequence at cycle n, T_o is the initial amount of target, and E is the efficiency of amplification. The maximum efficiency possible in PCR is 2– every PCR product is replicated every cycle. The minimum value is 1, corresponding to no amplification.

Real-Time Quantitative PCR

Typical real-time PCR curves monitored on the Lightcycler are shown in Fig. 1. The cycle where each reaction first rises above background is dependant on the amount of target present at the beginning of the reaction. This is easy to understand from Eq. 1. Let's say that it takes about 10^{10} copies of PCR product to produce a signal above background (a good estimate for real-time machines). Equation 1 predicts that if you have 1,000,000 copies at the beginning of PCR and an efficiency of 1.9, then you will first see a signal at cycle 14. If you start with 1000 copies you will first see a signal at cycle 25. If you start with a single copy you will have to wait until cycle 36. This is the basis of real-time quantitative PCR.

However, Eq. 1 does not describe the total PCR reaction; it describes an unsustainable exponential growth curve. Very few cycles of a PCR reaction actually fit this equation. In the early cycles the background dominates. In the later cycles the reaction plateaus.

There are multiple causes of the plateau:
1. PCR product re-annealing competing more and more effectively with hybridization of the primers.
2. Reaction by-products build up and inhibit the reaction.
3. The enzyme becomes limiting.

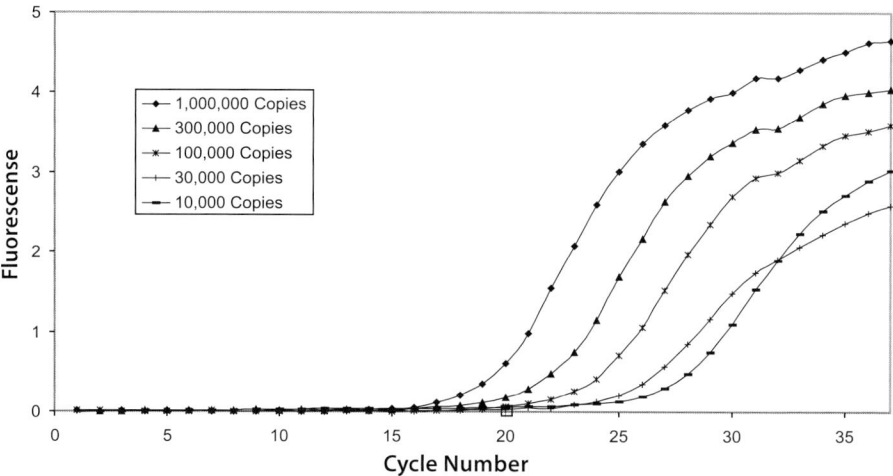

Fig. 1. Dilution series of purified PCR product from the human HER2/neu gene on the LightCycler. The reaction was monitored with SYBR Green I. Acquisitions were taken once per cycle after extension

These effects are PCR product concentration dependent, so a wide range of initial template concentrations all reach similar plateaus. This means that the quantitative information that was present during the log-linear phase is gone in the plateau phase. Equation 1 should not be applied to the PCR reaction in general, it only describes the exponential cycles.

Quantification with External Standards – "Kinetic PCR"

The idea behind kinetic PCR quantification is straightforward, but it was challenging to do using traditional end-point PCR methods. With all the quantitative information in just 4 or 5 exponential cycles out of 40, finding these cycles required either running many tubes or knowing the answer from the beginning.

With real-time PCR, the position of the exponential phase is easily identified. Higuchi et al. introduced fluorescence monitoring at each cycle for quantitative PCR, using ethidium bromide to monitor DNA synthesis [2]. With data on PCR product concentration from every cycle, the investigator can easily identify the exponential cycles after the reaction is over.

The Standard Curve

Figure 2 shows an example of a standard curve derived from the data in Fig. 1. How is this standard curve described and what can it tell us about the assumptions of the external standard method? Imagine that we can pick a point in the amplification curve that always represents the same amount of PCR product in every curve. Lets call that point in the amplification curve the crossing point. If the number of copies of PCR product present at the crossing point is K, and the cycle number is C_p, then by equation 1 above:

$$K = T_o(E)^{C_p} \tag{2}$$

This can be linearized to:

$$\log K = \log T_o + C_p * \log E \tag{3}$$

The initial concentration of the standard (T_o) is under the control of the investigator and the cycle number at the crossing point (C_p) is what we are measuring, so rearranging to the form y=mx+b gives:

$$C_p = -(1/\log E) * \log T_o + (\log K / \log E) \tag{4}$$

This is the equation of the standard curve with the log of the initial template copy number plotted horizontally and the cycle number at the crossing point plotted vertically. The slope of the line is the negative reciprocal of the log of the efficien-

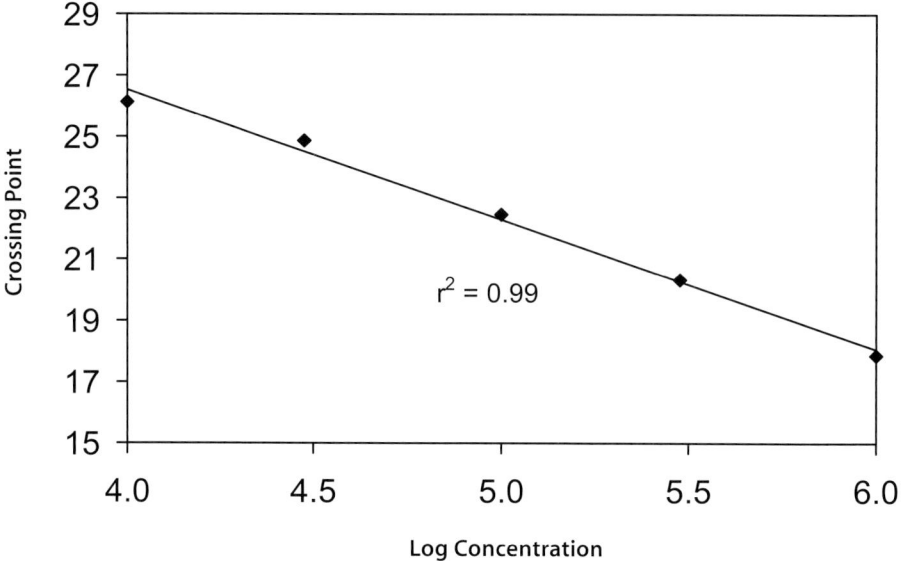

Fig. 2. A standard curve constructed from the data in Fig. 1. The slope of the line is −1/log (efficiency) giving an efficiency in this case of 1.73. The intercept is the log of the amount of DNA at threshold divided by the log of the efficiency

cy. The Y intercept is the log of the amount of PCR product at the crossing point divided by the log of efficiency. Looking at this equation gives insight into the assumptions of the method:

1. There is only one "E" in Eq. 4. Therefore the efficiency must be constant within each reaction from cycle one up to the crossing point.
2. A single number, the cycle number at the crossing point (C_p) defines the position of each amplification curve. C_p defines a "virtual cycle number", that is fractional (e. g., 29.72) where all the standards and unknowns have the same amount of PCR product.
3. The C_p values of the unknowns are converted to concentrations, using the equation derived from the standards. Therefore, the efficiency of the standards and unknowns must be the same and the amount of PCR product present at the crossing point must be the same for standards and unknowns.

Checking Efficiency

There is no simple way to check if the efficiency of the reaction is constant up to the crossing point, but it is possible to demonstrate that the efficiency of any given sample is constant over a range of initial template concentrations, and to compare the average efficiency of two samples. This is done by using the excellent dynamic range of PCR to analyze a dilution series of the sample by plotting C_p vs the log of the initial template concentration as shown in Fig. 2. If the efficiency is constant over the concentration range studied, the plot will be linear. If a quantification standard and an unknown are both studied, they have equal efficiencies if, and only if, they have the same slope. The actual concentrations of either are irrelevant to the determination of the slope.

How important are differences in efficiency? The effect is exponentially dependent on cycle number, but if the crossing point of an unknown is 25 cycles, and the efficiency of the standard is 1.9, then Table 1 gives the expected error for a range of efficiency differences between standard and unknown. Since the error can be calculated, in principle, the error can be corrected for in the analysis. How well this works in practice is under investigation.

Finding Your Way Through Cycle Space

To correlate initial template concentration to real-time fluorescence curves, a single number is needed that defines the position of the curve. This number is a virtual cycle number or crossing point, "C_p," where the amount of PCR product is the same in all curves that are being compared. There are a lot of methods for picking the crossing point, and happily, with good data, all of them give good answers.

The Threshold Method

To construct a standard curve, a single number is needed that defines the position of each amplification curve. A fluorescence threshold can be defined where the fluorescence signal exceeds background fluorescence [2]. The sensitivity of real-time instruments are about 10^{10} copies of PCR product, so the threshold is reached when about 10^{10} copies are formed. The threshold method assumes that all samples have the same concentration of DNA when the fluorescence significantly increases over background. Measuring "the level" of background fluorescence can be a challenge. It is no problem with nicely behaved reactions like Fig. 1, but too often the investigator has to deal with background drift over the course of the reaction. In real reactions, the baseline can drift up, drift down, or drift for some number of cycles and then level out. Averaging over a drifting background will give an overestimate of variance and thus increase the threshold level. In practice (as discussed below), it usually does not matter much where the threshold is set, as long as it is the same for all reactions that are compared.

Table 1. Effect of efficiency differences between standards and unknowns after 25 cycles

Efficiency of the Standards	Efficiency of the Unknowns	Efficiency Difference	Difference in Slope of the Standard Curve	Fold Underestimate
1.90	1.90	0.00	0.00	1.0
1.90	1.89	0.01	0.03	1.1
1.90	1.80	0.10	0.33	3.6
1.90	1.65	0.25	1.01	49

The "Fit Points" Method

The difficulty in defining the background and sample-to-sample differences in variance led us to develop a different method for defining the position of the curve in "cycle space". The fit points method has the user discard uninformative background points with a horizontal noise band, exclude plateau values by entering the number of log-linear points, and then fit a log-line to the exponential portion of the amplification curve. The intersection of these log lines with a horizontal threshold line, identifies the crossing points. The strength of this method is that it is extremely robust. The weakness is that it is not easily automated and requires user input.

Setting the Noise Band

In contrast to the threshold method, which uses the background fluorescence to establish a benchmark, the fit points method asks the user to throw away all the background fluorescence values with the noise band. Fig. 3 shows the experiment in Fig. 1 on a log scale, with a range of acceptable positions for the noise band (dotted lines). The optimal position of the noise band is as low as possible, without including any background points (solid line). When displayed on a log scale, the human eye is quite good at determining where the log-linearity is lost. However, it does not matter very much where the noise band is placed. In most experiments, the log-linear portion of the amplification is quite linear and any subset of points in this range can be used. It is not that the noise band position has no effect on the crossing points, but rather that, within a fairly wide range, the effect is small compared to other sources of variance in a quantitative PCR experiment.

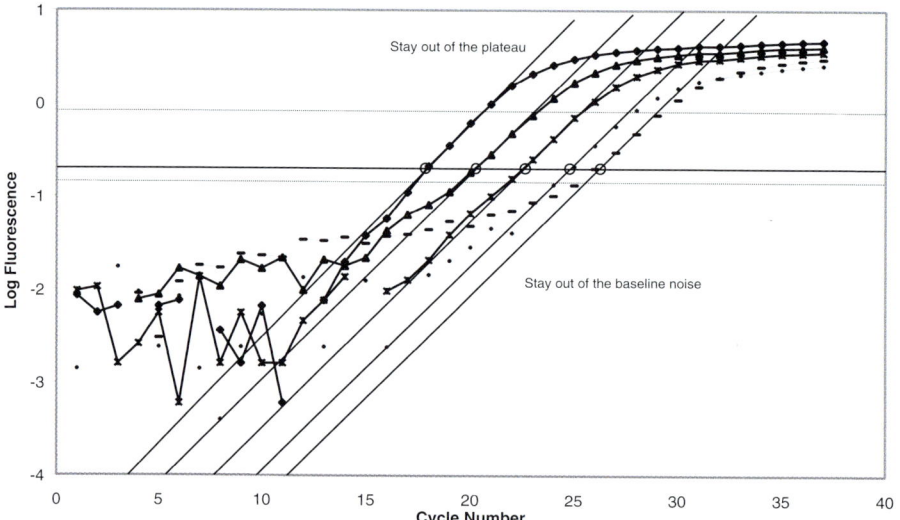

Fig. 3. The noise band should be set just above the background noise (*solid line*). However, there is a large range of acceptable levels for the noise band (bounded by the *dotted lines*) that will result in good standard curves

Picking the Number of Fit Points

The number of log-linear points varies from curve to curve. Fig. 4 shows the data from Fig. 1 after the background noise has been removed. In this data set, the number of log-linear cycles ranges from 2 in the low concentrations to 4 in the higher concentrations. It is best to be conservative and select the lowest value. Including plateau points in the line, can drastically decrease apparent efficiency, while excluding a log-linear point has little effect. The LightCycler software defaults to 2 points for this reason.

Setting the Threshold

If the position of the noise band doesn't matter very much, the position of the threshold barely matters at all. Once the log-lines have been calculated, any threshold position will give an acceptable standard curve. The only concern is extrapolation error, as the threshold gets further and further from the data. Fig. 5 shows two threshold band positions, separated by an order of magnitude in fluorescence. The inset graph in Fig. 5 shows that each gives an acceptable standard curve that differs only in Y intercept. Statistical theory says that the lowest extrapolation error will occur in the middle of the data used in the linear regressions. For this reason, the default position of the threshold line is at the same position as the noise band, insuring that the threshold line will be near the data points. The threshold is essentially an arbitrary benchmark for comparing curves. If the fluorescent signal depends on the amount of specific PCR product present, then it doesn't matter where you compare the amplification curves, it only matters that all the samples have the same threshold.

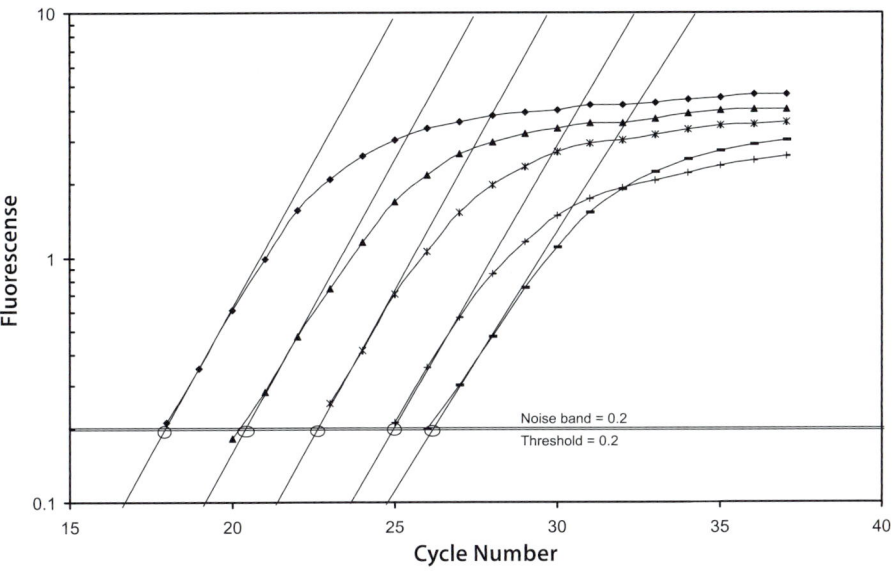

Fig. 4. After removing the background noise, log-lines are calculated for each curve and then extrapolated back to the threshold line. The software default for the threshold is the same position as the noise band

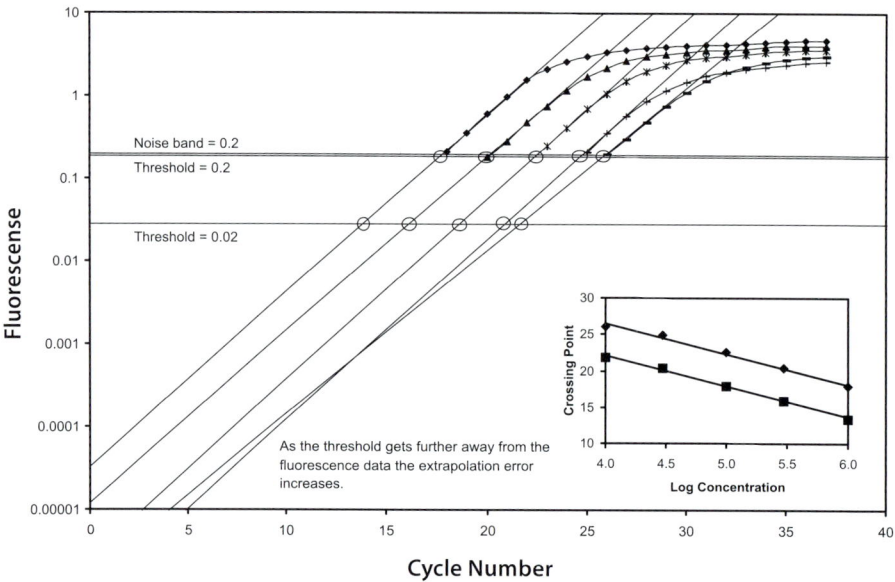

Fig. 5. A wide range of threshold positions give good standard curves. Two standard curves with thresholds of 0.2 and 0.02 are compared in the *inset*, and show identical slopes but different Y intercepts

The Second Derivative Maximum Method

Both threshold and fit points methods require that a human make decisions that will help the software find the log-linear portion of an amplification. Are there any characteristics of this part of the curve that software could automatically identify?

In the exponential part of the reaction, the fluorescence signal is increasing at an ever-increasing rate. This increase in the rate of increase (or acceleration) of the fluorescence signal slows as the reaction begins to enter the plateau. Therefore, the cycle where the second derivative is at its maximum should always be near the log-linear portion of the reaction. The amplification curve is not a function that is easily differentiable, but the derivatives can be estimated. Figure 6 shows how the second derivative maximum is usually found.

The fluorescence of the second derivative maximum is usually different from one sample to the next in any experiment. Unlike the threshold and the fit points methods, where the samples are compared at identical fluoresence levels, the second derivative maximum method rejects the idea that samples with the same fluorescence have the same DNA concentration. Instead, this method posits that the shape of the curve is a better guide to the concentration of PCR product, and the crossing point should be at the maximum acceleration, even if the fluorescence levels between curves are different.

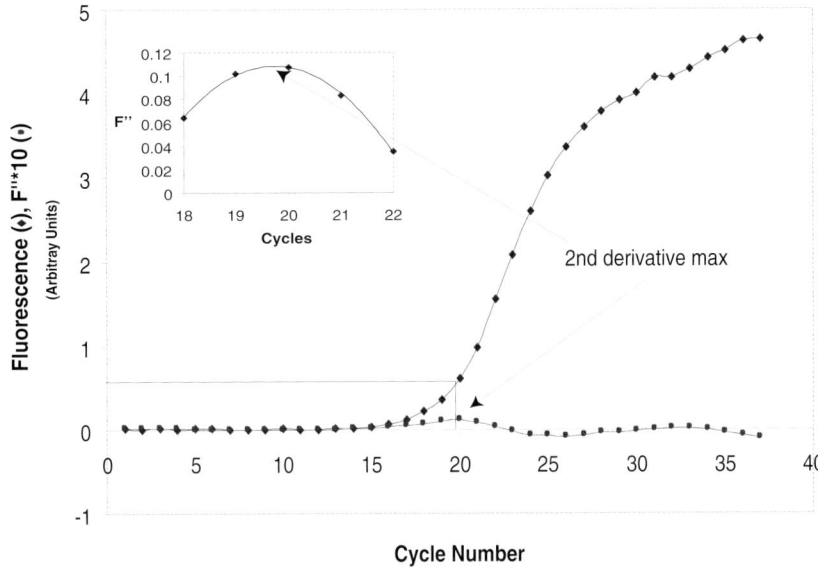

Fig. 6. A plot of the most concentrated sample of Fig. 1 and its estimated second derivative (×10). Once the cycle with the maximal second derivative is found, a parabola is fit to the region to determine the fractional cycle with the maximal second derivative (*inset*)

Reproducibility of Standard Curves

Interest in the reproducibility of standard curves usually takes the form of the question, "Do I need to run a standard curve with every run?" As always, it depends. In our experience, repeated runs of the same standard curve give variations around 2–3% in slope and 10% in Y-intercept. Because standard curves and unknowns vary together, this amount of variation allows twofold differences to be routinely detected. However, when unknowns from different runs are evaluated by a single standard curve, this correlation in variation is gone and larger errors will be seen. Five-fold changes in copy number can usually be discerned, but a twofold change might be missed. Is that good enough? The question "Do I need to run a standard curve every time?" is an experimental question that each investigator must answer based on the precision needed.

The number of points in the standard curves affects reproducibility. Varience in the standard curve will go with the square root of the number of points in the curve. Going from 3 points to 6 points will reduce varience by about 40%. Going from 6 points to 12 points will get you another 25% reduction. By increasing the number of points in the standard curve you can decrease varience, but there are dimishing returns to this strategy.

Obviously a line needs a minimum of 3 points. So we recommend a minimum of 3 points, or 1 point per log of concentration, whichever is the greater.

The absolute copy number of a sample also affects reproducibility. As you might expect, high copy samples are more reproducible than low copy samples. The most important reason for this is that, as the number of cycles increases, the affect of any variability in amplification efficiency increases: for example, a SYBR Green I-based quantification had a coefficient of variation of 6.4% at 10,000 copies per tube, 9.5% at 1,000 copies per tube, and 18% at 100 copies per tube. At very low copy numbers, a different source of error would become important. At 1–2 copies per tube, the random variation due to sampling error (also called Poisson error) becomes significant. Therefore, if, for example, your target averages 5 copies per reaction, only 18% of the reactions actually get 5 copies.

All Quantification Is Relative

Quantification by external standards is often referred to as "absolute" quantification, because an actual number of DNA molecules is directly obtained. However, the accuracy of external standard quantification is entirely dependent on the accuracy of the standards. It is easy to show a log linear relationship between target concentration and the position of amplification curves. These curves are also highly reproducible, which guarantees a precise answer, but not necessarily an accurate answer. As an example of how difficult it can be to achieve accuracy: the three most commonly used clinical HIV tests, PCR, bDNA and NASBA, all give extremely reproducible, but different answers [3]. All absolute copy number claims should be treated with a healthy skepticism.

Calibration of the Standards

The amount of care that needs to be taken in designing, making and calibrating standards should depend on the research question being asked. If only relative changes are important, say changes in gene expression with some kind of treatment, it is not important to know the actual copy number of the standards, it only matters that the standards are consistent. You may not even need RNA standards if you are measuring mRNA. Hayward et al. [4] showed that although the efficiency of reverse transcription (RT) differed greatly from target to target, it was consistent for a given target. This should allow the investigator to either ignore the RT efficiency or correct for it.

In our laboratory, we have found absorbance at 260 nm is a very consistent way to calibrate standards. However, absorbance may overestimate the amount of amplifiable DNA, sometimes by an order of magnitude or more, depending on the purification procedure used. For absolute quantification, standards can be quantified by limiting dilution [5,6].

The Denominator Problem

No matter how accurately the standard concentrations of the standards are known; final results are always reported relative to something. Copies per cell, copies per A_{260} unit of starting nucleic acid, copies per gram of tissue, copies per copy of some other target, copies per milliliter of blood are some examples. Again the choice here depends on the question being asked.

For systems where only relative changes are important, the denominator may be a housekeeping gene. For example, the number of mRNA copies of a particular gene relative to β actin mRNA. This sort of comparison of gene expression levels is probably the most common use of quantitative PCR in biological research. For this kind of study, there is little or nothing to be gained from accurately determining the absolute copy number.

If absolute changes in copy number are important, then the denominator must still be shown to be invariant across the comparison, but this sort of accuracy may only be needed in a few cases, like viral load determination.

In all cases, the quality of your quantitative data cannot be better than the quality of your denominator. Variation in your denominator will obscure real changes and produce artifactual changes. Careful use of controls is critical, in order to demonstrate that your choice of denominator was a wise one.

Options for Probes

SYBR Green I

SYBR Green I is a dsDNA-binding dye. It is thought to bind in the minor groove of dsDNA and upon binding increases in fluorescence over 100-fold (Figure 7a). It is compatible with PCR up to a point; at very high concentrations it starts to inhibit the PCR reaction. In the LightCycler, SYBR Green I is monitored in channel F1.

The biggest advantage of SYBR Green I is that it binds to any dsDNA; no design or optimization of probes is required. If you have a PCR that works, you can have a real-time quantitative assay working in about a day.

The biggest disadvantage of SYBR Green I is that it binds to any dsDNA; the specific product, non-specific products and primer dimers are detected equally well. There are a number of ways to handle this problem. Careful optimization of the PCR reaction can usually reduce primer dimers to a level that is only important for very low copy detection. Hot start techniques like TaqStart antibody [7] or Fast TaqKits (Roche) can be helpful in reducing primer dimers. The LightCycler allows melting curve analysis of the reaction [8]. This can help to determine the fraction of the signal coming from the desired product and the fraction coming from primer dimers [9]. Once the melting point of the product has been determined, fluorescence may be acquired above the melting temperature of the primer dimers, but below the melting temperature of the product.

Hybridization Probes

If sequence specific recognition is required, hybridization probes allow detection of only the specific product. Two probes are designed that hybridize side by side

I. SYBR Green I

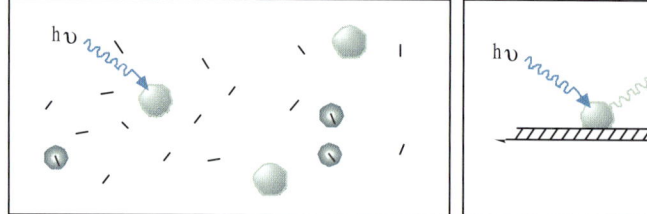

II. Hybridization Probes

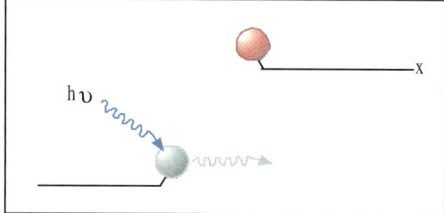
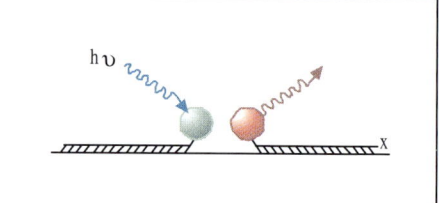

III. Taq Man Probes

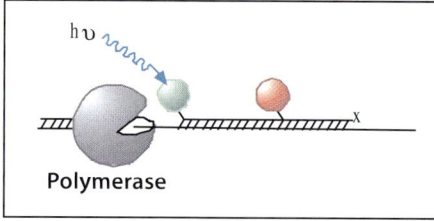
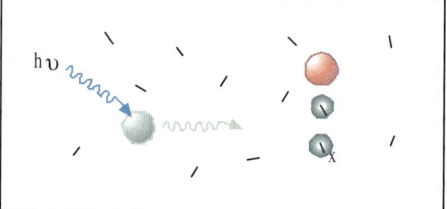

Fig. 7. Three probe systems compatible with the LightCycler

on the PCR product (Fig. 7b). The 3' end of the upstream probe is labeled with fluorescein, which acts as a fluorescence resonance energy transfer (FRET) donor [10]. The 5' end of the downstream probe is labeled with an acceptor dye, either LC Red 640, or LC Red 705. The FRET signal is seen only when two specific hybridization events occur. In the LightCycler, LCRed640 is monitored in channel F2, LCRed705 in channel F3. There may sometimes be an advantage to monitoring the ratio of the acceptor channel (where the signal goes up with increasing PCR product) to fluorescein in F1 (which goes down with increasing PCR product).

TaqMan Probes TaqMan probes derive their fluorescence signal from the hydrolysis of the probe by Taq's 5' to 3' exonuclease activity (Fig. 7c) [11]. The hydrolysis separates fluorescein from a quenching dye and results in an increased fluorescein signal. These probes can be used in the LightCycler and are monitored in F1 or F1/F2 [12].

Advantages of External Standard Quantification

External standard quantification with a real-time instrument is by far the easiest nucleic acid quantification system to set up and validate.

Once you have optimized primers you can make standards either directly by PCR, by cloning into a plasmid or by in vitro transcription. If you do not require complete sequence specificity, SYBR Green I can be used, eliminating the need to design, synthesize and optimize probes.

The large dynamic range of PCR quantification on the LightCycler allows accurate quantification in a single assay over a billion-fold range. When a wide variation in the amount of target DNA is expected, the external standard method is the method of choice for quantification. The LightCycler software completely supports external standard quantification, making data analysis significantly easier.

Disadvantages of External Standard Quantification

The main disadvantage of external standards is the lack of internal control for PCR inhibitors. All quantitative PCR methods assume that the target and the sample amplify with similar efficiency. The risk with external standards is that there is something different about the unknown samples that significantly reduces the efficiency of the PCR reaction in the unknowns. To test for PCR inhibitors, a different PCR product can be amplified from endogenous or spiked templates in the sample and the crossing point of the amplification compared to the control. Alternately, a dilution series can be run on the unknown sample. PCR inhibitors can usually be diluted out, resulting in a non-linear standard curve. Strong PCR inhibitors can sometimes be identified by using SYBR Green I and noting the absence of all products on melting curve analysis [8], i.e., neither specific amplicon nor primer dimer products are present. Even in negative samples, primer dimers are almost always produced if you cycle for long enough. Therefore, true negatives will have primer dimer peaks and negatives due to strong PCR inhibition will not.

Conclusion

The old "quantitative PCR is an oxymoron" joke isn't heard nearly as often these days. This is good evidence that the technique is gaining acceptance and credibility. But the reason for that old skepticism remains. Quantitative PCR is a powerful technique that can give answers that are difficult to find in any other way. Real-time PCR makes quantitative PCR much easier. However, careful experimental design, execution and analysis are still necessary if the investigator is to be rewarded.

References

1. Ferre F, Marchese A, Pezzoli P, Griffith S, Buxton E, Boyer V (1994) Quantative PCR: an overview. In: Mullis KB, Ferre F, Gibbs RA (eds) The polymerase chain reaction. Birkhauser, Boston 67-88
2. Higuchi R, Dollinger G, Walsh PS, Griffith R (1992) Simultaneous amplification and detection of specific DNA sequences. Biotechnology 10: 413–417
3. Schuurman R, Descamps S, Weverling G, Kaye S, Tijnagel J, Williams I, van Leeuwen R, Tedder R, Boucher C, Brun-Vezinet F, Loveday C (1996) Multicenter comparison of three commercial methods of quantification of human immunodeficiency virus type 1 RNA in plasma. J Clin Microbiol 34: 3016–3022
4. Hayward AL, Oefner PJ, Sabatini S, Kainer DB, Hinojos CA, Doris PA (1998) Modeling and analysis of competitive RT-PCR. Nucleic Acids Res 26(11): 2511–8
5. Wang Z, Spandoro J (1998) Determination of target copy number of quantitative standards used in PCR-based diagnostic assays. In: Ferre F (ed) Gene quantification. Birkhauser, Boston, pp 31–43
6. Morrison TB, Weis JJ, Wittwer CT (1998) Quantification of low copy transcripts by continuous SYBR Green I monitoring during amplification. Biotechniques 24: 954–958
7. Kellogg DE, Rybalkin I, Chen S, Mukhamedova N, Vlasik T, Siebert PD, Chenchik A (1994) Taqstart antibody:"hot start" PCR facilitated by a neutralizing monoclonal directed against Taq DNA polymerase. Biotechniques June 16(6): 1134–1137
8. Ririe KM, Rasmussen RP, Wittwer CT (1997) Product differentiation by analysis of DNA melting curves during the polymerase chain reaction. Anal Biochem 245: 154–160
9. Rasmussen RP, Morrison TB, Herrmann MG, WittwerCT (1998) Quantitative PCR by continuous flourescence monitoring of a double strand DNA specific binding dye. BioChemica 8–11
10. Wittwer CT, Herrmann MG, Moss AA, Rasmussen RP (1997) Continuous flourescence monitoring of rapid cycle DNA amplification. Biotechniques 22: 130–138
11. Holland PM, Abramson RD, Watson R, Gelfland DH (1991) Detection of specific polymerase chain reaction product by utilizing the 5' to 3' exonuclease activity of thermus aquaticius DNA polymerase. Proceedings of the National Academy of Science 88: 7276–80
12. Wittwer CT, Ririe KM, Rasmussen RP (1998) Flourescence monitoring of rapid cycle PCR for quantification. In: Ferre F (ed) Gene quantification. Birkhauser, Boston

Selection of Hybridization Probes for Real-Time Quantification and Genetic Analysis

Olfert Landt*

Introduction

The concept of LightCycler hybridization probes (HybProbes) is based on the detection of two adjacent oligonucleotide probes, whose fluorescent labels 'communicate' through fluorescence resonance energy transfer (FRET). The detection of a given nucleic acid is achieved by the near hybridization of two probes, or by an internal labeled PCR strand and a detection probe located on the opposite strand. The signal is dependent on the spatial approximation of the dyes, and is dependent on the amount of the target. Although hybridization is a very robust process and therefore most pairs of randomly chosen sequences will work satisfactorily, we believe that understanding how a probe finds its target will aid in the selection of probes with a better average performance. In the following text we will explain some rules for the selection of HybProbes. We will discuss their use for two possible applications, using HybProbes for the quantification of a target, or as probes for the detection of sequences variations. In both cases, effective competition of probe hybridization with rehybridization of the target and with elongation of the primers is important. In quantification, the probes can be placed anywhere within the amplicon. In contrast, the detection of sequence variants requires positioning the HybProbes over the variable site and difficult sequences cannot always be avoided. In the following text we will demonstrate how possible problems can be circumvented to increase the distinction of sequence variants. In any case, there may be a small number of molecular situations where no LightCycler-based assay can be constructed.

How Hybridization Works

Hybridization is influenced by many things, and only an 'efficient hybridization' will produce an intense signal. The hybridization is sensitive to the individual base composition and surrounding sequences. Therefore it is interesting to know how two complementary sequences find each other. The probes do not move directly together like magnets, they are instead driven by diffusion through the

---

* Olfert Landt (✉) (e-mail: Olandt@tib-molbiol.de)
  TIB MOLBIOL, Tempelhofer Weg 11–12, 10829 Berlin

reaction mix, and interact with many other nucleic acid molecules. They will 'sniffle' like a dog along all sequences that are present, pausing longer at sites with higher degrees of complementarity. The initial binding occurs at a 'crystallization point', possibly at one end, and continues like the closing of a zipper along the entire sequence. Since the process of binding is reversible, it can also be described in terms of probability. Most of the target will be hybridized to probe only below the melting temperature (T_m), the temperature where, at equilibrium, and containing equal amounts of probe and target, half of the probe is hybridized to the target. In order to promote saturation of the binding site, the concentration of probe is made much higher than the amplicon concentration produced during PCR.

HybProbes Have to Compete with Primer Extension and Product Re-Annealing

Access to a single stranded target is a prerequisite for specific HybProbe binding. The target is single stranded only during and directly after the thermal denaturation step. The binding competition between probes and primer extension starts as soon as the temperature begins to cool down to the annealing conditions, the competition with product re-annealing starts even earlier. The primers do not compete directly for the same site. However, the primer that binds to the same strand as the probes is immediately extended by the polymerase to transform the target sequence into a now inaccessible double stranded molecule. In order to ensure that FRET takes place, two labeled probes have to bind to the target at the same time. The probability of this occurring is the product of the probabilities of each individual binding event, compared to the probability of a single primer-binding event. Primer binding and extension is also favored because extension makes primer binding irreversible. If the hybridization rates of the probes and primers are about the same, only a minor fraction of the product will hybridize to both probes and contribute to fluorescence. Therefore, HybProbes should be designed so that they to bind to the target with greater affinity than the primers, especially than the primer that binds to the same strand. Re-annealing typically is only prominent in late cycles, where high concentrations of product are present. The effect of product re-annealing can sometimes be seen as a decrease in fluorescence after the logarithmic amplification phase.

Melting Temperature of HybProbes

Probe binding can be favored over primer binding by using HybProbes with a higher melting temperature than the primers, or at least higher than the competing primer on the same strand. We suggest using probes with a melting temperature 5–10°C higher than that of the primers. However, probe T_m should not be greater than 5°C above the extension temperature, that is, not over 75–80°C. Probe melting higher than 80°C may inhibit polymerase extension because the polymerase may not be able to displace the probes. If short or low T_m primers are used, the temperature difference between primers and probes may be higher than 10°C. In any case, the probe T_m needs to be higher than the annealing temperature if fluorescence is to be monitored each cycle.

Location of the HybProbe Target Site

The target site for HybProbes can be anywhere within the amplified sequence. However, because the primers will be extended and eventually block the target site, the probes are preferentially located far away from the primer on the same strand. That is, choose probes from the 'right half' of the amplicon if the probes are sequences in the upper strand, and from the 'left half' for probes in the lower strand.

HybProbes bind adjacent with facing fluorophores, leaving a gap of usually 1-5 bases in between, mainly to avoid collisional quenching between the fluorophores, and, to avoid any steric hindrance. This gap may even be bigger, since the FRET process will work within a range of about 50 Å – for comparison – a 10 bp full turn is about 34 Å.

HybProbes will bind to related targets at unexpected sites if the hybridization conditions are favorable. It is very unlikely that this would produce a fluorescent signal, because both probes must hybridize in the same vicinity and in the right orientation for FRET to occur. However, unexpected binding can reduce the effective concentration of available probes, and may be important if the alternative binding site is present at high concentrations. This can occur if the alternative binding site within an amplicon during the later cycles of PCR.

HybProbe Sequence

The sequence and secondary structure of probes affects probe binding. In general, probes should be "balanced" with an equal distribution of the four bases. The following situations can cause problems: Regions of very high binding strength within a probe, such as GC clusters, may focus the binding to a small motif and result in inappropriate binding to undesired sites. Repetitive motifs and stretches of the same base may result in "slippage" along the target. Complementary sequences, such as palindromes may result in dimer formation and lower the amount of free probe. Strong complementarity between donor and acceptor probes may result in binding between them, resulting in fluorescence independent of any target. Sequences which form a stem loop interfere with probe and target binding in two ways. They decrease the effective concentration of free probe and also the availability of the probe binding site in the target strand. Perhaps the most important rule is to avoid complementarity between probes and primers, especially to the 3'-termini of the primers, to prevent the production of probe-primer dimers during the early stages of PCR, when little amplifiable target is available. General rules for HybProbe design are summarized in Table 1.

Table 1. General rules for HybProbe design

The LightCycler HybProbe method uses two fluorescently-labeled oligonucleotide probes that hybridize, in a head-to-tail arrangement, to adjacent sequences on the target DNA. The interaction between the labels can only occur when both probes are bound

Design Rule	Explanation
Choose neighboring sequences on the same strand, leaving a gap of one to five bases	The energy transfer process is distance dependent. Donor and acceptor must be in close proximity (within 50 Å = 15 bases)
Block the 3'-termini of probes against polymerase extension, usually by a 3'-phosphate or a 3'-fluorescein	Probes must not amplify the target sequence
Label the adjacent ends of the probes; usually with a 5'-LightCycler Red label and a 3'-fluorescein label	The alternative, 5'-fluorescein and 3'-LightCycler Red labels are more difficult to obtain in pure form. Use FITC derivates of fluorescein for accurate color compensation of the different channels of the instrument with the Roche color compensation kit.
Use 'balanced' sequences with a normal distribution of the four bases	See following points.
Avoid same base stretches	Probes may slip and bind at different places
Avoid simple repetitive sequences	Probes may slip, that is, bind at different places
Avoid GC-rich sequences, especially G-rich sequences	Unexpected binding can occur at a short GC-rich motif
Avoid very purine-rich sequences	Tend to hybridize poorly.
Avoid palindromes and complementarity within each probe	Palindromes with 6 GC bases or more may cause probe-probe dimers. Internal probe complementarity may cause stem loops. Both palindromes and stem loops lower the concentration of "free" probe molecules and available single stranded target. This effect may be partly circumvented by using a higher probe concentration.
Avoid complementarity between probes	Stable dimers between probes may interact through resonance energy transfer, independent of the presence of PCR product.
Avoid complementarity between the 3'-termini of the primers and the probes	At the beginning of PCR there is very little target present, but high concentration of primers and probes may promote primer/probe dimers.
Avoid probe T_ms much greater than the extension temperature	Very stable probes may block polymerase extension and inhibit PCR because the polymerase will no longer be able to displace the probes.

Probes for Quantification

Because quantification requires monitoring fluorescence each cycle, the melting temperature of the probes must be higher than the temperature used for fluorescence acquisition. We recommend that the probe T_ms are within 2°C of each other and 5–10°C greater than the primer on the same strand. The target site can be located anywhere within the amplified sequence. However, the probes are preferably located far away from the primer that hybridizes to the same strand in order to minimize competition from primer extension. Therefore, to select probes for a given PCR fragment, analyze first the last part of the upper (sense) strand and the first part of the lower (anti-sense) strand. Special considerations for probes used in quantification are summarized in Table 2.

Probes for Mutation Analysis

The greatest potential for HybProbes is in the detection of known sequence variations, including mutations, polymorphisms, and other variant nucleic acid species, e.g., those due to translocations or alternative splicing. The molecular concept is rather simple: one of the probes is a tightly binding 'anchor probe' whereas the adjacent 'sensor probe' spans the region of sequence variation. During melting of the final PCR product, the sequence alteration is detected as a change in the melting temperature of the sensor probe. For a typical homozygous wild-type sample, a single melting peak is observed, for mixed alleles, two peaks, and for a homozygous mutated sample, a single peak at a temperature different from the wild-type allele. The temperature shift induced by one mismatched base is usually between 5–9°C and therefore easily observable.

Variants can be detected if they are present in comparable amounts, as in the case of two alleles in diploid cells. The method is generally not designed to detect minor variants that are present at relative concentrations less than 10%, for exam-

Table 2. Special considerations for probes used for quantification
Data is acquired every cycle. Therefore, the competition between primers and probes may be more important than the competition from product re-annealing

Design Rule	Explanation
Stay far away from the primer in the same orientation	One primer competes with the binding of two probes; if the primer is elongated, the target will be hidden and fluorescent cannot develop.
The T_m of the probes should be 5–10°C higher than the T_m of both primers, or at least the primer on the same strand	One primer competes with the binding of two probes. With identical binding properties, the primer will always hybridize faster than the two probes.
The T_m of the probes should be higher than the annealing temperature	The signal has to be obtained during annealing.
The T_m of the probes should not be much higher than the extension temperature	The probes might block primer extension.
To select probes for a given PCR fragment, analyze first the last part of the upper strand and the first part of the lower strand for suitable HybProbe pairs. Consider the general design rules above. If incompatibilities with primers are probable, consider using degenerate primers. For new PCR tests, look first for suitable probe sequences and then for compatible primers. Select two or three primers for each direction and select the best combination experimentally (running an agarose gel).	

ple, in minimal residual disease or antibiotic-resistant viral populations. In these cases, variant-specific primers should be used for analysis. The melting peak areas do not always accurately reflect the relative amount of the different alleles present. There are several reasons for this, including the fact that the probes may bind to the different alleles with different efficiencies. For example, errors might occur if the mismatched allele resulted in a stem-loop conformation in the target at the probe binding site that was not present in the matched allele. However, for systems with known properties, peak areas can be used to distinguish 1:2 or 2:3 ratios from 1:1, sufficient to see a missing allele or to identify a trisomy.

Because the sensor probe must overlay the region of sequence alteration, the possible positions for a sensor probe are limited. The location of the sensor probe is usually adjusted so that the single base mutation is located near the center of the probe. Placement nearer than 3 bases from either end should be avoided. The center of the probe is best defined by its binding properties, so a GC-rich end might be considerably "shorter" than a weak-binding end. One can choose the upper or the lower strand, and place the anchor left or right of the sensor probe. However, in some cases, it may be difficult to avoid troublesome sequences. For example, if the sequence alteration is located in a region that is not strongly involved in overall binding, the alternate sequences may not result in significantly different melting temperatures. Strong binding motifs can sometimes be avoided by moving the probe over the sequence and observing the resulting change in probe T_m with programs that provide thermodynamic analysis. Alternately, strong binding and complementary sequence motifs such as palindromes that can form stem-loop structures can be weakened or destroyed by inserting mismatches or base analogues, for example, deoxyinosine can be used to replace guanosine. However, changing the sequence of the probe will not change stem-loop formation in the target. Not all mismatches destabilize probe T_m by the same amount. The mismatch between guanosine and thymidine is relatively stable and causes a temperature decrease of only 2–4°C, which is more difficult to analyze in routine practice. In such cases, it is better to use the other strand and analyze the resultant A-C mismatch, or to use a mutation-specific probe.

Either the 3'- fluorescein or the 5'-LightCycler Red-labeled probe can be used as the sensor. A LightCycler Red-labeled sensor has the advantage of potential multiplexing with different LightCycler Red – labeled dyes. On the other hand, a fluorescein-labeled sensor can be synthesized faster, which is an advantage when analyzing multiple or daily new targets, as needed for the detection of patient-specific rearrangements in minimal residual disease.

One advantage of using HybProbes for mutation detection is that many different mutations can be detected under the sensor probe. In some cases, the different mutations can be differentiated and identified by characteristic T_m shifts. A potential disadvantage of HybProbes is that many different mutations can have approximately the same T_m. Unexpected rare mutations not previously described could result in the same T_m shift as the common mutation that a test was designed to detect. One way to avoid this possibility is to run a confirmatory test with a probe complementary to the mutation. The mutation probe alone may not be sufficient, because a double mutant might have the same T_m as a wild-type allele. For complete certainty, both probes are needed and can be multiplexed in the same sample, using the second LightCycler Red dye and the third detection channel of the instrument. In routine diagnostics, such certainty is seldom warranted because the probability of the expected mutation is high compared to the chance of new, rare mutations that happen to have the same T_m.

Finaly one should keep in mind that mutations at the primer sites may interfere with the assay, lowering or even inhibiting the amplification of one allele, suggesting a homozygous sample. Many primers are located in introns, which might not be as conserved as coding sequences. If the mutation at the primer site is known, degenerate primers can be used that amplify both sequences equally.

Special considerations for probes used in genotyping are summarized in Table 3.

Table 3. Special considerations for probes used for genotyping
The melting analysis is run after PCR. Therefore, the competition between primers and probes may be less important than the competition from product re-annealing

Design Rule	Explanation
Probe T_ms can be higher or lower than the primers.	If the probe T_m is lower than the annealing temperature, amplification cannot be followed during PCR.
Probes and primers do not need to be separated by many bases	Probe can be close to and even overlap with the primers (if they are in the same orientation). This is because extension is less prominent and the absolute fluorescence signal is less important than in quantification.
The anchor probe T_m must be higher than the sensor probe T_m	The melting transition of both probes should not overlap. To see only the melting of the sensor probe, the anchor probe must be annealed during the melting transition of the sensor probe.
The sequence variation can be anywhere under the sensor probe, but not closer than 3 bases from either end.	A mismatch separates the cooperativity between both parts of the sequence. Terminal positions have a smaller influence on the melting temperature.
Avoid G-T mismatches. Use instead the mutant-specific probe or the opposite strand (C-A mismatch)	G-T mismatches are better accepted and result in low T_m shifts (2–4°C instead of 5–10°C)
Avoid very strong-binding sequences within the sensor probe. Shift the probe position, introduce weak binding base analogues (like inosine), or introduce mismatches to weaken a strong-binding motif.	If the sequence alteration is near a tight-binding motif, smaller T_m shifts than expected may occur. The introduction of mismatches or base analogs will not change troublesome structures on the target strand.
Avoid stem-loop-forming sequences	Stem loops may preferentially affect binding to one allele, resulting in an unequal distribution of probe between wild type and mutant alleles.
If the fluorescence is low, asymmetric amplification may yield more single stranded target.	After symmetric amplification, product re-annealing can hide most of the target. An asymmetric amplification will produce excess target strands for probe binding.

Using the Nearest Neighbor Model for the Estimation of Matched and Mismatched Hybridization Probe Melting Points and Selection of Optimal Probes on the LightCycler

Nicolas von Ahsen*, Ekkehard Schütz

Introduction

The hybridization of oligonucleotide probes is a technique with widespread application in molecular biology. It has been used for the detection of immobilized nucleic acids in Northern or Southern blotting and for allele specific oligonucleotide hybridization. Provided stringency conditions are met, only perfectly matched probes will basepair with each other (for review see [1]). The method can therefore be used for mutation detection when the presence of a mismatch caused by a base exchange disrupts the Watson-Crick pairing and destabilizes the double strand. The LightCycler is a powerful tool for monitoring the hybridization of labeled oligonucleotides in a homogenous assay. Two different wavelengths are available for the detection of oligonucleotide hybridization when an appropriate fluorescence resonance energy transfer (FRET) pair is formed by the adjacent hybridization of two dye-labeled probes. General guidelines for the construction of hybridization probes for use in both quantification and mutation detection, have been given [2], but these depend on an accurate estimation of DNA melting temperatures. According to our experience, a 10% higher melting point (T_m) in °C of the anchor probe than that of the matched detection probe is sufficient. This ensures complete hybridization of the anchor during the analytical melting cycle and avoids the synthesis of unnecessarily long anchor probes. When a hybridization probe pair is used for quantification, the melting points of both probes should be equal, so that hybridization and energy transfer occur rapidly when the reaction is cooled below the T_m. An accurate estimate of T_m allows rapid optimization of reaction parameters by suggesting appropriate annealing and melting conditions. For mutation detection the T_m shift caused by the mutation should be as high as possible, to ensure good discrimination of heterozygotes [See also "Genotyping of the Most Common Thiopurine Methyltransferase Mutations (TPMT*3) with the LightCycler. Optimization of Hybridization Probe Assays for the Detection of Mutations Causing Stable Mismatches in DNA Duplexes" in this manual by E. Schütz and N. von Ahsen].

The melting temperature of short (<20 bp) oligonucleotides is often estimated with the Wallace/Ikatura rule (cited in [1]) T_m (°C)=2×(#A+#T)+4×(#G+#C). Although this approximation assumes a salt concentration of 0.9 M NaCl, typical

---

* Nicolas von Ahsen (✉) (e-mail: nahsen@gwdg.de)
  Department of Clinical Chemistry, Georg-August-University, Robert-Koch-Str. 40, 37075 Göttingen, Germany

for dot blots and other hybridizations, it can also be applied to PCR applications that are not very sensitive to different T_ms. However, for hybridization probes, a better estimation of T_m is required. The nearest neighbor (n-n) model is based on thermodynamic calculations and gives the most precise prediction of oligonucleotide stability. The model used throughout this chapter was proposed by Borer et al. [3] and modified by Allawi and SantaLucia [4] based on Gray's work [5] (see also [6] and references cited therein). This model assumes that the thermodynamic parameters for a given base pair depend on the identity of its adjacent base pairs and that these nearest neighbor parameters are additive. The stability of a given oligonucleotide sequence is a function of both, interstrand H bonding between Watson-Crick paired bases and the intrastrand base stacking. The application of this model for calculation of DNA duplex stability will be detailed in this chapter. The derivation of the formulas used can be found in the literature [4, 5] and will not be given here. Several data sets are now available that describe the ten n-n pairs that occur in double stranded DNA [4, 6–8]. The contribution of a mismatch to duplex stability depends on its location, orientation, and neighboring bases [4]. The destabilizing effect of the 48 possible single mismatches can also be taken into account if the n-n data of the respective mismatches are used in the calculation of the melting temperature [4, 9–12]. Oligonucleotides with repetitive sequences or strings of A-T base pairs may deviate from the n-n model as well as molecules which do not melt in a two state (helix versus random coil) manner such as long DNA duplexes [13]. Hybridization conditions (ionic strength and probe concentration) also influence the T_m and must be considered. We have recently shown that the n-n model is able to predict the melting points observed with hybridization probe assays with a standard error of less than 1°C [14]. The n-n model is most beneficial for the thermodynamic predictions of oligonucleotide DNA. In longer DNA strands, interactions that are independent of the neighboring bases become increasingly important. This reduces the utility of the n-n model for the prediction of longer (>50 bp) DNA duplexes (see below). For such polymer DNA calculations, simpler formulas give good T_m estimates (see Eq. 1a in [1]).

In this chapter we will:
1. Give a brief description of how thermodynamic data for an oligonucleotide are calculated with the n-n model.
2. Validate the application of the n-n model to hybridization probe assays on the LightCycler.
3. Use a thermodynamic model to calculate the extent to which various mutations destabilize various probes.

Methods

Formulas and Equations

The enthalpy (ΔH^0) and entropy (ΔS^0) of an oligonucleotide duplex is needed for the calculation of its melting temperature:

$$T_m = \frac{\Delta H°}{\Delta S°[Na+] + R \times \ln(CT)} - 273.15 \tag{1}$$

where R is the gas constant (1.987 cal K^{-1} mol^{-1}), CT is the strand concentration, ΔS^0 [Na] is the entropy at a given sodium equivalent concentration. For non-self-complementary sequences, CT is the concentration of oligonucleotides divided by 4. A sequence is considered to be self-complementary if two strands of the same sequence can form a duplex. Self-complementary oligonucleotides should never be used for PCR and hybridization applications. For this reason, only calculations for non-self-complementary oligonucleotides are performed in this chapter.

For the calculation of T_m we need to calculate ΔH^0 from Eq. 2 and ΔS^0 [Na] from Eqs. 3–5 and must know CT. According to the n-n model, energy for helix formation is the sum of three terms, one for helix propagation energy from the sum of every nearest neighbor pair, one for an energy change caused by helix initiation, and one to account for effects caused by duplex formation of self-complementary strands. The following formula applies for the calculation of ΔH^0 for an oligonucleotide:

$$\Delta H^0_{total} = \Sigma_i\, n_i\, \Delta H^0(i) + \Delta H^0(5'\text{init}) + \Delta H^0(3'\text{init}) \quad (2)$$

$\Delta H^0(i)$ is the enthalpy change for the ten possible n-n and n_i is the number of occurrences of each n-n.

For ΔS^0 a similar formula applies:

$$\Delta S^0_{total} = \Sigma_i\, n_i\, \Delta S^0(i) + \Delta S^0(5'\text{init}) + \Delta S^0(3'\text{init}) \quad (3)$$

For Eqs. 2 and 3 it should be noted that the symmetry correction is zero for non-self-complementary oligonucleotides and therefore omitted from the formula. The ΔH^0 and ΔS^0 values for Watson-Crick n-n with matched and mismatched base pairs and the initiation energy were taken from published data and are summarized in Table 1 and 2.

A standard PCR buffer contains 50 mM KCl and 1.5 mM Mg^{2+}; for rapid cycling applications, higher Mg concentrations are often used. Mg^{2+} cations have a 140-times higher stabilizing effect on duplexes compared to Na$^+$ [15]. The concentrations of the monovalent ions do not add much to this effect, at least under the conditions which apply for PCR [15]. We therefore calculate the sodium equivalents as follows:

$$\text{Na}^+ \text{ equivalent} = [\text{Mg}^{2+}] \times 140 \quad (4a)$$

Alternatively, a formula based on a 100-times higher DNA stabilizing effect of Mg^{2+} cations may be used. This also takes into account the monovalent ions [16]:

$$\text{Na}^+ \text{ equivalent} = [\text{Mg}^{2+}] \times 100 + \text{monovalent ions} \quad (4b)$$

The entropy is salt dependent and therefore ΔS^0 must be corrected if the sodium equivalent concentration is different from 1 M NaCl, the condition under which the parameters tabulated in Table 1 were derived [6].

$$\Delta S^0 [Na] = \Delta S^0 [1\ M\ Na] + 0.368 \times N \times \ln [Na^+] \tag{5}$$

where N is the total number of phosphates in the duplex divided by 2. This is equal to the oligonucleotide length minus 1.

If dimethylsulfoxide (DMSO) is included in a PCR reaction, it induces a T_m reduction of 0.6°C for each percentage of DMSO [17].

Calculation of T_m for an 18mer without Mismatches

As an example, the T_m of the following 18mer will be calculated:

    5'-ACG ATG GCA GTA GCA TGC-3'
    3'-TGC TAC CGT CAT CGT ACG-5'

This example oligonucleotide has 17 n-n doublets. These are AC/TG, CG/GC, GA/CT, AT/TA, TG/AC, GG/CC, GC/CG and so on (Fig. 1).

The first doublet (AC/TG) is identical with GT/CA and only the thermodynamic data for the ten unique n-n doublets are listed in Table 1. Equivalent sequences are not listed twice. The numbers of each n-n doublet found in the above 18mer oligonucleotide are as follows:

    0 AA/TT, 2 AT/TA, 1 TA/AT, 4 CA/GT, 2 GT/CA, 2 CT/GA, 1 GA/CT, 1 CG/GC, 3 GC/CG, 1 GG/CC.

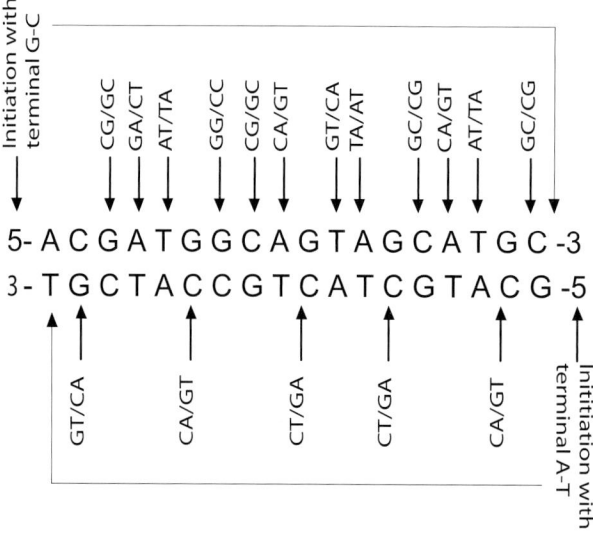

Fig. 1. The doublet nearest neighbor composition of an 18mer oligonucleotide. Seventeen n-n doublets and initiation energies for termination with G-C and A-T must be considered. The oligonucleotide is non-self-complementary. Doublets that match the tabulated items are shown above the strand while doublets shown below the strand are equivalent to others listed in the table

Table 1. Unified parameters for oligonucleotide ΔH^0 and ΔS^0 in 1 M NaCl [6]. Nearest neighbor dimer duplexes are presented with a slash separating strands in antiparallel orientation, e.g., AT/TA means 5'-AT-3' Watson-Crick paired with 3'-TA-5'

Sequence	ΔH^0 kcal/mol	ΔS^0 cal K^{-1} mol^{-1}
AA/TT	−7.9	−22.2
AT/TA	−7.2	−20.4
TA/AT	−7.2	−21.3
CA/GT	−8.5	−22.7
GT/CA	−8.4	−22.4
CT/GA	−7.8	−21.0
GA/CT	−8.2	−22.2
CG/GC	−10.6	−27.2
GC/CG	−9.8	−24.4
GG/CC	−8.0	−19.9
Initiation with terminal G-C	0.1	−2.8
Initiation with terminal A-T	2.3	4.1

The ΔH^0 calculation is according to Eq. 2:

$\Sigma_i\, n_i\, \Delta H^0(i)$ results in 0×(−7.9)+2×(−7.2)+1×(−7.2)+4×(−8.5)+2×(−8.4)+2×(−7.8) +1×(−8.2)+1×(−10.6)+3×(−9.8)+1×(−8.0)=−144.2 kcal/mol

The initiation terms are 0.1 kcal/mol for the terminal G-C and 2.3 kcal/mol for the terminal A-T. Therefore, ΔH^0_{total}=−144.2+0.1+2.3 kcal/mol=−141.8 kcal/mol.
The ΔS^0 calculation is according to Eq. 3:

$\Sigma_i\, n_i\, \Delta S^0(i)$ results in 0×(−22.2)+2×(−20.4)+1×(−21.3)+4×(−22.7)+2×(−22.4)+ 2×(−21.0)+1×(−22.2)+1×(−27.2)+3×(−24.4)+1×(−19.9)=−382.2 cal K^{-1} mol^{-1}

The initiation terms are −2.8 cal K^{-1} mol^{-1} for the terminal G-C and 4.1 cal K^{-1} mol^{-1} for the terminal A-T, total +1.3 cal K^{-1} mol^{-1}.

ΔS^0_{total}=−380.9 cal K^{-1} mol^{-1}

The sodium equivalents are calculated according to Eq. 4a. Assuming the PCR contains 2.5 mM Mg^{2+}, the sodium equivalent is 2.5 mM×140=350 mM=0.35 M. Then, ΔS for a given sodium equivalent is calculated according to Eq. 5. For a 18-bp duplex, N=17 and ΔS^0 [Na]=−380.9 cal K^{-1} mol^{-1}+0.368×17×ln[0.35]=−387.5 cal K^{-1} mol^{-1}. Finally, T_m is calculated according to Eq. 1:

$$T_m = \frac{-141800 \; cal/mol}{-387.5 \; cal/K \times mol + 1.987 \; cal/K \times mol \times \ln\left(\frac{0.1 \times 10^{-6}}{4}\right)} - 273.15 = 62.7°C$$

Because of the different units, ΔH is multiplied by 1000 to give cal/mol. The concentration of the probe is 0.1 μM=0.1×10⁻⁶ M and this is divided by 4 because the sequence is non-self-complementary. Finally, subtract 273.15 to convert from Kelvin to Celsius.

Calculation of T_m for an 18mer with Mismatches

As an example, the already known 18mer 5'-ACG ATG GCA GTA GCA TGC-3' oligonucleotide as above, is used and a G→A mutation at the third base is assumed. Hybridization of the wild type probe with the mutated strand leads to the following duplex with a G-T mismatch in the trimer CGA/GTT:

```
5'-ACG ATG GCA GTA GCA TGC-3'
3'-TGT TAC CGT CAT CGT ACG-5'
```

The oligonucleotide example has 15 Watson-Crick paired doublets (Table 1) and 2 mismatched doublets (Table 2). These are AC/TG, CG/GT, GA/TT, AT/TA, TG/AC, GG/CC, GC/CG and so on. The number of different doublets are:
 0 AA/TT, 2 AT/TA, 1 TA/AT, 4 CA/GT, 2 GT/CA, 1 CT/GA, 1 GA/CT, 0 CG/GC, 3 GC/CG, 1 GG/CC.

Table 2. Parameters for mismatched oligonucleotide ΔH⁰ and ΔS⁰ in 1 M NaCl. References are given in the table. Nearest neighbor dimer duplexes are presented with a slash separating strands in antiparallel orientation, e.g., AT/TG means 5'-AT-3' mismatched with 3'-TG-5'. Mismatches are in the second position of the listed doublet except for double mismatches. These parameters are not valid for the calculation of mismatches in the terminal or penultimate position

Sequence	ΔH⁰ kcal/mol	ΔS⁰ cal K⁻¹ mol⁻¹
Internal G-A mismatches [11]		
AA/TG	−0.6	−2.3
AG/TA	−0.7	−2.3
CA/GG	−0.7	−2.3
CG/GA	−4.0	−13.2
GA/CG	−0.6	−1.0
GG/CA	0.5	3.2
TA/AG	0.7	0.7
TG/AA	3.0	7.4
Internal C-T mismatches [10]		
AC/TT	0.7	0.2
AT/TC	−1.2	−6.2
CC/GT	−0.8	−4.5
CT/GC	−1.5	−6.1
GC/CT	2.3	5.4
GT/CC	5.2	13.5
TC/AT	1.2	0.7
TT/AC	1.0	0.7

Sequence	ΔH⁰ kcal/mol	ΔS⁰ cal K⁻¹ mol⁻¹
Internal A-C mismatches [9]		
AA/TC	2.3	4.6
AC/TA	5.3	14.6
CA/GC	1.9	3.7
CC/GA	0.6	−0.6
GA/CC	5.2	14.2
GC/CA	−0.7	−3.8
TA/AC	3.4	8.0
TC/AA	7.6	20.2
Internal G-T mismatches [4]		
AG/TT	1.0	0.9
AT/TG	−2.5	−8.3
CG/GT	−4.1	−11.7
CT/GG	−2.8	−8.0
GG/CT	3.3	10.4
GT/CG	−4.4	−12.3
TG/AT	−0.1	−1.7
TT/AG	−1.3	−5.3
Internal A-A mismatches [12]		
AA/TA	1.2	1.7
CA/GA	−0.9	−4.2
GA/CA	−2.9	−9.8
TA/AA	4.7	12.9
Internal C-C mismatches [12]		
AC/TC	0.0	−4.4
CC/GC	−1.5	−7.2
GC/CC	3.6	8.9
TC/AC	6.1	16.4
Internal G-G mismatches [12]		
AG/TG	−3.1	−9.5
CG/GG	−4.9	−15.3
GG/CG	−6.0	−15.8
TG/AG	1.6	3.6
Internal T-T mismatches [12]		
AT/TT	−2.7	−10.8
CT/GT	−5.0	−15.8
GT/CT	−2.2	−8.4
TT/AT	0.2	−1.5
Double mismatches [4]		
GG/TT	5.8	16.3
GT/TG	4.1	9.5
TG/GT	−1.4	−6.2

The two doublets affected by the mutation are: CG/GT and GA/TT. Each of these doublets occurs once.

The ΔH^0 calculation is according to Eq. 2.

For the Watson-Crick paired bases $\Sigma_i\ n_i\ \Delta H^0(i)$ results in 0×(−7.9)+2× (−7.2)+1×(−7.2)+4×(−8.5)+2×(−8.4)+2×(−7.8)+0×(−8.2)+0×(−10.6)+3×(−9.8)+1

×(−8.0)=−125.4 kcal/mol. Two doublets are affected by mismatches: for the CG/GT doublet, ΔH^0=−4.1 kcal/mol, and for the GA/TT (TT/GA in Table 2) ΔH^0=−1.3 kcal/mol.

The initiation terms are 0.1 kcal/mol for the terminal G-C and 2.3 kcal/mol for the terminal A-T. Therefore, ΔH^0_{total}=−125.4−4.1−1.3+0.1+2.3=−128.4 kcal/mol. The ΔS^0 calculation is according to Eq. 3.

For the Watson-Crick paired bases $\Sigma_i\ n_i\ \Delta S^0(i)$ results in 0×(−22.2)+2×(−20.4)+1×(−21.3)+4×(−22.7)+2×(−22.4)+2×(−21.0)+0×(−22.2)+0×(−27.2)+3×(−24.4)+1×(−19.9)=−332.8 cal K^{-1} mol^{-1}. Two doublets are affected by mismatches and their thermodynamic properties are again found in Table 2. For the CG/GT doublet ΔS^0=−11.7 kcal K^{-1} mol^{-1}, for GA/TT (look up TT/GA in Table 2, this is the same as GA/TT) ΔS^0=−5.3 kcal K^{-1} mol^{-1}.

The initiation terms are −2.8 cal K^{-1} mol^{-1} for the terminal G-C and 4.1 cal K^{-1} mol^{-1} for the terminal A-T, total+1.3 cal K^{-1} mol^{-1}.

ΔS^0_{total}=−348.5 cal K^{-1} mol^{-1}

The sodium equivalents are calculated according to Eq. 4a. With 2.5 mM Mg^{2+}, the sodium equivalent is 2.5 mM×140=350 mM=0.35 M. Then, ΔS for a given sodium equivalent is calculated according to Eq. 5.

For a 18-bp duplex, N=17 and ΔS^0 [Na]=−348.5 cal K^{-1} mol^{-1}+0.368×17×ln[0.35]=−355.1 cal K^{-1} mol^{-1}.

T_m is calculated according to Eq. 1:

$$T_m = \frac{-128400\ cal/mol}{-355.1\ cal/K \times mol + 1.987\ cal/K \times mol \times \ln\left(\frac{0.1 \times 10^{-6}}{4}\right)} - 273.15 = 56.2°C$$

Again, ΔH is multiplied by 1000 to change to cal/mol and 273.15 is subtracted to convert from Kelvin to Celsius. The concentration of the probe is 0.1 μM= 0.1×10^{-6} M and is divided by 4 because the sequence is non-self-complementary.

Calculation of ΔT_m for a Hybridization Probe Mutation Detection Assay

In the first example, we have calculated a T_m of 62.7°C for an 18mer probe. In the second calculation, a lower T_m of 56.2°C was calculated due to a single mismatch caused by a G→A mutation. When the probe hybridizes to the mutated strand, a calculated reduction in T_m of 6.5°C results.

Limitations of the Model

The thermodynamic calculations assume that the hybridization occurs at pH 7.0. This is not the case in standard PCR buffer systems, but the resulting error seems to be small. It is more important to note that mismatches in the terminal and penultimate positions are not properly reflected by the parameters in Table 2 [4, 12]. The thermodynamic data set for tandem mismatches is still incomplete. A reliable prediction is only possible for the few tandem mismatches listed in Table 2. Thermodynamic calculations are impossible if probe hybridization leads to the formation of bulges or loops because ΔH and ΔS changes of such secondary structures are as yet unpublished. The influence of the salt concentration on oligonucleotide stability is also still under investigation [6, 7, 15]. Nevertheless, the method presented here will allow a reliable estimate of hybridization probe T_m on the LightCycler, but may require correction of the sodium equivalents according to individual PCR conditions. For the T_m calculation of polymer DNA, formulas are available that are easier to use [1]. We have calculated the T_m of seven PCR amplicons up to 250 bp and compared the observed T_m with those from the polymer DNA formula (Eq. 1a in [1]) or the n-n model. The mean difference of observed versus predicted T_m was 1.1±0.70°C using the polymer formula, and 0.9±0.46°C using the n-n model.

Results

Validation of the Nearest Neighbor Model on the LightCycler

We have incorporated the formulas described above into a Microsoft Excel-spreadsheet which is capable of calculating the thermodynamic data for matched and mismatched oligonucleotides under various concentrations and ionic strengths [18]. This program is available for download via http://www.meltcalc.de. A total of 89 melting points resulting from 12 different published [14, 19–30] and 12 unpublished (N. von Ahsen, data on file) hybridization probe assays correlated with the predictions from the n-n model (Fig. 2). Eq. 4b was used to calculate the Na$^+$ equivalents in Fig. 2, because it gave a better fit (n=89, y=1.07x−3.95, $S_{y|x}$ 2.04, r=0.959) than Eq. 4a (n=89, y=1.12x−6.09, $S_{y|x}$ 2.52, r=0.943) when published T_m data from various laboratories were used. If only the data from our own laboratory were used, we found a better overall model fit and not much difference between Eq. 4b (n=54, y=1.01x−1.07, $S_{y|x}$ 1.54, r=0.964) and Eq. 4a (n=54, y=1.01x−1.53, $S_{y|x}$ 1.53, r=0.964). It is difficult to make a comparison with published data, because certain technical aspects are not always mentioned in the publications. These include the ramp rates used for melting curve acquisition and the report of average T_m data versus those of single experiments.

Sensitivity and Specificity of Hybridization Probe Genotyping

We have developed genotyping assays for the apolipoprotein B-3500 and factor V Leiden mutation. For both loci, different mutations have been described that occur under the detection probe. These are the G10699A and C10698T mutations of apolipoprotein B [31] and the G1691A, A1692C, and G1689A mutations for factor V [24, 32]. Increasing knowledge of the prevalence of single nucleotide polymorphisms in the human genome [33, 34] raises the important question about the sensitivity of hybridization probe assays towards various

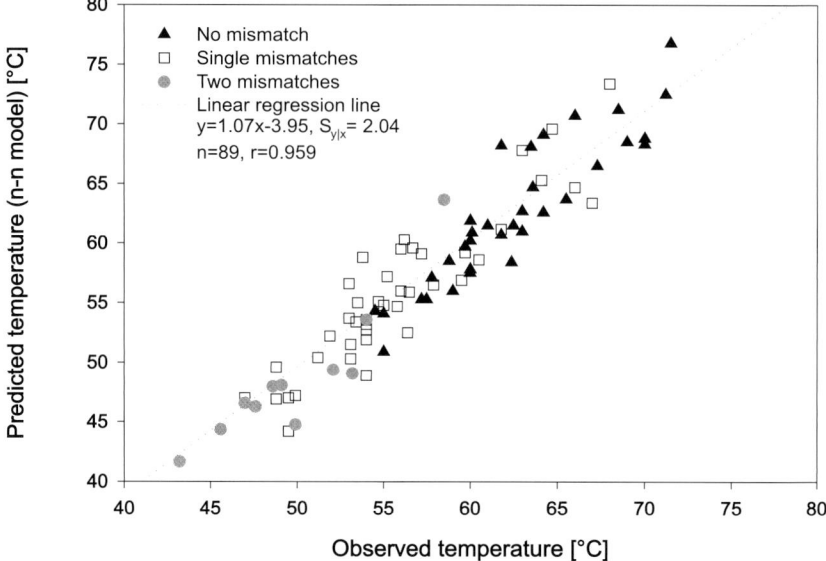

Fig. 2. Experimentally observed melting points versus melting points calculated with the nearest neighbor model. The linear regression line is shown, with its standard error of the estimate ($S_{y|x}$) for all samples in the plot. The mean difference for observed versus predicted temperatures is 0.2± 2.1°C. Note the predictive accuracy, no matter whether none, one or two mismatches are present under the probe

mutations under a probe. Melting point calculations for various possible mutations under a 19mer probe suggest that no mutation under a properly designed probe will remained undetected. However, there are some mutations that destabilize the probe binding to the same extent and will not be detected [14]. Figure 3 shows how 19mer probes with a variable GC content are destabilized by a single base mutation. Melting point calculations were performed for all possible 3×10^6 mismatches (excluding ultimate and penultimate positions) under 60,000 randomly chosen 19mers. For this purpose, the sequence of a starting probe was randomly altered by exchanging nucleotides using a computer program. The randomizer algorithm was programmed to produce a Gaussian shaped distribution with respect to probe GC content and a preponderance of probes with 50–60% of GC. Only 0.055% of mismatches will cause a melting shift lower than 1.25°C. The most important trend is that probes with higher GC content show less T_m shift due to a single mutation. Another principal point is that shorter probes show higher T_m shifts because the destabilizing impact of the mutation on the total stability is higher in shorter duplexes as evident from the n-n model.

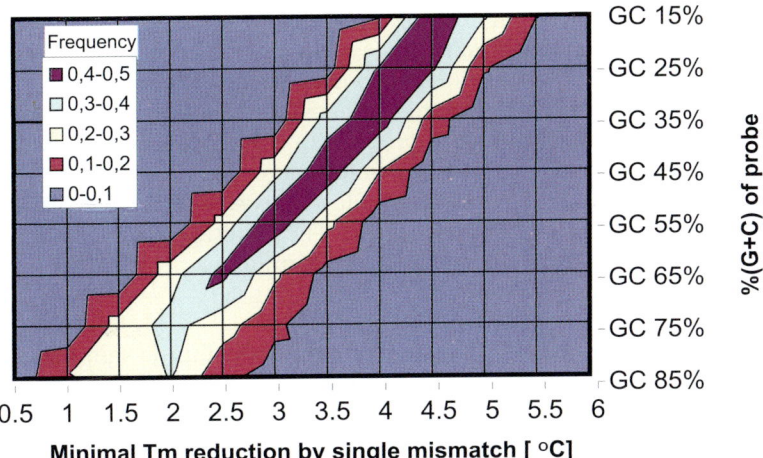

Fig. 3. The interdependence of the melting temperature shift of 19mer probes versus their GC content is shown with a contour plot. The various frequencies are shown in *colors*. There is a trend for higher destabilization in probes with a lower GC content. Melting point calculations were performed for all possible 3×10^6 mismatches (excluding ultimate and penultimate positions) under 60,000 randomly chosen 19mers (see text)

Comments

The thermodynamic n-n model provides the theoretical basis for the stability of matched and mismatched oligonucleotides. The comparison of experimentally derived melting points with those calculated from the n-n model shows very good agreement. Using a detection probe that has been properly designed according to this model, we speculate that the detection of every point mutation is possible. However, various mutations under a probe may result in the same T_m reduction.

An important factor for the design of hybridization probes is the absence of significant self- or cross-dimerization of probes and amplification primers. The gap between the fluorophores should be as small as possible since the FRET efficiency is inversely proportional to the sixth power of the distance between the donor and the acceptor dye [35]. Nucleotide gaps between one and five bases give good results [2]. For quantification, equal probe melting points should be used and there should not be any false priming sites on the amplicon [2]. For mutation detection, the detection probe should have the maximum T_m shift for the investigated mutation (see below) and the anchor probe should have a T_m approximately 10% higher than that of the detection probe. Probe placement is uncritical if the investigated mutation is of the deletion/insertion type. This type of polymorphism leads to bulge loop formation of the non-complementary base within the DNA duplex. The destabilization of the n-n doublet affected by the bulge formation will in most cases be high enough for easy detection.

Provided it is not in the three outer positions, the place where the mutation is located under the probe is not important. The resulting melting point shift will be affected by the type of mutation and the neighboring bases but not significantly by its position [12]. The most common mutations are G→A mutations [36]. However, the resulting G:T mismatch under a probe with the sequence of the sense strand is among the most stable mismatches which can occur, together with G:A and G:G mismatches [11]. It is therefore advisable in this case to use a detection probe with the sequence of the antisense strand. The resulting complementary C:A mismatch has a higher destabilizing effect and is more readily detected.

Note Added in Proof:
The accuracy of the prediction of DNA double strand stability is significantly improved by the inclusion of dangling end parameters. This was recently shown in a paper from SantaLucia's group which includes the 32 dangling end parameters to perform these calculations (S. Bommarito, N. Peyret, and J. SantaLucia, Jr. Thermodynamic parameters for DNA sequences with dangling ends. Nucleic Acids Res 38:1929–1934, 2000).

References

1. Wetmur JG (1991) DNA probes: applications of the principles of nucleic acid hybridization. Crit Rev Biochem Mol Biol 26: 227–259
2. Caplin BE, Rasmussen A, Bernard PS, Wittwer CT (1999) LightCycler hybridization probes. Biochemica 1: 5–8
3. Borer PN, Dengler B, Tinoco I, Uhlenbeck OC (1974) Stability of ribonucleic acid double-stranded helices. J Mol Biol 86: 843–853
4. Allawi HT, SantaLucia J, Jr. (1997) Thermodynamics and NMR of internal G.T mismatches in DNA. Biochemistry 36: 10581–10594
5. Gray DM (1997) Derivation of nearest-neighbor properties from data on nucleic acid oligomers. I. Simple sets of independent sequences and the influence of absent nearest neighbors. Biopolymers 42: 783–793
6. SantaLucia J, Jr. (1998) A unified view of polymer, dumbbell, and oligonucleotide DNA nearest- neighbor thermodynamics. Proc Natl Acad Sci U S A 95: 1460–1465
7. SantaLucia J, Jr., Allawi HT, Seneviratne PA (1996) Improved nearest-neighbor parameters for predicting DNA duplex stability. Biochemistry 35: 3555–3562
8. Sugimoto N, Nakano S, Yoneyama M, Honda K (1996) Improved thermodynamic parameters and helix initiation factor to predict stability of DNA duplexes. Nucleic Acids Res 24: 4501–4505
9. Allawi HT, SantaLucia J, Jr. (1998) Nearest-neighbor thermodynamics of internal A.C mismatches in DNA: sequence dependence and pH effects. Biochemistry 37: 9435–9444
10. Allawi HT, SantaLucia J, Jr. (1998) Thermodynamics of internal C.T mismatches in DNA. Nucleic Acids Res 26: 2694–2701
11. Allawi HT, SantaLucia J, Jr. (1998) Nearest neighbor thermodynamic parameters for internal G.A mismatches in DNA. Biochemistry 37: 2170–2179
12. Peyret N, Seneviratne PA, Allawi HT, SantaLucia J, Jr. (1999) Nearest-neighbor thermodynamics and NMR of DNA sequences with internal A.A, C.C, G.G, and T.T mismatches. Biochemistry 38: 3468–3477
13. Owczarzy R, Vallone PM, Gallo FJ, Paner TM, Lane MJ, Benight AS (1997) Predicting sequence-dependent melting stability of short duplex DNA oligomers. Biopolymers 44: 217–239
14. von Ahsen N, Oellerich M, Armstrong VW, Schütz E (1999) Application of a thermodynamic nearest-neighbor model to estimate nucleic acid stability and optimize probe

design: Prediction of melting points of different mutations of apolipoprotein B 3500 and factor V Leiden with a hybridization probe genotyping assay on the LightCycler. Clin Chem 45: 2094–2101
15. Nakano S, Fujimoto M, Hara H, Sugimoto N (1999) Nucleic acid duplex stability: influence of base composition on cation effects. Nucleic Acids Res 27: 2957–2965
16. Chester N, Marshak DR (1993) Dimethyl sulfoxide-mediated primer Tm reduction: a method for analyzing the role of renaturation temperature in the polymerase chain reaction. Anal Biochem 209: 284–290
17. Musielski H, Mann W, Laue R, Michel S (1981) Influence of dimethylsulfoxide on transcription by bacteriophage T3- induced RNA polymerase. Z Allg Mikrobiol 21: 447–456
18. Schütz E, von Ahsen N (1999) Spreadsheet software for thermodynamic melting point prediction of oligonucleotide hybridization with and without mismatches. Biotechniques 27: 1218–1224
19. Bernard PS, Lay MJ, Wittwer CT (1998) Integrated amplification and detection of the C677 T point mutation in the methylenetetrahydrofolate reductase gene by fluorescence resonance energy transfer and probe melting curves. Anal Biochem 255: 101–107
20. von Ahsen N, Oellerich M, Schütz E (2000) Using two reporter dyes without interference in a single tube rapid cycle PCR: alpha1-antitrypsin genotyping by multiplex real time fluorescence PCR with the LightCycler. Clin Chem 46: 156–161
21. Bernard PS, Ajioka RS, Kushner JP, Wittwer CT (1998) Homogeneous multiplex genotyping of hemochromatosis mutations with fluorescent hybridization probes. Am J Pathol 153: 1055–1061
22. Bernard PS, Pritham GH, Wittwer CT (1999) Color multiplexing hybridization probes using the apolipoprotein E locus as a model system for genotyping. Anal Biochem 273: 221–228
23. Lay MJ, Wittwer CT (1997) Real-time fluorescence genotyping of factor V Leiden during rapid-cycle PCR. Clin Chem 43: 2262–2267
24. Lyon E, Millson A, Phan T, Wittwer CT (1998) Detection and Identification of Base Alterations Within the Region of Factor V Leiden by Fluorescent Melting Curves. Mol Diagn 3: 203–209
25. Aslanidis C, Schmitz G (1999) High-speed apolipoprotein E genotyping and apolipoprotein B3500 mutation detection using real-time fluorescence PCR and melting curves. Clin Chem 45: 1094–1097
26. Aslanidis C, Nauck M, Schmitz G (1999) High-speed prothrombin G–>A 20210 and methylenetetrahydrofolate reductase C–>T 677 mutation detection using real-time fluorescence PCR and melting curves. Biotechniques 27: 234–6, 238
27. Gaffney D, Reid JM, Cameron IM, Vass K, Caslake MJ, Shepherd J, Packard CJ (1995) Independent mutations at codon 3500 of the apolipoprotein B gene are associated with hyperlipidemia. Arterioscler Thromb Vasc Biol 15: 1025–1029
28. von Ahsen N, Schütz E, Armstrong VW, Oellerich M (1999) Rapid detection of prothrombotic mutations of prothrombin (G20210 A), factor V (G1691 A) and methylenetetrahydrofolate reductase (C677 T) by real time fluorescence PCR with the LightCycler. Clin Chem 45: 694–696
29. Nauck M, Wieland H, März W (1999) Rapid, Homogeneous Genotyping of the 4G/5G Polymorphism in the Promoter Region of the PAI1 Gene by Fluorescence Resonance Energy Transfer and Probe Melting Curves. Clin Chem 45: 1141–1147
30. Nauck MS, Gierens H, Nauck MA, Marz W, Wieland H (1999) Rapid genotyping of human platelet antigen 1 (HPA-1) with fluorophore- labelled hybridization probes on the LightCycler. Br J Haematol 105: 803–810
31. Tybjaerg-Hansen A, Steffensen R, Meinertz H, Schnohr P, Nordestgaard BG (1998) Association of mutations in the apolipoprotein B gene with hypercholesterolemia and the risk of ischemic heart disease. N Engl J Med 338: 1577–1584
32. Liebman HA, Sutherland D, Bacon R, McGehee W (1996) Evaluation of a tissue factor dependent factor V assay to detect factor V Leiden: demonstration of high sensitivity and specificity for a generally applicable assay for activated protein C resistance. Br J Haematol 95: 550–553
33. Cargill M, Altshuler D, Ireland J, Sklar P, Ardlie K, Patil N, Lane CR, Lim EP, Kalayanaraman N, Nemesh J, Ziaugra L, Friedland L, Rolfe A, Warrington J, Lipshutz R, Daley GQ, Lander ES

(1999) Characterization of single-nucleotide polymorphisms in coding regions of human genes. Nat Genet 22: 231–238
34. Brown SM (1999) Snapping up SNPs. Biotechniques 26: 1090–1093
35. Clegg RM (1995) Fluorescence resonance energy transfer. Curr Opin Biotechnol 6: 103–110
36. Cooper,D.N., Krawczak,M., and Antonarakis,S.E. (1995): The nature and mechanism of human gene mutation. In: Scriver CR et al (eds) The metabolic and molecular bases of inherited disease pp. 259–291. McGraw-Hill, New York

Quantification of Human Papilloma Virus Type 16 Using Quantitative Competitive PCR on the LightCycler

Brian Erich Caplin*

Introduction

Human papillomavirus type 16 (HPV 16) is one of the most common high-risk serotypes of human papilloma viruses. Clinical samples containing HPV 16 typically have approximately 1300–53,000 viral genome equivalents per cell [1]; a 40-fold range of viral load. Accurate quantification over this range of initial copy numbers can easily be achieved with several dilutions of the unknown samples and multiple quantification reactions. However, a cost effective means of applying PCR to diagnostic work requires a broad dynamic-range for the assay.

The broadest dynamic-range for quantitative PCR can be produced by using external standards. Unfortunately, the results from external standard reactions are difficult to interpret when unknown samples appear negative. Internal reaction controls of quantitative competitive PCR (QC-PCR) provide a convenient means of ensuring that each PCR sample is amplifying.

Performing QC-PCR on the LightCycler is simplified by the real-time detection of the accumulating PCR product. Real-time detection during PCR provides a rapid method for the determination of crossing threshold for both the competitor and the target amplicon.

Described here is a method for quantifying HPV-16 virus, using hybridization probes with two acceptor colors, one for the target (HPV 16) and one for the competitor. The method described uses internal reaction controls and competitive PCR. The dynamic range of the assay covers a 50-fold target-to-competitor ratio.

Materials

Agarose Gel DNA Extraction Columns (Amicon, Beverly, MA) **Equipment**
LightCycler Instrument (Roche Molecular Biochemicals, Indianapolis, IN)

Plasmid DNA purification kit (5′-3′ Inc., Boulder, CO) **Reagents**
KlenTaq (AB Peptides, St. Louis, MO)
TaqStart (Clontech, Inc. Palo Alto, CA)

---

* Brian Erich Caplin (✉) (e-mail: brianc@idahotech.com)
  390 Wakara Way, Salt Lake City, Utah 84108, USA

Oligonucleotides (IT Biochem, Salt Lake City, UT)
Restriction Endonucleases (Roche Molecular Biochemicals, Indianapolis, IN)
10× Reaction Buffer (Idaho Technology, Inc., Salt Lake City, UT) [500 mM Tris pH 8.3 (25°C), 40 or 32.5 mM $MgCl_2$, 2.5 mg/ml BSA]
1× Enzyme diluent (Idaho Technology, Inc., Salt Lake City, UT) [10 mM Tris pH 8.3 (25°C), 2.5 mg/ml BSA]

Procedure

Sample Preparation

Plasmid pUC18 DNA was purified from *E. coli* DH5αa cells. Plasmid DNA was restriction endonuclease digested with Eco RI and Bam HI enzymes. Digested DNA was purified by agarose gel electrophoresis and DNA gel extraction columns. HPV-16 E1 gene nucleotides 862–1271 were amplified with Eco RI and Bam HI tailed primers (Set 1) and subcloned into the plasmid.

Mutagenic primers were used to scramble a 27 base pair region of the HPV-16 E1 gene, to produce a competitor template (Set 2). The competitor template was also subcloned using Eco RI and Bam HI tailed primers (Set 1). To ensure high purity of the DNA, both plasmids were purified from *E. coli*, and super-coiled plasmid DNA was extracted from the agarose gel. Excised gel slices were electroeluted to produce the final purified DNA template. The initial copy number for each template was determined by spectrophotometry as well as limiting dilution assays. A detailed description of this method and analysis software is available at the following web site: http://ubik.microbiol.washington.edu/cbu/quality/jquality.html.[2]

Primer Design

Primers for the amplification of the HPV-16 E1 gene were designed with the use of DNAstar- PrimerSelect software. The primers were tested versus the BLAST algorithm to ensure that combined, these primers would only amplify HPV 16 DNA (Table 1).

Green I Amplification: Plasmid Inserts and Competitor Template

SYBR Green I Master Mix for each 20 µl reaction:

	Volume [µl]	[Final]
10X 40 mM $MgCl_2$ Reaction Buffer	2	4.0 mM
dNTPs (8.0 mM)	2	0.8 mM
Taq DNA polymerase	2	0.2 units
Forward Primer (1.0 µM)	2	0.1 µM
Reverse Primer (1.0 µM)	2	0.1 µM
H_2O (PCR Grade)	8	
Total volume	18	

In all, 18 µl of reaction mixture and 2 µl of either HPV 16 or Competitor DNA were added to each capillary. Sealed capillaries were centrifuged in a Microcentrifuge and placed into the LightCycler rotor.

Table 1. Oligonucleotides

Human papilloma virus 16 (Genebank Accession # K02718)					
	Position	DNA	Length	GC (%)	T_m (°C)
Set 1					
GGGGATCCACTTCAGTATTGC		BamHI HPV16	21	52.4	63.4
GGGAATTCCATGGCTGATCC-TGCAGG		EcoRI HPV16	26	57.7	71.9
Product	862-1271	Tailed Insert	422		
Set 2					
GATCCTGCAGGTACCGATGGA-TAGTGAGCGAGAGATAGGTAG-GGATGGTTTTATGTAG	870	HPV 16	58	48.8	78.9
CCACTTCAGTATTGCCATACCC	1271R	HPV 16	22	50.0	63.6
Product	870-1271	Competitor	401		
Set 3					
CCATGGCTGATCCTGCAGGTAC	862	HPV 16 and Competitor	22	59.1	67.8
CCACTTCAGTATTGCCATACCC	1271 R	HPV 16 and Competitor	22	50.0	62.1
Product	862-1271	Quantified Products	409		
Hybridization Probes					
CTCGTCATCTGATATAGCATCCC-CTGTTTTTTTTTCCACTACAGCC-TCTACATAAAACC-F	916	HPV 16 and Competitor	58	39.7	74.7
LCRed705-ATTACATCCCGTACCC-TCTTCCCCATT-P	885	HPV 16	27	48.1	68.1
LCRed640-CTACCTATCTCTCGCT-CACTATCCATC-P	885	Competitor	27	48.1	64.4

The following PCR amplification protocol was used for SYBR Green I amplification of the plasmid inserts using primers from Set 1, and the competitor amplicon, using primers from Set 2:

Parameter	Value			
Cycles	30			
Type	Quantification			
	Segment 1	Segment 2	Segment 3	
Incubation time [s]	0	2	18	
Temperature transition rate [°C/s]	20	20	10	
Acquisition mode	None	None	Single	
Gains	F1=1	F2=8	F3=22	

Quantitative Competitive PCR with Hybridization Probes:

Hybridization Probe Master Mix for each 20-μl Reaction:

	Volume [μl]	[Final]
10X 32.5 mM MgCl$_2$ Reaction Buffer	2	3.25 mM
dNTPs (8.0 mM)	2	0.8 mM
KlenTaq DNA Polymerase	2	0.4 units
Forward Primer (5.0 μM)	2	0.5 μM
Reverse Primer (1.25 μM)	2	0.125 μM
Fluorescein Probe (3.0 μM)	2	0.3 μM
LCRed640Probe (1.0 μM)	2	0.1 μM
LCRed705Probe (1.0 μM)	2	0.1 μM
H$_2$O (PCR Grade)	2	
Total volume	18	

For hot start, premix the KlenTaq DNA polymerase with TaqStart Antibody in a 1-to-1 mixture. Incubate at room temperature prior to dilution of the enzyme. Add polymerase to reaction as usual.

In all, 18 μl of reaction mixture and 1 μl of HPV-16 template, and 1 μl of competitor template were added to each capillary. HPV-16 template was held constant at a single initial copy number, and the competitor template was varied over a range of 100-fold greater to 100-fold less than the target initial copy number. Sealed capillaries were centrifuged in a Microcentrifuge and placed into the LightCycler rotor.

The following PCR protocol was used for the hybridization probe detection during QC-PCR using primers from Set 3:

- Denaturation for 15 s at 95°C.
- Amplification with LightCycler Hybridization Probes:

Parameter	Value			
Cycles	35			
Type	Quantification			
	Segment 1	Segment 2	Segment 3	Segment 4
Target temperature [°C]	92	47	51	78
Incubation time [s]	0	0	6	10
Temperature transition rate [°C/s]	20	20	0.4	10
Acquisition mode	None	None	Single	None
Gains		F1=1	F2=8	F3=22

Data Analysis of Two-Color Competitive PCR

A color-compensation file was created according to the instructions provided in the LightCycler software. The PCR results were color compensated and analyzed, using the fit points method in the Quantification Module of the LightCycler Data Analysis (LCDA) software that is provided with the instrument.

The LCDA software produces the crossing threshold data for both the competitor and target. These results were transferred to a spreadsheet file for further analysis.

Derivation of quantification based on the difference in crossing thresholds has been previously reported.[3]

Results

Comparison of competitor and target amplification

To ensure that the amplification of the competitor is comparable with that of the HPV-16 target four criteria needed to be met:
- Competitor and target probes should have high specificity.
- Amplicon detection sensitivity should be equivalent for both probes.
- Crossing thresholds of equal dilutions for each template should be equivalent.
- Detection of the target and competitor must occur with the same reaction conditions.

Amplification of both the competitor and the HPV-16 target templates is with the same primer set, and these templates differ by only 27 scrambled base pairs. The 27 scrambled base pairs have the same G::C content, and the LCRed labeled detection probes have similar predicted T_ms.

Figure 1 demonstrates the crossing threshold comparison between the competitor detected with the LCRed640 probe and the target detected with the LCRed705 probe. The two detection probes do not demonstrate any cross-reactivity (data not shown). Both templates can be detected over the same range of

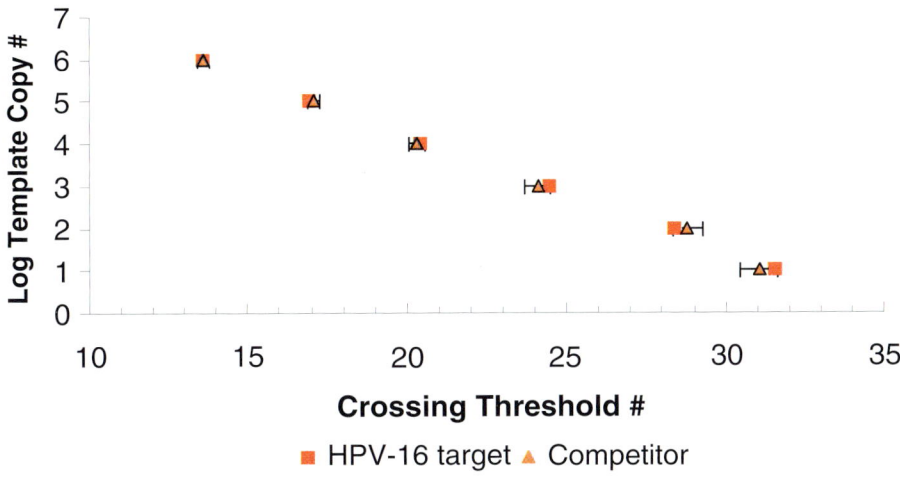

Fig. 1. Crossing thresholds of competitor and target templates. Ten-fold serial dilutions, a range of 1×10^1 to 1×10^6, of either competitor template or HPV-16 target template were amplified. Only the hybridization probes specific to the template being amplified was included with each reaction. Reaction efficiencies were determined for each template/probe system. Both the target and the competitor were found to be amplified and detected with similar efficiencies. The 95% confidence intervals are presented as x-axis error bars

initial template copy number. The reaction efficiency is determined by the slope of the lines produced from the crossing threshold data. Both the target and the competitor have a reaction efficiency of 1.9, as calculated from the following equation. Efficiency $=10^{-slope}$

Crossing Threshold Analysis

Under circumstances where two templates have equivalent reaction efficiencies, the crossing threshold for one template will be earlier in the reaction if it has a higher initial copy number than the second template. The shift in crossing thresholds and the quantification of this shift is the basis for quantitative-competitive PCR [3].

A method for quantifying the difference between the two crossing thresholds was devised which was based on previous QC-PCR analyses. A linear plot of the data can be represented by the equation:

$$\text{Log } C_0 = \log E(\Delta C.T.) + \log T_0$$

Where C_0 is the initial competitor copy number, T_0 is the initial HPV-16 target copy number, E is the reaction efficiency, and $\Delta C.T.$ is the shift in crossing threshold between the competitor and the HPV-16 target. Log T_0 represents the y-intercept of the line produced by this equation (Fig. 2), and is the value of the HPV-16 target initial copy number. The linear fit to these data produce results that are within 30% of the actual initial copy number.

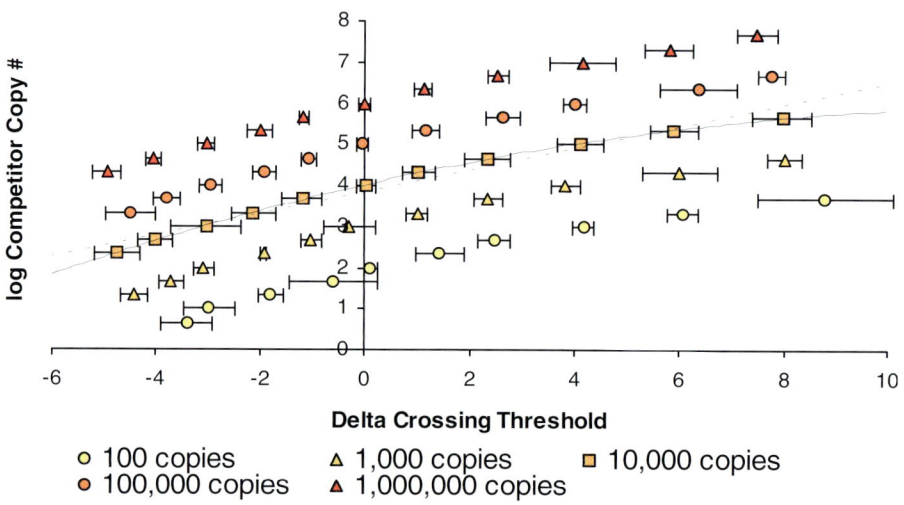

Fig. 2. Dynamic range of the Δ C.T. with linear and non-linear analysis. Both linear and non-linear "best-fit" analysis are shown for the 1×10^4 HPV-16 target reactions. Competitor-to-target ratio ranged from 50:1 to 0.02:1, ratios were varied in 1/3 log steps. Copy numbers are for the HPV-16 target DNA templates as determined by limiting dilution assay. Analysis of the two fit methods is presented in table 2

The linear "best-fit" line (Fig. 2, *dashed lined*) consistently underestimates the amount of HPV-16 initial template copy number. A quadratic equation, however, produces a better fit to the data (Fig. 2, *solid line*). In Table 2 the two methods of data fitting are compared. A quadratic fit to the data in Fig. 2 consistently results in better accuracy. A non-linear analysis of the PCR allows for a broader range of template-to-competitor ratios. Non-linear relations between competitor and target during the PCR has been considered previously [4].

Maximizing the Dynamic-Range of the Assay

Comments

QC-PCR relies on there being minimal differences between the target and competitor templates; closely matched amplicons, using the same primers, will behave similarly during PCR. To effectively perform QC-PCR, it must be possible to distinguish the competitor from the target. Thus, QC-PCR with post reaction analysis requires either complex competitor designs or complex detection methods.

Template and Probe Design

Real-time PCR and hybridization probes are a logical tool for distinguishing the competitor from the target. The difficulty here is not in designing the template specific probes, but rather in designing two template specific detection probes with similar thermodynamic characteristics. T_m calculators specific to the LightCycler (T_m Utility from Idaho Technology, Inc.) have recently, become available. This reduces the challenge of this task. Once similar hybridization probes are selected for QC-PCR, one must simply apply the standard LightCycler hybridization probe criteria, as previously described [5].

The preparation and purification of the template DNA for use in these studies proved to be one of the most time consuming efforts of the entire method. The plasmid DNA was purified, using several methods in tandem to ensure the high-

Sample Preparation and Template Quantification

Table 2. Linear and Quadratic analysis of the data in Figure 2 is presented for comparison of the two methods of data analysis. Accuracy (%) is the variation of the calculated HPV-16 target copy number from the actual target copy number

HPV-16 copy number	Equation	Linear Fit (R^2)	Quadratic Fit (R^2)	Accuracy (%)
10^6	y = 0.269x + 5.86	0.985		71.8
	y = -0.009x^2 + 0.290x + 5.99		0.999	96.9
10^5	y = 0.266x + 4.81	0.974		65.2
	y = -0.012x^2 + 0.303x + 4.98		0.999	95.3
10^4	y = 0.263x + 3.85	0.975		70.8
	y = -0.011x^2 + 0.295x + 4.01		0.999	96.8
10^3	y = 0.267x + 2.83	0.965		67.6
	y = -0.014x^2 + 0.312x + 3.02		0.999	94.9
10^2	y = 0.248x + 1.81	0.963		64.7
	y = -0.013x^2 + 0.309x + 1.93		0.997	85.7

est purity that could be achieved without introducing large quantities of various salts, which may interfere with the PCR.

Following purification of the template DNA, a reliable determination of the initial copy number was needed. Using standard spectrophotometric methods, the two templates produced results similar to those shown in Fig. 1. However, the observed crossing thresholds did not correspond with the calculated reaction efficiencies. Using a limiting dilution assay to calculate the initial copy number of each template, produced results that were more consistent with the observed crossing thresholds. This method, although time consuming, proved to be an effective method for determining the starting copy numbers of each template: target and competitor.

References

1. Wickenden C, Malcolm A, Byrne M, Smith C, Anderson M, Coleman D (1987) Prevalence of HPV DNA and viral copy numbers in cervical scrapes from women with normal and abnormal cervices. J Pathol 153: 127–135
2. Rodrigo A, Goracke P, Rowhanian K, Mullins J (1997) Quantitation of target molecules from polymerase chain reaction-based limiting dilution assays. Aids Res Hum Retroviruses 13: 737–742
3. Fink L, Seefer W, Ermert L, Stahl U, Grimminger F, Kummer W, Bohle R (1998) Real-time quantitative RT-PCR after laser-assisted cell picking. Nat Med 4: 1329–1333
4. Rawymaekers L, (1993) Quantitative PCR: theoretical considerations with practical implications. Analytical Biochemistry 214: 582–585
5. Caplin BE, Rasmussen RP, Bernard PS, Wittwer CT (1999) LightCycler Hybridization Probes. Biochemica 1: 5–8

Use of TaqStart Antibody to Increase the Sensitivity of Herpesvirus Quantitative PCR on the LightCycler

Karen Brengel-Pesce*, Gérard Bargues, Patrice Morand, Jean-Marie Seigneurin

Introduction

Human viral load monitoring needs to be accurate and sensitive. Quantitative PCR assays are therefore useful. The development of real-time quantitative PCR on the LightCycler combines speed and accuracy. However, during PCR, non-specific amplification may compete with formation of specific products, leading to a reduced PCR efficiency and lower sensitivity. In this study, we demonstrate that an anti-Taq DNA polymerase antibody (TaqStart Antibody, Clontech) avoids the formation of primer-dimers and non-specific products in LightCycler PCR and improves the sensitivity of Epstein-Barr virus (EBV) and Human Herpesvirus type 8 (HHV8) DNA quantification.

Materials

Equipment

LightCycler Instrument
Oligo Primer Analysis Software

Reagents

QIamp blood extraction kit (Qiagen)
Amplification primers (Genset SA)
TaqStart Antibody (Clontech)
Heat-labile uracil DNA glycosylase (UNG, Roche Diagnostics)
LightCycler-DNA Master SYBR Green I (Roche Diagnostics)
TA-cloning Kit (Invitrogen)
B95–8 cell line: an EBV productive B lymphoid cell line [1]
BBG1 cell line: a malignant cell line containing HHV8 DNA [2]

Procedure

Template Preparation

A 169 bp PCR product in the BXLF1 gene encoding EBV thymidine kinase was cloned into a plasmid to define an external standard for quantification. Briefly, viral DNA from the B95–8 EBV-positive cell line was extracted with the QIamp

---

* Karen Brengel-Pesce (✉) (e-mail: Karen.Brengel-Pesce@ujf-grenoble.fr)
  Laboratoire de Virologie, Faculté de Médecine de Grenoble, 38706 La Tronche, France

Blood kit. The PCR fragment was amplified and cloned into the PCR II plasmid of the TA-cloning kit. Dilution series of the plasmid DNA allowed the measurement in a single run over a range of concentrations from 20 to 2×10^5 copies.

The HHV8 standard curve was derived from a 10-fold serial dilution of BBG1 cell line DNA, after extraction with the QIamp Blood kit. A linear range, beginning at 10 pg (300 copies) and extending through 100 ng (3×10^6 copies) of DNA was established.

Primer Design

For EBV, amplification primers were designed using Oligo Primer analysis software. For HHV-8, the PCR primer pair was previously described [3]. Oligonucleotide primers are shown in Table 1.

Table 1. EBV and HHV8 PCR primer pairs

EBV (Genebank Accession # V01555)				
	Position	Length	GC (%)	T_m (°C)
GGGGCAAAATACTGTGTTAG	143411	20	45	58.3
CGGGGGACACCATAGT	143579 R	16	62.5	59.6
PCR product	169			
HHV8 (Genbank Accession # U75698)				
AGCCGAAAGGATTCCACCAT	47287	20	50	63.6
TCCGTGTTGTCTACGTCCAG	47519 R	20	55	63.5
PCR product	233			

LC-PCR

The procedure below describes the amplification of either herpesvirus, EBV or HHV8, in the presence of the fluorescent dye SYBR Green I.

SYBR Green I Master Mix for each 20 µl reaction:

	Volume [µl]	[Final]
LightCycler-DNA Master SYBR Green I	2	1×
MgCl$_2$ (25 mM)	1.6	3 mM
Primers (10 µM each)	1	0.5 µM
UNG	1	
H$_2$O (PCR grade)	12.4	
Total volume	18	

If TaqStart Antibody is used, 0.2 µl of the antibody is added to each 2 µl of LightCycler-DNA Master SYBR Green I and incubated for 10 min at room temperature, before the addition of other reagents.

A total of 18 µl of master mix are placed into each precooled capillary and 2 µl of the appropriate DNA template are added. A negative control replaces the template DNA with PCR grade H$_2$O. Sealed capillaries are centrifuged in a microcentrifuge and placed into the LightCycler rotor.

The following experimental PCR protocol was used with the SYBR Green detection:
- Denaturation at 95°C for 2 min
- Amplification by using the following cycle program data for EBV and HHV8 (in parentheses when different from EBV):

Parameter	Value			
Cycles	45			
Type	Quantification			
	Segment 1	Segment 2	Segment 3	Segment 4
Target temperature (°C)	95	58 (55)	72	81 (78)
Incubation time [s]	5	10	15	2
Temperature transition rate [°C/s]	20	20	5	5
Acquisition mode	None	None	None	Single
Gains	F1=6			

- Melting curve program

Parameter	Value		
Cycles	1		
Type	Quantification		
	Segment 1	Segment 2	Segment 3
Target temperature (°C)	95	55 (60)	95
Incubation time [s]	0	30	0
Temperature transition rate [°C/s]	20	20	0.2
Acquisition mode	none	none	Cont
Gains	F1=6		

Results

PCR amplification was first performed without TaqStart Antibody and low melting products were present on the melting curves of both herpesviruses (Figs. 1a,2a). To eliminate the detection of non-specific products, the fluorescence signal was measured during each cycle at a temperature above the melting temperature of the primer-dimers, e.g., 81°C for EBV (Fig. 1c) [4]. In the case of HHV8, however, the melting temperature of non-specific products was close to that of the specific PCR product. Thus, it was difficult to increase the temperature of fluorescence measurement in order to avoid non-specific products without losing specific signal. Quantification data shows that the position of fluorescence traces from samples more diluted than 10^{-3} do not correlate with copy number. (Fig. 2b).

Quantification of Viral DNA Using an External Standard Without TaqStart Antibody

a/ melting curve analysis

Without TaqStart Antibody

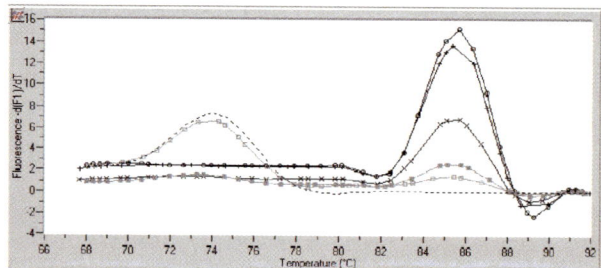

b/ quantification

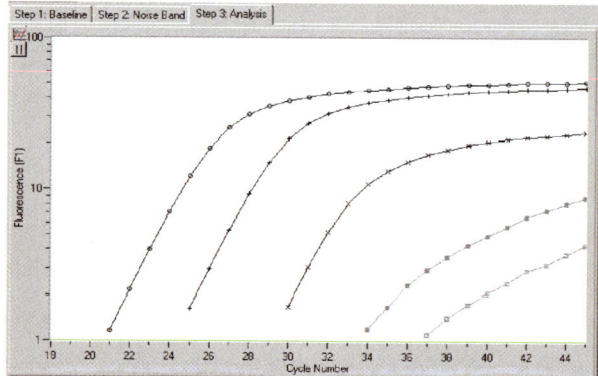

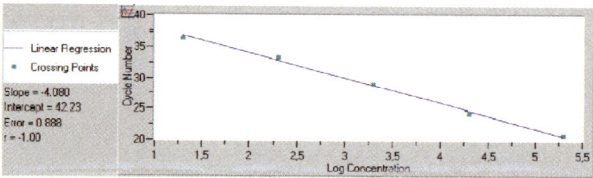

Fig. 1. EBV DNA quantification (SYBR Green I detection) with and without TaqStart Antibody

Quantification of Viral DNA Using an External Standard with TaqStart Antibody

To increase quantification sensitivity, the reduction of non-specific products is indispensable. Addition of TaqStart Antibody inhibits Taq DNA polymerase prior to amplification and decreases primer-dimers and other non-specific amplification products, thereby reducing non-specific peaks on melting curve graphs (Figs. 1c,2c). Consequently, the profile of quantification curves is better with TaqStart Antibody, showing parallel and evenly-spaced titration curves (Figs. 1d, 2d).

With

TaqStart Antibody

c/ melting curve analysis

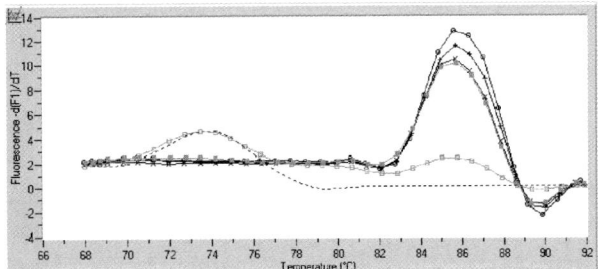

d/ quantification

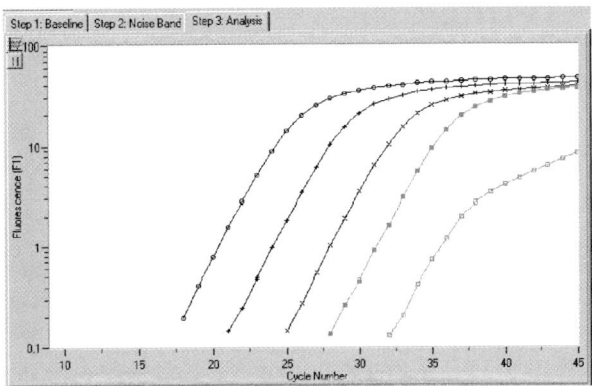

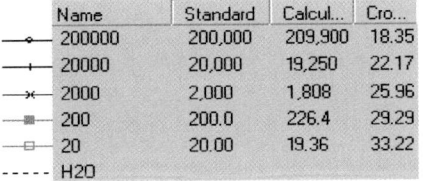

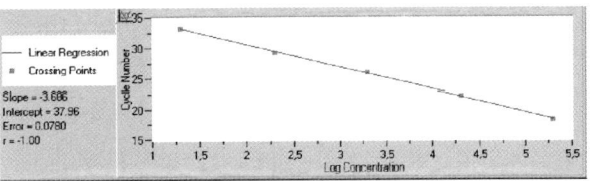

Fig. 1. *Continued.*

EBV DNA amplification induced primer-dimer formation in both the H_2O control and the 20 copy plasmid reaction, even when TaqStart Antibody was used. However, the PCR sensitivity was clearly improved with TaqStart Antibody, since the curves are more parallel at low copy numbers. Furthermore, "crossing points vs log concentration" graphs show the difference in sensitivity with and without TaqStart Antibody: the crossing points of each dilution and the intercept value are lower with TaqStart Antibody (Fig. 1).

For HHV8 DNA amplification, we observed that TaqStart Antibody abolished low melting product formation. While the high copy number samples were not

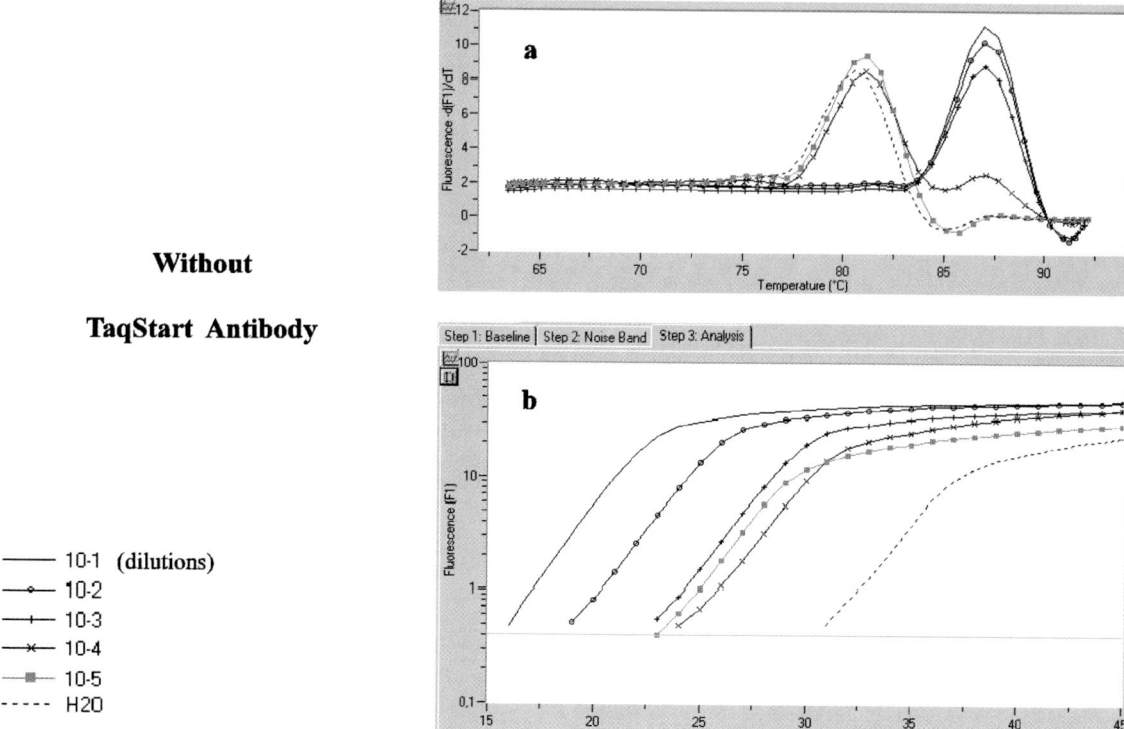

Fig. 2. HHV8 DNA quantification (SYBR Green I detection) with and without TaqStart Antibody

affected much by TaqStart Antibody addition, the 10^{-4} and 10^{-5} dilution templates became quantifiable with a linear external standard curve. The fluorescence vs cycle number plots are parallel and evenly-spaced, even down to a 10^{-5} dilution.

Comments

Quantitative PCR on the LightCycler depends on good primer design, and computer software can facilitate analysis of criteria such as melting temperature, GC content, PCR product length and sequence. However, undesired "non-specific" amplification can still occur, even with computer-assisted primer design. The use of TaqStart Antibody improves the amplification efficiency of specific products and therefore the sensitivity of quantitative PCR. We tested TaqStart Antibody at various concentrations (from 0.16 µl to 0.35 µl per tube) and various incubation times (5, 10 and 15 min) with the PCR mix. Optimal conditions appeared to be a volume of 0.2 µl per tube (i.e., 0.2 µl in 2 µl of mix) and an incubation period of 10 min.

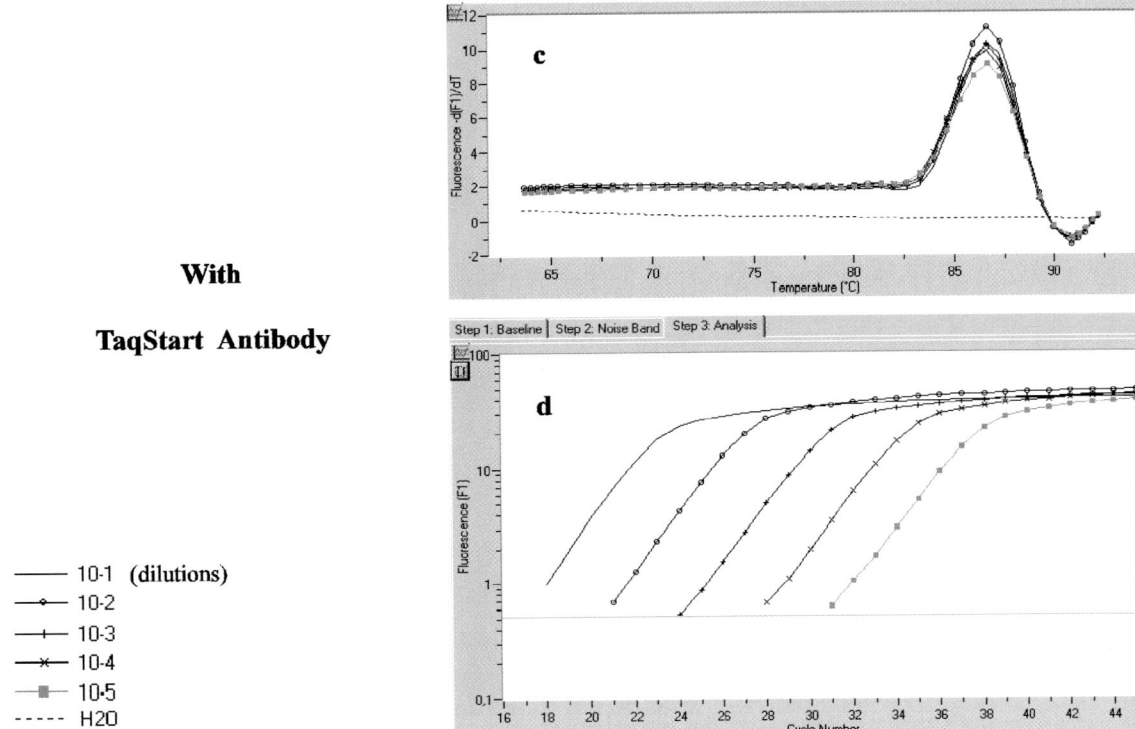

Fig. 2. *Continued.*

References

1. Miller G, Lipman M (1973) Release of infectious Epstein-Barr virus by transformed marmoset leukocytes. Proc Natl Acad Sci USA 70: 190–194
2. Morand P, Buisson M, Collandre H, Chanzy B, Genoulaz O, Bourgeat MJ, Pinel N, Leclercq P, Leroux D, Marechal V, Fritsch L, Ruigrok R, Seigneurin JM (1999) Human herpesvirus 8 and Epstein-Barr virus in a cutaneous B-cell lymphoma and a malignant cell line established from the blood of an AIDS patient. Leuk Lymphoma 35: 379–387
3. Chang Y, Cesarman E, Pessin MS, Lee F, Culpepper J, Knowles DM, Moore PS (1994) Identification of herpesvirus-like DNA sequences in AIDS-associated Kaposi's sarcoma. Science 266: 1865–1869
4. Morrison TB, Weiss JJ, Wittwer CT (1998) Quantification of low-copy transcripts by continuous SYBR Green I monitoring during amplification. Biotechniques 24 (6): 954–962

Genotyping of Human Germline Variations

High-Speed Detection of α_1-Antitrypsin Deficiency Alleles Pi*S and Pi*Z on the LightCycler 75
CHARALAMPOS ASLANIDIS, GERD SCHMITZ

High-Speed Methylenetetrahydrofolate Reductase C $\rightarrow$ T 677 Mutation Detection on the LightCycler 83
CHARALAMPOS ASLANIDIS, GERD SCHMITZ

Dual Color Detection of Splice Variants of the c-erbA α (Thyroid Hormone Receptor α) Gene 91
ONNO BAKKER

Detection of Three Major Polymorphisms in the *N*-Acetyltransferase 2 Gene by Melting Peak Analysis Using Fluorogenic Hybridization Probes 97
BRUNHILDE BLÖMEKE

Genotyping of Cytochrome P450 2D6*4 Mutation with Fluorescent Hybridization Probes Using LightCycler 105
JEANETTE BJERKE, CHUNG-CHE CHANG, CHUCK SCHUR, STEVEN WONG, NAZIHA NUWAYHID

Fluorescent Hybridization Probe Detection of the F508del Cystic Fibrosis Allele on the LightCycler 111
CAMERON N. GUNDRY

Genotyping β-globin Mutations (Hb S, Hb C, Hb E) by Multiplexing Probe Color and Melting Temperature 119
MARK G. HERRMANN

Simultaneous Detection of C282Y and H63D Hemochromatosis Mutations Using LCRed 640 and LCRed 705 Labeled Hybridization Probes 127
CINDY A. MEADOWS, BS CLSP (MB), MAREC PHILLIPS, MING Y. HUANG, MS and ELAINE LYON, PH.D

**Genotyping of Angiotensin-Converting Enzyme
and Angiotensinogen Polymorphisms with the LightCycler System** 135
Eiichi Sakai, Minori Tajima, Mitsuko Mori, Reiko Inage,
Manabu Fukumoto, Kan-ichi Nakagawara

**Genotyping of the Most Common Thiopurine Methyltransferase Mutations
with the LightCycler** 143
Ekkehard Schütz, Nicolas von Ahsen

**Detection of the Mitochondrial DNA Mutation MELAS3243
Using Hybridization Probes** 153
Stephanie Kleinle, Sabina Gallati

High-Speed Detection of α_1-Antitrypsin Deficiency Alleles Pi*S and Pi*Z on the LightCycler

CHARALAMPOS ASLANIDIS*, GERD SCHMITZ

Introduction

Protease inhibitor 1 (α_1-antitrypsin, AT) is the main serum inhibitor of proteolytic enzymes. In AT-deficiency, enzymes such as neutrophil elastase can damage the lung tissues, leading to pulmonary emphysema. More than 90 different alleles have been identified so far for the protease inhibitor 1 (*PI*) gene. The three most important variants are type M (90% of population), type S (Pi*S) and type Z (Pi*Z). Homozygotes of type Z have a considerable reduction in the serum α_1-antitrypsin concentration and may develop pulmonary emphysema or hepatic cirrhosis. SZ-heterozygotes are less severely affected [1–3].

Pi*S and Pi*Z genotyping is recommended in the following cases: molecular diagnosis in patients with AT-deficiency, hepatitis of unknown origin in newborns and the young, icterus prolongatus in newborns, hepatitis or liver cirrhosis of unknown origin in the adult, pulmonary emphysema and prenatal diagnosis in families with AT-deficiency.

PCR-based technologies are now widely used for the identification of the mutations underlying the Pi*Z allele (Glu342Lys, GAG to AAG) and the Pi*S allele (Glu264Val, GAA to GTA) [4]. Even though the results of these methods are unequivocal, they are time consuming and require optimization of the PCR reaction to eliminate unspecific PCR products. The newly available LightCycler system is perfectly suitable for mutation detection. This technology allows the real time detection of a specific PCR product followed by melting curve analysis of a hybridization oligonucleotide [5, 6].

Here we describe a procedure for high-speed mutation detection for the protease inhibitor 1 gene, in order to identify the Pi*Z and Pi*S alleles using the LightCycler [7]. In this homogeneous system, contamination of the PCR reaction due to sample handling is minimized.

Materials

LightCycler Instrument (Roche Diagnostics, Mannheim, Germany)   **Equipment**
LightCycler Capillaries (Roche Diagnostics)

---

* C. Aslanidis (✉) (e-mail: Charalampos.Aslanidis@klinik.uni-regensburg.de)
  Klinikum der Universität Regensburg, Institut für Klinische Chemie und Laboratoriumsmedizin, Franz-Josef-Strauß-Allee 11, 93053 Regensburg, Germany

Kits	LightCycler-DNA Master Hybridization Probes (Roche Diagnostics)
	QIAamp DNA Blood Midi Kit (Qiagen, Hilden, Germany)
Reagents	Amplification primers (Metabion, Munich, Germany)
	Hybridization probes (TIB MOLBIOL, Berlin, Germany)

Procedure

Sample Preparation

DNA was isolated from human blood (EDTA -blood). In general, freshly drawn blood was available, although blood stored for 1 week in the cold was equally good for DNA-extractions and amplification on the LightCycler. High quality DNA was extracted from 2 ml of blood using the QIAamp DNA Blood Midi Kit from Qiagen and eluted in 300 µl buffer. Depending on the number of leukocytes, approximately 40–120 µg of DNA was extracted from 2 ml of blood from healthy individuals (130–400 ng/µl). DNA was stored at 4°C or was frozen at –20°C.

Oligonucleotide Design

A pair of primers flanking each of the mutations in the Pi*S and Pi*Z alleles was designed according to the genomic DNA sequence of the human AT-gene [7]. The oligonucleotides used for the Pi*S and Pi*Z PCR are shown in Table 1. A 238-bp PCR fragment of the Pi*S-locus and a 253-bp PCR product of the Pi*Z-locus were amplified (Fig. 1 C, F). The 3'-phosphorylated detection probe for the Pi*S allele is located 2 bp downstream of the anchor probe. The detection probe for the Pi*Z allele is located 3 bp downstream of the anchor probe. Both detection probes are homologous to the wild type sequence.

Table 1. Primer sequences and hybridization probes for Pi*S and Pi*Z typing

α-Antitrypsin (Genebank Accession # K02212)					
	Position	Exon	Length	GC (%)	T_m (°C)
Pi*S locus primers					
GGTGCCTATGATGAAGCGTTTAGGC	9488	3	25	52.0	66.2
AGGTGTGGGCAGCTTCTTGGTCA	9725 R	3	23	56.5	68.8
Product	9488-9725		238		
Pi*S locus probes					
TTCTTCCTGCCTGATGAGGGGAAACTA-F	9591	3	27	48.1	66.6
LCRed640-GCACCTGGAAAATGAAC-P	9620	3	17	47.1	53.3
PI*Z locus primers					
TCCACGTGAGCCTTGCTCGAGGCCTG	11843	5	26	65.4	70.7
TTGGGTGGGATTCACCACTTTTC	12095 R	5	23	47.8	60.3
Product	11843-12095		253	52.2	
Pi*Z locus probes					
CTCCAGGCCGTGCATAAGGCTGT-F	11904	5	23	70.6	65.6
LCRed640-GACCATCGACGAGAAAGGG-P	11930	5	19	57.9	56.8

Variable nucleotides in the detection probes are underlined.

Pi*S Hybridization Probe Master Mix for each 20 µl reaction:

LightCycler-PCR

	Volume (µl)	Final
LightCycler-DNA Master Hybridization Probes	2	1×
MgCl$_2$ (25 mM)	1.6	3.0 mM
Primers (5 µM each)	1+1	0.25 µM
Probes (4 µM each)	1+1	0.2 µM
H$_2$O (PCR grade)	10.4	
Total volume	18	

Pi*Z Hybridization Probe Master Mix for each 20 µl reaction:

	Volume (µl)	Final
LightCycler-DNA Master Hybridization Probes	2	1×
MgCl$_2$ (25 mM)	0.8	2.0 mM
Primers (5 µM each)	1+1	0.25 µM
Probes (4 µM each)	1+1	0.2 µM
H$_2$O (PCR grade)	11.2	
Total volume	18	

A total of 18 µl of master mix and 2 µl of genomic DNA (40–100 ng) were added to each glass capillary placed in adapters. Sealed capillaries were centrifuged briefly with the adapters in a microcentrifuge and put in the LightCycler rotor.

The following PCR protocol was used for amplification and melting curve analysis:
- Denaturation at 94°C for 2 min
- Amplification

Parameter	Value		
Cycles	40		
Type	Quantification		
	Segment 1	Segment 2	Segment 3
Target temperature (°C)	95	55	72
Incubation time (s)	0	10	15
Temperature transition rate (°C/s)	20	20	20
Acquisition mode	None	Single	None
Gains	F1=10 ; F2=10		

- Melting Curve Analysis

Parameter	Value		
Cycles	1		
Type	Melting Curves		
	Segment 1	Segment 2	Segment 3
Target temperature (°C)	94	40	70
Incubation time (s)	0	5	0
Temperature transition rate (°C/s)	20	20	0.2
Acquisition mode	None	None	Step
Gains	F1=10; F2=10		

- Cooling was for 30 s at 40°C (Temperature transition rate 20°C/s)

Results

Optimization of PCR in Glass Capillaries

Both primer sets resulted in specific PCR products without the need for laborious optimizations. Unique bands of 238 bp (Pi*S) and 253 bp (Pi*Z) were generated when the $MgCl_2$ concentration was adjusted to 3 mM (Fig. 1C) and 2 mM (Fig. 1F), respectively. In contrast, the high GC-content of apoE required several primer combinations to be tested in the presence of dimethyl sulfoxide for successful genotyping on the LightCycler [8].

On-line PCR and Pi*S Mutation Detection with Hybridization Probes

The fluorescence profiles generated from DNA samples and negative controls are shown in Fig. 1. Using hybridization probes for the Pi*S allele (Fig. 1A), fluorescence increased in the samples with DNA (samples 2 and 3), whereas no fluorescence was detected in the H_2O control (sample 1). When PCR products were analyzed on agarose gels, a specific 238 bp PCR product of equal intensity in samples 2 and 3 was detected (Fig. 1C). The melting curves of the same samples are shown in Fig. 1B. The melting point (T_m) of the wild type sample (curve 3) is at 55.7°C, whereas the heterozygous sample (curve 2) shows melting peaks at 48.6°C and 55.7°C. It is interesting to note that, despite the equal amounts of PCR product in the capillaries, the fluorescence intensity in sample 2 (Fig. 1A) is considerably lower than in sample 3. This is because the T_m of the mismatched probe in half of the PCR products from the heterozygotes (sample 2) is below the annealing temperature in the PCR reaction. Consequently, when analyzing DNA from homozygous mutant individuals, no fluorescence will be monitored during PCR. However, melting peaks will reveal the proper genotype due to the low temperature starting point of the melting curve.

On-line PCR and Pi*Z Mutation Detection with Hybridization Probes

Fluorescence monitoring with the hybridization probes specific for the Pi*Z allele is illustrated in Fig. 1. As can be seen in Fig. 1D, the three different genotypes result in different fluorescence signal intensities, even though analysis of PCR products on agarose gels revealed equal amounts of PCR product (Fig. 1F). The

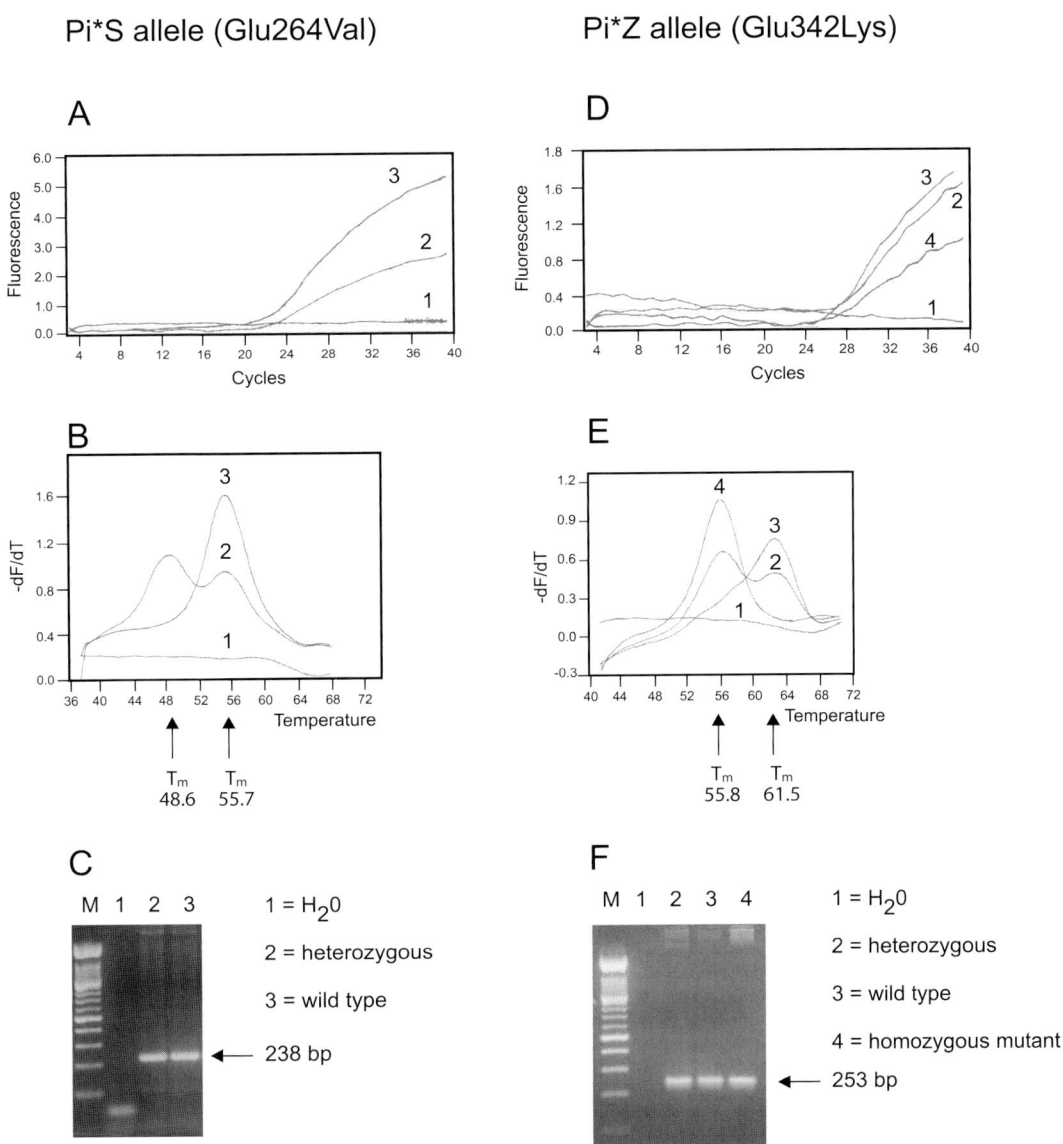

Fig. 1 A–F. Fluorescence vs cycle number, melting peaks, and agarose gel electrophoresis of PCR products for Pi*S (A–C) and Pi*Z (D–F) alleles. DNAs of various genotypes were amplified in glass capillaries (LightCycler) with primers derived from the regions of the Pi*S and Pi*Z mutations, and fluorescence was monitored using Pi*S- and Pi*Z-specific hybridization probes. The assignment of samples to curves is shown by *numbers*. Melting curves were converted to melting peaks by plotting the negative derivative of the fluorescence with respect to temperature (-dF/dT) against temperature, and are shown on the right. The melting point (T_m) of the individual detection probes is shown by an arrow. PCR products from glass capillaries were analyzed in 3% agarose gels. M; 100 bp molecular weight marker. (Figure reprinted from [7])

detection probe has a T_m of 61.5°C with the wild type sequence (curve 3) and is stable at the PCR annealing temperature (55.0°C), whereas basepairing with PCR products from a homozygous mutant individual (curve 4) is impaired due to a relatively low T_m (55.8°C) (Fig. 1E). This apparent variability in signal intensity does not affect the quality of the melting curves because monitoring of the melting behavior starts at 40°C.

Determination of Pi*S and Pi*Z Genotypes

Individuals with two Pi*S alleles exhibit a single peak at 48.6°C, heterozygous individuals have two peaks, and individuals with the wild type sequence on both alleles have a single peak at 55.7°C. Individuals with two Pi*Z alleles exhibit a single peak at 55.8°C, in heterozygous individuals two peaks are detected, and individuals with the wild type sequence on both alleles result in a single peak at 61.5°C (Table 2).

Comments

Sample Preparation

Sample preparation is not a major concern in this qualitative assay. We did not experience any difficulties in the amplification and detection process on the LightCycler when we used DNA isolated from different sources (blood, fibroblasts, buccal swabs). DNA that was isolated from blood kept cold for 1 week or at room temperature for several days was equivalent to DNA isolated from fresh samples.

Primer Design

We have used the LightCycler successfully for apoE genotyping, mutation detection in methylenetetrahydrofolate reductase, prothrombin, factor V-Leiden, hemochromatosis, apolipoprotein B3500 and in several other polymorphism analyses in various research projects [7–9]. The design requirements of the PCR primers are the same as for traditional thermal cyclers and protocols. Extensive optimization is not required. Variation of the $MgCl_2$ concentration in initial experiments is recommended. The PCR primers for the Pi*S and Pi*Z determinations generated specific PCR products. The hybridization probes for the Pi*S allele (anchor and detector probes) were at a distance of 2 bp, whereas the probes for the Pi*Z genotyping were at a distance of 3 bp. It is possible that the difference

Table 2. Melting temperatures of Pi*S and Pi*Z alleles

Locus	Allele	Pairing	T_m (observed)
Pi*S	WT	A-T match	55.7°C
	Pi*S	A-A mismatch	48.6°C
Pi*Z	WT	G-C match	61.5°C
	Pi*Z	G-T mismatch	55.8°C

WT, wild type sequence

in signal generation seen in Fig. 1 A, D (5 vs 1.7) results from less efficient energy transfer at a distance of 3 bases as compared to 2. If monitoring amplification during cycling is desired, the melting point of the detector probe should be higher than the annealing temperature in the PCR reaction. The Tm shifts that result from the Pi*S and Pi*Z mutations are 7.1°C and 5.7°C, respectively. The different genotypes are readily apparent by melting curve analysis and are reproducible and unambiguous.

References

1. Crystal RG (1989) The alpha-1-antitrypsin gene and its deficiency states. Trends Genet 5:411–417
2. Crystal RG (1990) Alpha-1-antitrypsin deficiency, emphysema, and liver disease. Genetic basis and strategies of therapy. J Clin Invest 85:1343–1352
3. Hutchison DCS (1998) α_1-Antitrypsin deficiency in Europe: geographical distribution of Pi types S and Z. Respir. Med. 92:367–377
4. Braun A, Meyer P, Cleve H, Roscher AR (1996) Rapid and simple diagnosis of the two common α_1-proteinase inhibitor deficiency alleles Pi*Z and Pi*S by DNA analysis. Eur J Clin Chem Clin Biochem 34:761–764
5. Wittwer C, Ririe K, Andrew R, David D, Gundry R, Balis U (1997) The LightCycler: A microvolume multisample fluorimeter with rapid temperature control. Biotechniques 22:176–81
6. Lay M, Wittwer C (1997) Real time fluorescence genotyping of factor V Leiden during rapid cycle PCR. Clin Chem 43:2262–2267
7. Aslanidis C, Nauck M, Schmitz G (1999) High-speed detection of the two common α_1-antitrypsin deficiency alleles Pi*Z and Pi*S by real-time fluorescence PCR and melting curves. Clin Chem 45:1872–1875
8. Aslanidis C, Schmitz G (1999) High-speed apolipoprotein E genotyping and apolipoprotein B3500 mutation detection using real-time fluorescence PCR and melting curves. Clin Chem 45:1094–1097
9. Aslanidis C, Nauck M, Schmitz G (1999) High-speed prothrombin G→A 20210 and methylenetetrahydrofolate reductase C→T 677 mutation detection using real-time fluorescence PCR and melting curves. Biotechniques 27:234–238

High-Speed Methylenetetrahydrofolate Reductase C → T 677 Mutation Detection on the LightCycler

CHARALAMPOS ASLANIDIS*, GERD SCHMITZ

Introduction

Venous thrombosis is a common cardiovascular disease in Caucasians, affecting 1 out of 1000 individuals, and strongly associates with pulmonary embolism [1]. Genetic factors have been found to play a crucial role in disease development. Among these, mutations in the Factor V gene (FV Leiden) and in the prothrombin gene have been shown to account for a large number of thromboembolisms [2]. An additional independent risk factor is hyperhomocysteinemia. Increased plasma homocysteine is partly due to a thermolabile variant of methylenetetrahydrofolate reductase (MTHFR) resulting from a C to T transition at nucleotide 677 (Ala222Val) and is implicated as a risk factor for vascular and thromboembolic disease and neural tube defects [3].

MTHFR C677T mutation analysis is recommended in the following cases: familial occurrence of thromboembolisms at young age, prescription of oral contraceptives in women with predisposition to thromboembolism, repeated thromboembolisms of unknown origin, and relatives of index patients.

To date, mutation detection for MTHFR C677T is mainly performed by PCR, followed by digestion with restriction enzymes and restriction fragment length polymorphism (RFLP) analysis, and separation of the resulting DNA-fragments on agarose or acrylamide gels [3]. However, these methods, even though they are robust and lead to unequivocal results, are time consuming and do require optimization of the PCR reaction in order to eliminate unspecific PCR products that would disturb the genetic analysis. Contamination is of major concern because of the non-homogeneous nature of the methodology.

The newly available LightCycler system is perfectly suitable for mutation detection. This technology allows the real-time detection of the specific amplicon followed by detection of the mutation by identification of the melting behavior of one of the two hybridization oligonucleotides [4–6].

We describe here a high-speed and easy to perform mutation detection for the methylenetetrahydrofolate reductase C677T mutation using the LightCycler technology and melting curves [7]. This homogeneous system minimizes PCR con-

---

* Charalampos Aslanidis (✉) (e-mail: Charalampos.Aslanidis@klinik.uni-regensburg.de)
Institute for Clinical Chemistry and Laboratory Medicine, University of Regensburg, Franz-Josef Strauß Allee 11, 93042 Regensburg, Germany

tamination concerns related to sample handling and does not require digestion of PCR products with restriction enzymes and fragment separation on gels.

Materials

Equipment LightCycler Instrument (Roche Diagnostics, Mannheim, Germany)
LightCycler Capillaries (RMB)

Kits LightCycler-DNA Master Hybridization Probes (RMB)
QIAamp DNA Blood Midi Kit (Qiagen, Hilden, Germany)

Reagents Amplification primers (Metabion, Munich, Germany)
Hybridization probes (TIB MOLBIOL, Berlin, Germany)

Procedure

Sample Preparation Human blood with anticoagulant (EDTA blood) was used for genomic DNA isolation. In general, freshly drawn blood was preferable, but material that was kept for no longer than 1 week in the cold was equally good for DNA extractions and amplification on the LightCycler. DNA was isolated from 2 ml of blood using the QIAamp DNA Blood Midi Kit from Qiagen and eluted in 300 µl buffer, according to the recommendations of the manufacturer. Depending on the particular number of leukocytes, approximately 40–120 µg of DNA was extracted from 2 ml of blood of healthy individuals (130–400 ng/µl). DNA was stored at 4°C or was frozen at −20°C for long term storage.

Oligonucleotide Design The primers for the MTHFR PCR produced a 233-bp PCR product from genomic DNA (Fig. 1C and Table 1). The 3′-phosphorylated detection primer is located downstream of the anchor primer at a distance of 2 nucleotides.

Table 1. Oligonucleotides.

Methylenetetrahydrofolate Reductase, exon 4 (Genebank Accession # AF105980)				
	Position	Length	GC (%)	T_m (°C)
Primers				
CGAAGCAGGGAGCTTTGAGGCTG	70	23	60.9	66.9
AGGACGGTGCGGTGAGAGTG[1]	302R	20	65.0	65.7
Product		233		
Probes				
TGACCTGAAGCACTTGAAGGAGAAGGTGTC-F	91	30	50.0	67.7
LCRed640-CGGGAGCCGATTTCATCAT-P[2]	123	19	52.6	58.7

[1] Reverse primer derived from intron 4 as per [3] and AF105980.
[2] Variable nucleotide in the detection probe is underlined.

High-Speed Methylenetetrahydrofolate Reductase C→T 677 Mutation Detection on the LightCycler

LightCycler-PCR

MTHFR C677T Hybridization Probe Master Mix for each 20 µl reaction:

	Volume (µl)	Final
LightCycler-DNA Master Hybridization	2	1×
$MgCl_2$ (25 mM)	1.2	2.5 mM
Primers (5 µM)	1+1	0.25 µM
Probes (4 µM)	1+1	0.2 µM
H_2O (PCR grade)	10.8	
Total volume	18	

A total of 18 µl of master mix and 2 µl of genomic DNA (40–100 ng) were added to each glass capillary placed in adapters. Sealed capillaries were centrifuged briefly in a microcentrifuge and put in the LightCycler rotor.

The following PCR protocol was used for MTHFR C677T amplification and melting curve analysis:

- Denaturation at 94°C for 2 min
- Amplification

Parameter	Value		
Cycles	40		
Type	Quantification		
	Segment 1	Segment 2	Segment 3
Target temperature (°C)	94	55	72
Incubation time [s]	0	10	15
Temperature transition rate [°C/s]	20	20	20
Acquisition mode	None	Single	None
Gains	F1=1; F=10		

- Melting Curve Analysis

Parameter	Value		
Cycles	1		
Type	Melting Curves		
	Segment 1	Segment 2	Segment 3
Target temperature (°C)	94	40	80
Incubation time [s]	0	5	0
Temperature transition rate [°C/s]	20	20	0.6
Acquisition mode	None	None	Step
Gains	F1=1; F=10		

- Cooling was for 30 s at 40°C (Temp. Transition Rate 20°C/s)

Results

Optimization of PCR in Glass Capillaries

The primers generated specific PCR products without the need for laborious optimization, as was also the case in our α_1-antitrypsin deficiency protocol [8], the prothrombin G20210A protocol [7] and apoB3500 mutation detection (unpublished). As can be seen on the agarose gel in Fig. 1C, a unique band of 233 bp is generated. In another application (apoE-genotyping), due to high GC content, several primer combinations had to be tested in the presence of dimethyl sulfoxide in order to establish a robust protocol [9].

On-line MTHFR-PCR and Mutation Detection with Hybridization Probes

The fluorescence profiles generated from selected DNA samples and negative controls are shown in Fig. 1. When the hybridization probes for MTHFR were used (Fig. 1A), fluorescence increased constantly in the samples with DNA (samples 2, 3 and 4), whereas no fluorescence was detected in the control sample (H$_2$0, sample 1). As can be seen in Fig. 1C, equal amounts of the 233 bp PCR product were generated in the various samples. The melting curves of the same samples are shown in Figs. 1B and 1D. The melting point (T_m) of the "wild-type" sample (curve 4) is at 63.1 °C, whereas the T_m of the homozygous "mutant" sample (curve 2) is at 55.2 °C. The differences in the fluorescence signal intensities in Fig. 1A result from the fact that the detection probe is derived from the wild-type sequence leading to stable hybridization at the annealing temperature in the PCR (55.0 °C), while base pairing with PCR products from homozygous mutant (curve 2) and heterozygous individuals (curve 3) is impaired due to a lower T_m (55.2 °C), compared to the melting point in the wild-type sample (curve 4), which is 63.1 °C. The T_ms are also affected by the temperature transition rate during the melting process (Figs. 1B and 1D).

Determination of the MTHFR Genotypes

Individuals with two copies of the wild-type sequence (C/C) show a single melting peak at 63.1°C, homozygous mutant individuals (T/T) show a single peak at 55.2 °C and heterozygous individuals (C/T) result in two peaks in this analysis (Table 2).

Table 2. Melting temperatures of MTHFR C677T alleles

Locus	Allele	Pairing	T_m (observed)
MTHFR C677T	WT	C-G match	63.1°C
	C677T	C-A mismatch	55.2 °C
WT, wild-type sequence			

Comments

Sample Preparation

In this qualitative assay, sample preparation is not of major concern. We have used DNA isolated from different sources (blood, fibroblasts, buccal swabs) and

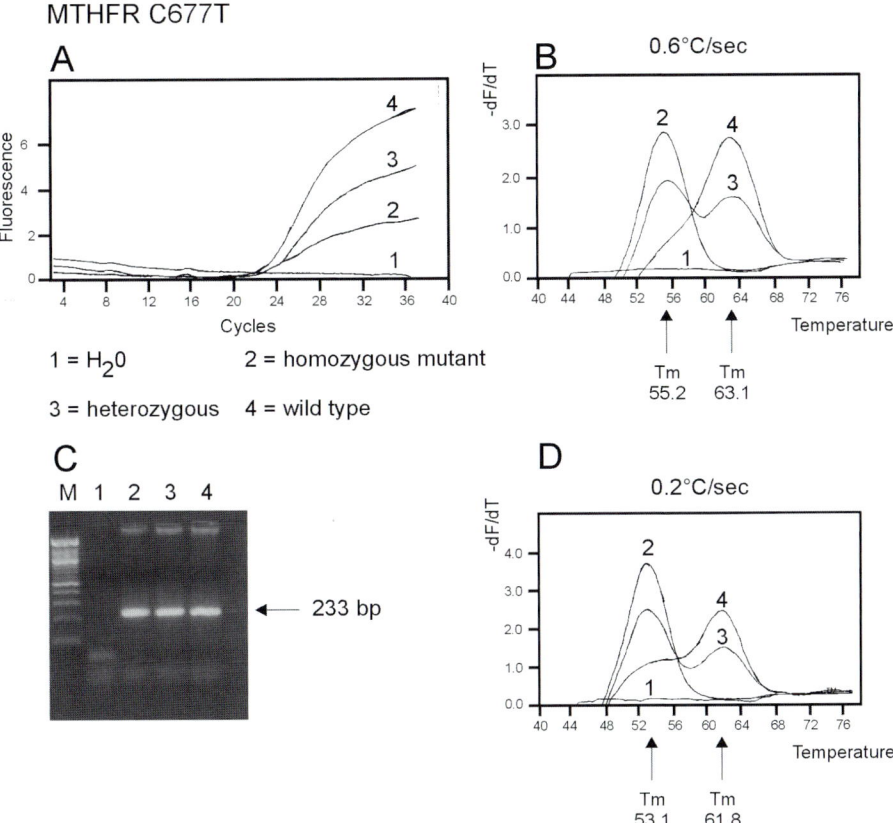

Fig. 1. Fluorescence vs. cycle number, melting peaks, and agarose gel electrophoresis of PCR products for MTHFR C677T mutation detection. **A** DNAs from various individuals were amplified with MTHFR-specific primers in glass capillaries (LightCycler), and fluorescence was monitored using MTHFR-specific hybridization probes. The assignment of samples to curves is shown by numbers. **B, D** Melting curves were converted to melting peaks by plotting the negative derivative of the fluorescence with respect to temperature (-dF/dT) against temperature. The melting point (T_m) of the individual detection probes is shown by arrows. The melting peaks generated by two different melting protocols (0.6 °C/s vs. 0.2 °C/s) are shown. **C** Agarose gel electrophoresis of PCR products from glass capillaries in 3% NuSieve GTG agarose. M, 100-bp molecular weight marker. (Figure reprinted from [7])

did not experience any difficulties in the amplification and detection on the LightCycler. DNA that was isolated from blood that has been kept in the cold for 1 week or at room temperature for several days, was as good as DNA isolated from fresh samples.

We have established additional applications on the LightCycler, e.g., apolipoprotein E genotyping and apolipoprotein B3500 mutation detection [9], α_1-anti-

Primer Design

trypsin Pi*Z and Pi*S allele determinations [8], prothrombin G20210 A mutation detection [7], and several polymorphism analyses for scientific purposes, and have optimized detection probes for apoB3500 (unpublished). In our experience, the basis for the design of the PCR primers is the same as that required for traditional thermal cyclers and protocols. Extensive optimization is not required. In one application (apo E genotyping) we had to optimize the primers and the reaction conditions because of a very high GC-content of the region of interest [9]. With respect to the hybridization probes, a distance of 2 nucleotides between anchor primer and detection primer was ideal for efficient energy transfer and signal generation. It is also important that the annealing temperature in the PCR reaction is kept below the melting point of the detection probe, to allow monitoring of the fluorescence during the cycling. The T_ms in the heterozygous individuals are well separated (7.9 °C). In our experience, a minimum distance of 5°C-6°C between melting peaks is necessary in order to generate reproducible and unambiguous results.

Bernard et al. [6] have published a similar protocol for MTHFR C677T typing. Unlike here, one of the primers for PCR was labeled internally with Cy5 and the detection probe was derived from the antisense strand. Thus, a G-T mismatch was generated, leading to a T_m shift of 3°C. In our case, the detection probe was derived from the sense strand and generated a C-A mismatch. It is known that C-A mismatches have a higher destabilizing effect than G-T mismatches [6]. Thus the combination of the C-A mismatch with a short detection probe (19-mer here, 26-mer at Bernard et al.) results in a higher T_m shift (7.9°C).

References

1. Kniffin WD, Baron JA, Barret J, Birkmeyer JD, Anderson FA Jr (1994) The epidemiology of diagnosed pulmonary embolism and deep venous thrombosis in the elderly. Arch Intern Med 154: 861–866
2. Poort SR, Rosendaal FR, Reitsma P.H, Bertina RM (1996) A common genetic variation in the 3'-untranslated region of the prothrombin G20210A protocol [7] and an optimized apoB3500 mutation detection protocol (unpublished).
3. Frosst P, Blom HJ, Milos R, Goyette P, Sheppard CA, Matthews RG et al (1995) A candidate genetic risk factor for vascular disease: a common mutation in methylenetetrahydrofolate reductase. Nat Genet 10: 111–113
4. Wittwer C, Ririe K, Andrew R, David D, Gundry R, Balis U (1997) The LightCycler: A microvolume multisample fluorimeter with rapid temperature control. Biotechniques 22: 176–81
5. Lay M, Wittwer C (1997) Real time fluorescence genotyping of factor V Leiden during rapid cycle PCR. Clin Chem 43: 2262–2267
6. Bernard PS, Lay MJ, Wittwer C (1998) Integrated amplification and detection of the C677 T point mutation in the methylenetetrahydrofolate reductase gene by fluorescence resonance energy transfer and probe melting curves. Anal Biocem 255: 101–107
7. Aslanidis C, Nauck M, Schmitz G (1999) High-speed prothrombin G→A 20210 and methylenetetrahydrofolate reductase C→T 677 mutation detection using real-time fluorescence PCR and melting curves. Biotechniques 27:234–238
8. Aslanidis C, Nauck M, Schmitz G (1999) High-speed detection of the two common α_1-antitrypsin deficiency alleles Pi*Z and Pi*S by real-time fluorescence PCR and melting curves. Clin Chem 45: 1872–1875

9. Aslanidis C, Schmitz G (1999) High-speed apolipoprotein E genotyping and apolipoprotein B3500 mutation detection using real-time fluorescence PCR and melting curves. Clin Chem 45: 1094–1097

Dual Color Detection of Splice Variants of the c-erbA α (Thyroid Hormone Receptor α) Gene

Onno Bakker*

Introduction

Thyroid hormone (T_3) is important for many metabolic processes and signals its presence to the cell via nuclear thyroid hormone receptors (TR) [1]. There are two genes encoding for these receptors giving rise to four isoforms of the receptor (α1, α2, β1 and β 2). The TRα1 and TRα2 isoforms are the result of alternative splicing from the primary transcript of the c-erbA-α gene. The expression of these receptors changes as a result of various developmental and metabolic conditions. One of these is non-thyroidal illness, also called the sick-euthyroid syndrome (SES). It is characterized by an aberrant thyroid hormone metabolism in which the levels of the active hormone T_3 in the blood are lowered without an accompanying rise in TSH. The pathophysiology of this phenomenon is not clear, but cytokines are known to play a role. Furthermore, some liver genes (like malic enzyme) become refractory to exogenous T_3 as a result of the SES. This could be the consequence of an increased expression of the TRα2 isoform which can act as a dominant negative of the other three receptors. In the past, we have studied the expression of the isoform mRNAs in the liver and heart of fasting rats (a model of SES) and found that the level of the liver TRα2 mRNA indeed increases approximately threefold [2]. These results were obtained by using a semi-quantitative PCR technique based on competition between a standard and the target. With this technique the expression levels of the TRα1 and -α2 mRNAs in one sample have to be measured in separate reactions, thereby introducing the possibility of errors. With the advent of the LightCyler we are now able to measure the TRα1 and -α2 mRNA levels (and/or the ratio) directly and specifically in one reaction, using hybridization probes and the dual color detection format. This chapter describes the approach we have developed to accomplish this.

Materials

RNase-free glassware and disposables **Equipment**
LightCycler Instrument

---

* Onno Bakker (✉) (e-mail: o.bakker@amc.uva.nl)
  Endocrinology, F5-171, Academic Medical Centre, Meibergdreef 9, 1105 AZ Amsterdam, The Netherlands

Reagents   Amplification Primers (TIB MOLBIOL, Berlin, Germany)
Hybridization Probes (TIB MOLBIOL, Berlin, Germany)
TriPure (Roche Diagnostics, Mannheim, Germany)
First strand cDNA synthesis kit (Roche Diagnostics, Mannheim)
LightCycler-DNA Master Hybridization Probes (RocheDiagnostics, Mannheim)
LightCycler-Color Compensation set (Roche Diagnostics, Mannheim)

Procedure

Primer Design   Primers were designed on the basis of those already published [2] with some changes so that T_ms were comparable and about 5–10°C below the calculated T_m of the hybridization probes (Table 1). Specificity of the sequence was checked by comparing the sequence to the Genebank data base.

The basic design is such that there is a common forward primer with two different isoform specific reverse primers (Fig. 1, *black bars*). The same principle has guided our hybridization probe design (Fig. 1) where we have one common 5′ fluorescein (*F*) labeled probe and two isoform specific probes labeled with *LC-640* ($\alpha1$) and *LC-705* ($\alpha2$). The hybridization probes were designed to anneal on the 3′ half of the amplicon to allow enough detection time before the polymerase displaces the probes.

Sample Preparation   Total RNA was purified from liver with the use of TriPure reagent (Roche Diagnostics, Mannheim, Germany). From this, cDNA was prepared by using the First strand cDNA synthesis kit with random primers (AMV reverse transcriptase; Roche Diagnostics, Mannheim). These cDNAs were also used to prepare standards by cloning the fragments into Bluesript SK+ (Stratagene, La Jolla, USA), allowing expression of the standards using T_7-polymerase.

PCR with the LightCycler   Hybridization Probe Mastermix for a 20-µl reaction:

	Volume [µl]	[Final]
LightCycler-DNA Master Hybridization Probes	2	1×
MgCl$_2$ (25 mM)	2.4	4 mM
Primers (TRα1) (5 µM each)	2	0.5 µM
Probes (TRα1) (2 µM each)	1	0.1 µM
Primers (TRα2) (5 µM each)	2	0.5 µM
Probes (TRα2) (2 µM each)	1	0.1 µM
H$_2$O (PCR grade)	7.6	

In order to minimize the risk of primer precipitation due to MgCl$_2$ at low temperatures, the mastermix was not placed in the cooling block until the DNA Master was added.

A total of 18 µl of master mix and 2 µl of cDNA template was added in each capillary. It is necessary to make sure that the liquid is at the bottom of the plas-

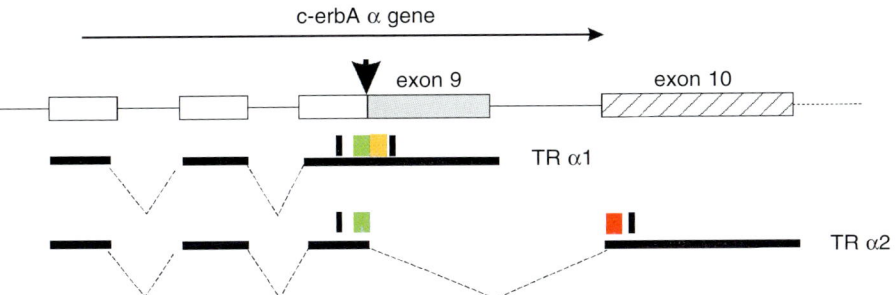

Fig. 1. Schematic representation of the c-erbAα gene. The figure shows the last exons of the c-erbAa gene which encodes both the TRα1 and TRα2 as a result of alternative splicing. The alternative splice site in exon 9 is indicated by an *arrow*. When the site is skipped, TRα1 mRNA is formed and when it is used the TRα2 mRNA is formed by splicing exon 10 to part of exon 9. The forward and reverse primers are indicated by *vertical black lines*. The common fluorescein labeled 5′-probe is indicated by the *green block*. The 3′-probes for TRα1 and TRα2 are indicated by the *yellow and red blocks* respectively

Table 1. Oligonucleotides. Sequences of the primers and hybridization probes used for the dual color detection of the TRα1 and TRα2 isoform mRNAs

c-erbA-1 gene (Genebank Accession # X55005 (α1) and J03239 (α2))					
	Position	Exon	Length	GC (%)	T_m (°C)
TRα1					
CATCTTTGAACTGGGCAAGT	1350	8	20	45	63.0
CTGAGGCTTTAGACTTCCTGATC	1707R	9	23	48	63.8
Product	1350–1707		358		
TRα2					
CATCTTTGAACTGGGCAAGT	951	8	20	45	63.0
GACCCTGAACAACATGCATT	1288R	10	20	45	62.7
Product	951–1288		338		
Hybridization Probes					
GGCCCAAGCTGCTGATGAAG-F (TRα1 and TRα2)	1557/1148	8	20	60	66.4
LCRed640-TGACTGACCTCCGCAT-GATCG-P (TRα1)	1578	9	21	57	66.7
LCRed704-GAGAAGTGCAGAGTTCG-ATTCTGTACAAG-P (TRα2)	1172	10	29	45	67.4

tic reservoir to avoid getting it caught between the stopper and the rim. Sealed capillaries were centrifuged briefly in an Eppendorf microcentrifuge (a short blast at full speed) and placed carefully into the LightCycler rotor.

Note: if the capillary breaks when placing it in the rotor, just take the glass part containing the sample (it usually breaks above the liquid level) and place it upside down in a new capillary and centrifuge for a few seconds in an Eppendorf microcentrifuge (check that it fits when the lid is closed). It is also helpful to clean the capillary positions in the rotor with a small brush before placing the capillaries.

The LightCycler was programmed as follows:
- Denaturation for 30 s at 95°C
- Amplification

Parameter	Value		
Cycles	45		
Type	Quantification		
	Segment 1	Segment 2	Segment 3
Target temperature [°C]	95	50	72
Incubation time [s]	0	10	15
Temperature transition rate [°C/s]	20	20	20
Acquisition mode	None	Single	None
Gains	F1=1, F2=15, F3=30		

Note: since the primers and their optimal annealing conditions were known from previous experiments, these were taken as a starting point for the PCR in the LightCycler.
- Cooling to 40°C.
- Color compensation was used to correct signal overflow between channels 2 and 3. The color compensation file was generated with the use of the Color Compensation set.

Results and Discussion

With the LightCycler, we are able to detect both the mRNA for TRα1 and for TRα2 in the same capillary. As can be seen in the typical example in Fig. 2, the dual color detection with the LightCycler allows for quantification over a range of at least 10000-fold for both splice variants.

Reproducibility appears to be good. We also performed the assays for both TR isoforms separately in SYBR Green I format, but due to the presence of primer-dimers, quantification was not possible over such a large range as when we used dual color detection. The standard curves are slighty more shallow when dual color detection is compared to single color format, which may be due to the fact that we used the same forward primer for both the TRα1 and TRα2 mRNAs (Fig. 1). Since this can give rise to competition resulting in a less efficient PCR, we used a final forward primer concentration double the recommended strength [3], in order to prevent depletion of this primer. A comparison of the characteristics of the standard curves obtained using either single or dual color detection is shown in Table 2.

We previously developed a semiquantitative competitive PCR technique to measure the relative levels of the splicing variants of the TRα gene [2]. With this

Dual Color Detection of Splice Variants of the c-erbA α (Thyroid Hormone Receptor α) Gene

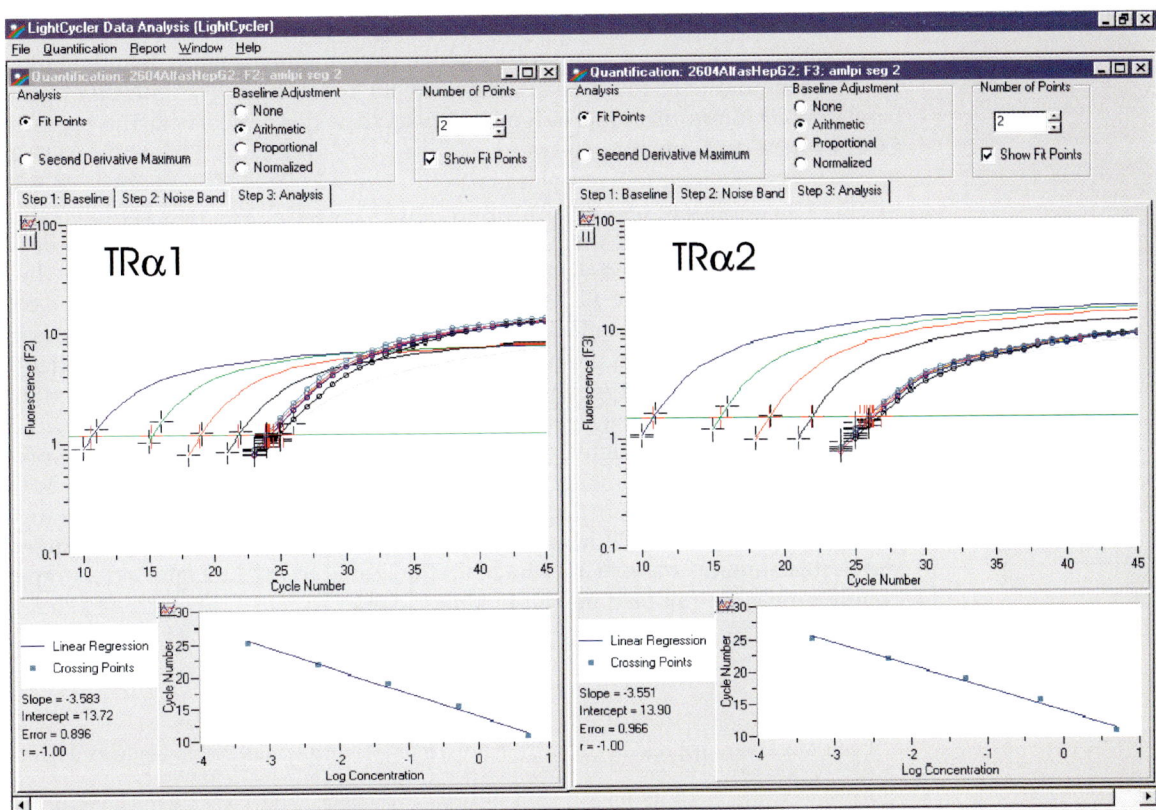

Fig. 2. The LightCycler Data Analysis screen of the dual color detection of TRα1 and TRα2. The figure shows the quantification screen with standard and sample curves of the TRα1, detected in channel F2 (*left hand side*, F2) and the TRα2 detected in channel F3 (*right hand side*, F3)

Table 2. Comparison of the characteristics of the standard curves obtained using either single or dual color detection

| | Single color | | Dual color | |
| | α1 | α2 | α1 | α2 |
	Mean of 2 experiments		Mean±SD of 5 experiments	
Slope	−4.87	−5.33	−3.68±0.13	−3.70±0.13
Cv	nd	nd	3.6%	3.7%
R	−0.97	−1.00	−1.00	−0.99

method, TRα1 and TRα2 mRNA levels had to be determined in separate tubes and it relied on the use of competitors which were slightly smaller then the target fragment so that they could be separated on an agarose gel. The intensity of the bands on ethidium bromide stained gels was then quantified with the use of a CCD camera and suitable software. Because our average experimental set up involves 6 points with 6–8 samples per point, around 600 PCRs and about 20 gels would be necessary to measure "absolute" mRNA levels. Since we were interested in the relative changes of the TRα1 and α2 mRNA, we used a fixed amount of competitor to calculate a competitor/target ratio per sample instead, thus decreasing the number of PCRs fourfold. Changes in TRα1 and α2 levels were then calculated from the individual competitor/target ratios. The disadvantages of this method are that in practice the maximum difference that can be reliably measured between the target and competitor is about tenfold and that the amplification efficiencies of the target and the competitor should be the same [4].

The advantage of the LightCycler method described here over the one we previously developed is that we are now able to quantify the levels (or ratio) of both splice variant mRNAs in one capillary at the same time over a much wider range then previously possible. With the same average experimental set up mentioned above, this amounts to 96 PCRs which, taking into account the LightCyclers rapid cycling programs, can be done "all in a day's work".

References

1. Lazar MA (1993) Thyroid hormone receptors: multiple forms, multiple possibilities. Endocr Rev 14:184–193
2. Bakker O, Razaki H, de Jong J, Ris-Stalpers C, Wiersinga WM (1998) Expression of the alpha 1, alpha 2, and beta 1 T3-receptor mRNAs in the fasted rat measured using competitive PCR. Biochem Biophys Res Comm 242: 492–496
3. LightCycler Operators Manual (1999), p D107 Roche Diagnostics, Mannheim
4. Raeymakers L (1993) Quantitative PCR: theoretical considerations with practical implications. Anal Biochem 214: 582–585

Detection of Three Major Polymorphisms in the *N*-Acetyltransferase 2 Gene by Melting Peak Analysis Using Fluorogenic Hybridization Probes

BRUNHILDE BLÖMEKE*

Introduction

Variability in drug acetylation was discovered in 1953 when the fate of isoniazid was studied in patients who had tuberculosis. Family studies revealed that the marked interindividual variation in the elimination of isoniazid was due to the genetically controlled expression of the *N*-Acetyltransferase 2 enzyme (NAT2). The so-called slow acetylators of isoniazid elimination were homozygous for a recessive gene, whereas rapid acetylators were either homozygous or heterozygous for the normal or wild type gene. In the following years, numerous pharmaceutical drugs and some carcinogens, such as heterocyclic amines and aromatic amines, were found to be metabolized by this enzyme. Furthermore, the slow acetylator phenotype has been correlated with several diseases, including some types of cancer and adverse drug reactions [1]. The proportions of rapid and slow phenotypes vary in different ethnic groups. In Caucasians, 40–70% have the slow acetylator phenotype, whereas Asian populations have only 10–30% slow acetylators. Several polymorphisms in the NAT2 gene are responsible for the slow acetylator phenotype. Disease and pharmacologic associations are the basis for increasing interest in polymorphism detection in the new research field of molecular epidemiology. Major goals of this research area are to better understand individual disease susceptibility and risk assessment in order to develop prevention strategies. To that end, assays that allow for the investigation of large sample sets in as little time as possible are desirable.

Traditional methods for the detection of these polymorphisms rely on conventional PCR amplification and subsequent detection of restriction fragment polymorphisms. These approaches have several drawbacks. First of all, they are lengthy, labor-intensive procedures, which limit the number of samples that can be analyzed in a reasonable time. Secondly, unambigous genotype identification is dependent on complete restriction digestion, which cannot always be achieved when samples from difficult sources like serum or formaldehyde fixed paraffin-embedded tissues are used. Thirdly, this kind of analysis is confined to polymorphisms in sequence motifs that either contain a restriction enzyme cutting site or

---

* Brunhilde Blömeke (✉) (e-mail: b.bloemeke@rwth-aachen.de)
  Department of Dermatology, University Hospital RWTH Aachen, Pauwelsstr. 30, 52074 Aachen, Germany

allow the introduction of such sites via modified PCR primers. And, last but not least, one persisting problem with these methods is the cross-contamination that might occur during one of the many manual steps eventually leading to incorrect typing. To overcome these problems, we developed a simple and very rapid assay, using the new LightCycler technology. This assay is based on hybridization probes labeled with fluorescent dyes that allow fluorescence resonance energy transfer. The probes were designed for the detection of the NAT2 wild type allele, NAT2*4, and three polymorphic forms [NAT2*5A ($C^{481}T$), NAT2*6A ($G^{590}A$), NAT2*7A ($G^{857}A$)] [1]. This technique allows for easy, unambigous and rapid detection of polymorphisms with DNA from various sources (whole blood, serum, and formalin fixed paraffin-embedded tissue sections).

Materials

Equipment LightCycler Instrument

Reagents QIAmp blood kit (Qiagen)
Proteinase K (Roche Diagnostics)
RNase (Roche Diagnostics)
Buffer-saturated Phenol (Life Technologies)
Glycogen (Roche Diagnostics)
Amplification primers (Ark Scientific, Germany)
Hybridization probes (TIB MOLBIOL, Berlin)
LightCycler–DNA Master Hybridization Probes

Procedure

Sample Preparation DNA from whole blood was extracted with the QIAmp blood kit (Qiagen) according to the manufacturer's instructions. DNA from serum and paraffin-embedded tissues was extracted as described by Blömeke et al. [2]

Serum Two hundred and fifty microliters of serum were mixed with 3 vol buffer (50 mM Tris-Cl, pH 8.0; 1 mM EDTA; 100 mM NaCl). The samples were centrifuged for 12 min at 90,000 × g (or 15 min at 20,000 × g). The resulting pellets were resuspended in 200 µl of an aqueous solution of proteinase K (1 mg/ml) containing 0.1% SDS and incubated for 6 h at 65° C. The samples were extracted with 1 vol phenol/chloroform/isoamylalcohol (25/24/1). One hundred microliters of 10 M ammonium-acetate, $MgCl_2$ to give a final concentration of 0.01 M and 20 µg glycogen were added to the top phase. The DNA was precipitated with 2.5 vol absolute ethanol, the pellet collected by centrifugation, air dried and resuspended in 50 µl distilled water.

Formalin-Fixed Paraffin-Embedded Tissues DNA from formalin-fixed and paraffin-embedded tissue sections was extracted according to Goelz et al. [5]. Tissue slides were deparaffinized by two washing steps in xylene (15 min each) and three washing steps in ethanol (100%, 90% and

70%, 15 min each) and the tissues were subsequently air dried. The tissue material was then scraped off the slides into Eppendorf tubes and rehydrated with 250 µl TE (10 mM Tris, 1 mM EDTA, pH 8.0). The samples were then incubated at 37°C after the addition of proteinase K (30 µl of 5 mg/ml in 10% SDS). The digestion was performed until no remaining tissue pieces were visible (48–72 h). If necessary, additional aliqouts of proteinase K were added after several hours. The samples were then further processed as described above for serum extraction.

Primer Design

The amplification primers used for polymorphism detection in NAT 2 are unique based on published sequences [1]. Hybridization probes to distinguish four alleles [NAT2*4, NAT2*5A ($C^{481}T$), NAT2*6A ($G^{590}A$), NAT2*7A ($G^{857}A$)] were designed and synthesized by TIB MOLBIOL (Berlin). Details are shown in Table 1.

Table 1. Oligonucleotides

LC-PCR

N-Acetyltransferase 2 (Genebank Accession # AF055875 and D10872)				
	Position	Length	GC (%)	T_m (°C)
Primers				
TGCATTTTCTGCTTGACA	442	18	38.9	56.2
GTTGGGTGATACATACACAA	914 R	20	40.0	56.1
Product	442–914	473		
Hybridization probes (NAT2*5A)				
GCATTTTCTGCTTGACAGAAGAGAGAGGA-F	443	29	44.8	66.3
LCRed640-TCTGGTACCTGGACCAAAT-CAGGA-P	473	24	50.0	64.7
Hybridization probes (NAT2*6A)				
ACGTCTGCAGGTATGTATTCATAGACTCAAAAT-F	634R	33	36.4	65.5
LCRed640-TCAATTGTTCGAGGTTCAAGCGT-P	599R	23	43.5	63.1
Hybridization probes (NAT2*7A)				
TTCCTTGGGGAGAAATCTCGTGC-F	819	23	52.2	64.6
LCRed640-CAAACCTGGTGATGGATCCCT-P	843	21	52.4	61.8

PCR was performed with the following reaction mix:

	Volume [µl]	[Final]
LightCycler-DNA master mix	2	1×
MgCl$_2$ (20 mM)	2	3.0 mM
NAT2 primers (3.0 µM)	2+2	0.3 µM
Sensor probe (2.0 µM)	2	0.2 µM
Anchor probe (2.0 µM)	2	0.2 µM
H$_2$O (PCR grade)	7	
Total volume	19	

19 μl master mix and 1μl DNA (20–50 ng/μl) were added to each capillary. Sealed capillaries were spun (1000 x g for 15 s) and placed into the LightCycler.

The following PCR protocol was used for amplification:

- Denaturation for 2 min at 95°C.
- Amplification

Parameter	Value		
Cycles	1		
Type	Melting curve		
	Segment 1	Segment 2	Segment 3
Target temperature (°C)	95	40	94
Incubation time (s)	0	30	0
Temperature transition rate (°C/s)	20	20	0.2
Acquisition mode	None	None	Cont.
Gains	F1=1; F2=10		

- Melting curve analysis and cooling was done according to manufacturers instructions with a target temperature in segment 2 of 40°C and a temperature transition rate in segment 3 of 0.2°C with continuous acquisition.

Results

Figures 1–3 show representative examples of melting curves for genotyping analysis. A 473 bp NAT2 fragment was amplified separately in the presence of NAT2*5, NAT2*6 and NAT2*7 hybridization probes and melting curve analysis were performed. A sudden decrease in fluorescence indicates sensor probe melting. When samples are taken from either homozygous wild type or homozygous mutant individuals, melting curves with a single transition are obtained that may differ in their magnitude of fluorescence. If the sample is heterozygous, a bimodal melting curve results from the presence of two alleles with different melting points (Figs. 1a, 2a, 3a). A mathematical transformation of the data – the negative first derivative of fluorescence with respect to temperature (-dF/dT) vs T – allows for an easier interpretation of the results where the maximum represents the melting point or T_m (Fig. 1b, 2b, 3b).

Melting temperatures for the probes at all alleles are given in Table 2. Homozygous donors displayed one melting peak at T=65.7°C (WW) or T=58.6°C (NAT2*5A), and samples from heterozygous donors displayed two melting peaks at the same temperatures as the peaks from the homozygous samples. The shift in T_m between wild type and mutant alleles is due to a mismatch between the PCR fragment and the sensor hybridization probe, causing decreased thermal stability of the heteroduplex.

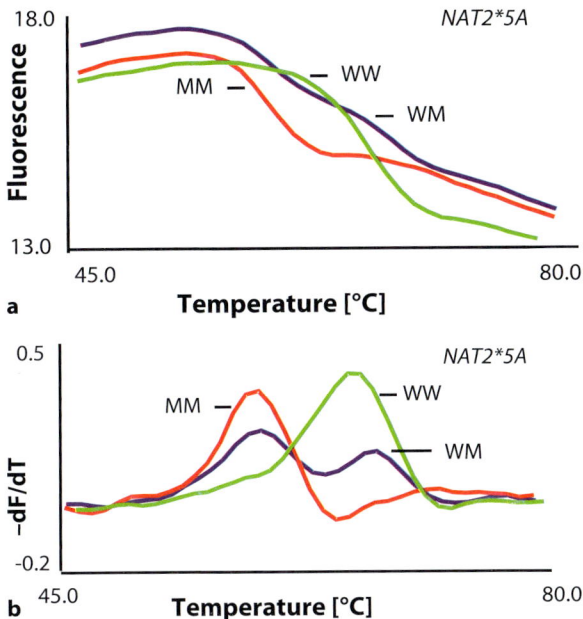

Fig. 1. Melting curves for different amplification products using NAT2*5A probes (**a**), and melting peaks (**b**). Data were acquired after 45 cycles of amplification with NAT 2 primers in the presence of wild type alleles, mutant alleles, and both alleles. The data from (**a**) were mathematically transformed to the negative first derivative of fluorescene with respect to temperature (–dF/dT) and displayed as melting peaks (**b**). Alleles with a perfect match to the hyprization probes have higher T_ms than alleles with a mismatch. Heterozygotes displayed two melting peaks at exactly the same temperatures as the respective homozygous samples

A lower absolute fluorescence level was observed with the NAT2*5 hybridization probe (Fig. 1). This is most likely due to overlapping of one of the hybridization probes with the forward PCR primer. However, fluorescence levels were high enough to allow unambigous product identification. A general finding was that higher fluorescence levels are observed with perfect matching hybridization probes than with probes containing one mismatched base (Table 2). Negative controls without added template did not show any signal (not shown).

The results of this facile approach were compared to our validated PCR-RFLP assay using in a total of 155 samples. These samples included DNA from whole blood (n=120), serum (n=25), and DNA from paraffin-embedded tissue sections (n=10). The concordance rate was 100% for each polymorphic site [4].

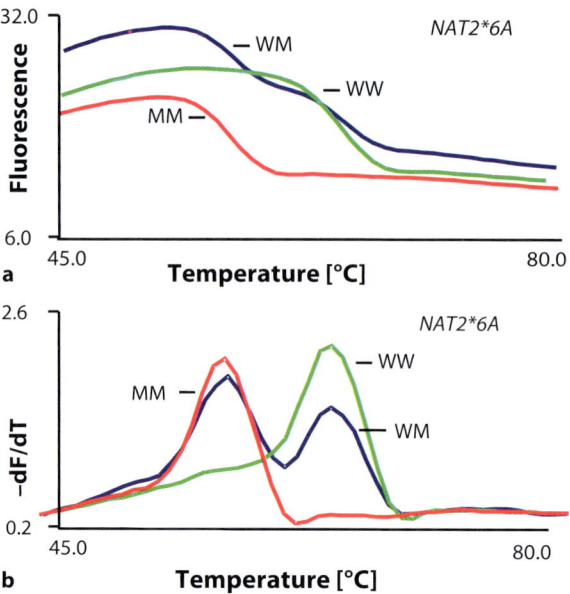

Fig. 2. Melting curves for different amplification products using NAT2*6A probes (**a**), and melting peaks (**b**)

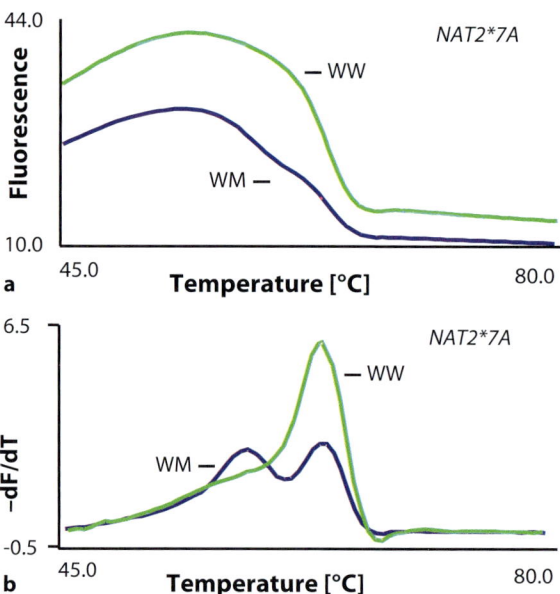

Fig. 3. Melting curves for different amplification products using NAT2*7A probes (**a**), and melting peaks (**b**)

Table 2. Melting temperatures at NAT2*5A, NAT2*6A, and NAT2*7A loci

Locus	Allele	Pairing with sensor probe	T_m (observed)
NAT2*5A	WW	C-G match	65.7
	NAT2*5A	C-A mismatch	58.6
NAT2*6A	WW	G-C match	64.5
	NAT2*6A	G-T mismatch	57.1
NAT2*7A	WW	G-C match	63.4
	NAT2*7A	G-T mismatch	58.3

Comments

Maintenance of Hybridization Probes

An important component of successful genotyping lies in the storage of the hybridization probes. We prefer preparing multiple aliquots, in order to avoid repetitive freezing and thawing. We dilute them to our working concentration and store them in aliquots for 10 or 30 samples at –20°C. We do not refreeze them. For long term storage, -70°C is optimal.

Controls

The analysis of samples for epidemiology or disease diagnosis requires specific and consistent quality control measures [3]. We include a known standard for each informative genotype in each experiment to verify assay performance. We do repeats on 20% of the samples to verify the results.

References

1. Blum M, Demierre A, Grant DM, Heim M, Meyer UA (1991) Molecular mechanism of slow acetylation of drugs and carcinogens in humans. Proc Natl Acad Sci USA 88: 5237–5241
2. Blömeke B, Bennet WP, Harris CC, Shields PG (1997) Serum, plasma and paraffin-embedded tissues as sources of DNA for studying cancer susceptibility genes. Carcinogenesis 18: 1271–1275
3. Blömeke B, Shields PG (1999) Laboratory methods for the determination of genetic polymorphisms in humans. IARC Sci Publ148 (13): 135–149
4. Blömeke B, Sieben S, Spötter D, Landt O, Merk HF (1999) Identification of N-Acetyltransferase 2 genotypes by continuous monitoring of fluorogenic hybridizytion probes. Anal Biochem 275: 93–97
5. Goelz SE, Hamilton SR, Vogelstein B (1985) Purification of DNA from formaldehyde fixed and paraffin embedded human tissue. Biochem Biophys Res Commun 130: 118–126

Genotyping of Cytochrome P450 2D6*4 Mutation with Fluorescent Hybridization Probes Using LightCycler

JEANETTE BJERKE, CHUNG-CHE CHANG*, CHUCK SCHUR, STEVEN WONG, NAZIHA NUWAYHID

Introduction

The family of cytochrome P450 (CYP) is an important oxidative enzyme system involved in the metabolism of important drugs. More than 30 isoenzymes of CYP have been characterized in humans [1–4]. One of these, CYP2D6 (debrisoquine hydroxylase) is responsible for the metabolism of anti-depressants and neuroleptic agents.

The CYP2D6 gene (GeneBank index number: 181303) resides in the CYP2D6 cluster on chromosome 22 in association with the CYP2D7P and CYP2D8P pseudogenes. The genetic variants of the CYP2D6 are grouped into three classes of phenotypes based on the extent of drug metabolism: extensive (EM), poor (PM), and ultraextensive (UEM) metabolizers resulting in normal, high and low blood levels of these drugs. The most common mutation CYP2D6*4, (allelic frequency: 28.6%) involves a base substitution of G_{1934} with A. The homozygous mutants lead to PM, while the wild types are EM [1–3]. Heterozygous individuals have partially impaired metabolic activity and have been designated intermediate metabolizers. Furthermore, the clinical significance of heterozygosity for one inactive allele has not been clearly demonstrated [3]. Ultra-extensive metabolism (UEM) is an autosomal dominant trait arising from gene amplification [1].

Current methods for genotyping point mutations, such as CYP2D6*4, include oligonucleotide ligation, allele-specific oligonucleotide hybridization, and PCR restriction fragment length analysis. These methods require multiple manual steps and are time consuming, labor intensive, and difficult for routine clinical application.

We describe here a strategy for developing an assay using fluorescent hybridization probes with LightCycler to detect the mutation at CYP2D6*4 site. This assay provides a short turn around time, about 40 min after DNA preparation, and is suitable for routine clinical application.

---

* Chung-Che Chang (✉) (e-mail: jeffchang@pol.net)
  Department of Pathology, Medical College of Wisconsin, Milwaukee, WI 53226, USA

Materials

Equipment
LightCycler Instrument
Primer Designer 4 software (Scientific and Education Software)

Reagents
Puregene DNA Isolation Kit (Gentra Systems Inc.)
Amplification Primers (Operon Technologies)
Hybridization Probes (Operon Technologies)
Opti Prime 10 × Buffer #1 (Stratagene Inc.)
$MgCl_2$ Stock Solution (Idaho Technology)
dNTPs-dUTP (1.25 mM Stock containing dATP, dCTP, dGTP, dTTP, dUTP in a ratio of 1:1:1:1:1) (Amersham Pharmacia Biotech)
AmpliTaq Gold polymerase (Perkin Elmer)
Uracil-N-Glycosylase (RocheDiagnostics, Mannheim, Germany)

Procedure

Sample Preparation
DNA was extracted from whole blood samples using the Puregene DNA Isolation Kit (Gentra Systems Inc.). The DNA was quantified spectrophotometrically and stored at –20°C.

Primer Design
The forward (5′ CCA ACC ACT CCG GTG GG 3′) and the reverse (5′ AAT CCT GCT CTT CCG AGG C 3′) primers were modified from published data using Primer Design 4 Software to produce a 347 bp PCR product. Hybridization probes, (anchor probe: 5′ GTC CAA GAG ACC GTT GGG GCG A FITC 3′; mutant probe: 5′ LAG GGG CGT CCT GGG CP 3′ (L=Amino Linker/LCRed640, P=Phosphate)) were designed according to published guidelines [5]. Briefly, the two probes were designed to have different melting temperatures (T_m). The T_ms differed by about 6°C, which allowed the mutant probe to "melt off" the target at a lower temperature than the anchor probe during melting curve analysis. The disassociation of the mutant probe was monitored by a change in fluorescence intensity at a real-time fashion (please refer to the result section for details).

Table 1. Oligonucleotides

Cytochrome P450 2D6 (Genebank Accession # 181303)				
	Region	Length	GC (%)	T_m (°C)
Primers				
CCAACCACTCCGGTGGG	3366–3383	17	71	64.4
AATCCTGCTCTTCCGAGGC	3693–3712	19	58	63.7
Mutation Probe				
LCRed640AGGGGCGTCCTGGGGP	3461–3476	15	80	65.1
Anchor Probe				
GTCCAAGAGACCGTTGGGGCGAF	3475–3497	22	64	68.4
Product		347		

Genotyping of Cytochrome P450 2D6*4 Mutation with Fluorescent Hybridization Probes Using LightCycler

LC-PCR

Hybridization Probe Master Mix:

	Volume [µl]	[Final]
Opti Prime Buffer #1 (10×)	2.0	1×
dNTPs (1.25 mM Stock)	3.2	200 µM
$MgCl_2$ (40 mM)	0.5	2.5 mM
Primers (10 µM each)	1.0 + 1.0	0.5 µM
Mutation probe (10 µM)	0.6	0.3 µM
Anchor probe (10 µM)	0.4	0.2 µM
AmpliTaq Gold DNA Polymerase (5 U/µl)	0.6	3 units
Heat-Labile Uracil-DNA Glycosylase (5 U/µl)	1.0	1 unit/reaction
Sample DNA	2.0	100 ng/reaction
Sterile glass distilled H_2O	7.7	
Total volume	20	

The reaction mix was added to the capillary tubes which were sealed, centrifuged in a microcentrifuge, and placed in the LightCycler rotor.

The following PCR protocol was used:

- Initial Denaturation and Activation of AmpliTaq Gold Polymerase
- Denaturation and activation of AmpliTaq Gold polymerase for 10 min at 95°C.
- Amplification

Parameter	Value			
Cycles	45			
Type	Quantification			
	Segment 1	Segment 2	Segment 3	
Target temperature (°C)	95	60	75	
Incubation time [s]	0	10	15	
Temperature transition rate [°C/s]	20	20	20	
Acquisition mode	None	Single	None	
Gains	F1=2	F2=25	F3=10	

- Melting Curve Analysis

Parameter	Value				
Cycles	1				
Type	Melting Curve				
	Segment 1	Segment 2	Segment 3	Segment 4	Segment 5
Target temperature (°C)	95	50	45	40	82
Incubation time [s]	0	20	20	60	0
Temperature transition rate [°C/s]	20	20	20	20	0.2
Acquisition mode	None	None	None	None	Step

Genotyping by Restriction Enzyme Analysis

After the melting curves were completed, the samples were cooled to 40°C for 60 s.

CYP2D6*4 mutations can be detected by treating the 347-bp amplified product with Mva I restriction enzyme. The PCR products were incubated with Mva I at 37°C for 4 h. The fragments were visualized on a 3% agarose gel after digestion. The wild-type allele contains the restriction site, 5'CCCCAGGA 3', and results in two fragments, 242 and 105 bps, following digestion. The mutant allele contains a point mutation, 5'CCCCAAGA 3', which leads to the deletion of the restriction site for Mva I. This results in an undigested band at 347 bp. Heterozygotes result in three bands: one band at 347 bp (from the mutant allele) and bands at 242 bp and 105 bp (from the wild-type allele) (Fig. 1).

Results

Genotyping by Fluorescent Hybridization Probes

During the hybridization steps in the quantification cycles, the anchor and the mutant probes were brought close together by annealing to the amplification product, and fluorescence resonance energy transfer (FRET) occurred. The FRET led to the increase of fluorescent intensity with advancing cycles.

During fluorescence, melting curve analysis after the completion of the amplification, the probe then "melts-off" at a characteristic temperature (T_m). The T_m

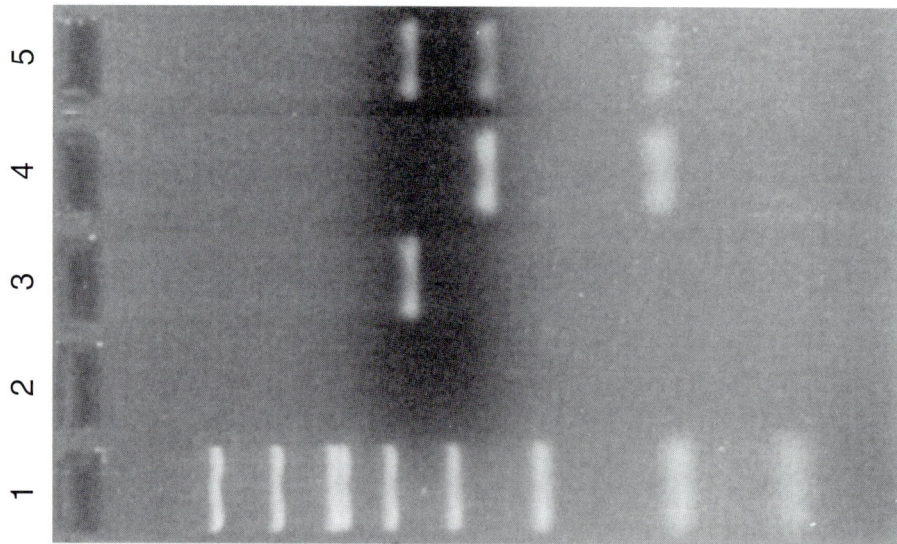

Fig. 1. Ethidium bromide stained 3% agarose gel illustrating the Mva I restriction fragments of the Cytochrome P450 2D6*4 PCR products. Lane 1: Molecular weight marker (Gel marker I, Research Genetics). Lane 2: Water Control. Lane 3: Homozygote- 347 bp fragment. Lane 4: Wild-type- 242 bp and 105 bp fragments. Lane 5: Heterozygote 347 bp, 242 bp, and 105 bp fragments

depends on the stability of the probe-allele hybridized double strand structure. The probes were designed so that during hybridization, the LCRed640-labeled probe formed an A:C mismatch with the mutant-type allele and a complete match with the wild-type allele extended by the forward primer. The single base mismatch in the mutant allele decreases the stability of the probe-allele double strand structure, resulting in a lower T_m (T_m=51.2°C) than that of the wild-type allele (T_m=63.2°C). The heterozygote has T_ms of the wild-type and mutant allele (Fig. 2). We studied 22 samples from healthy volunteers using the above assay. One of them was mutant homozygote, two were mutant heterozygotes, and the rest were wild-type.

Comparison of the Genotyping by Fluorescent Hybridization Probes and by Restriction Enzyme Analysis

There was a 100% match in genotyping results of homozygotes, heterozygotes, and wild-types among the samples studied using the two different methods. However, the turn around time was much shorter with fluorescent hybridization probes than that with restriction enzyme analysis (40 min vs 6–8 h, respectively). The technician time was also substantially reduced by the former method.

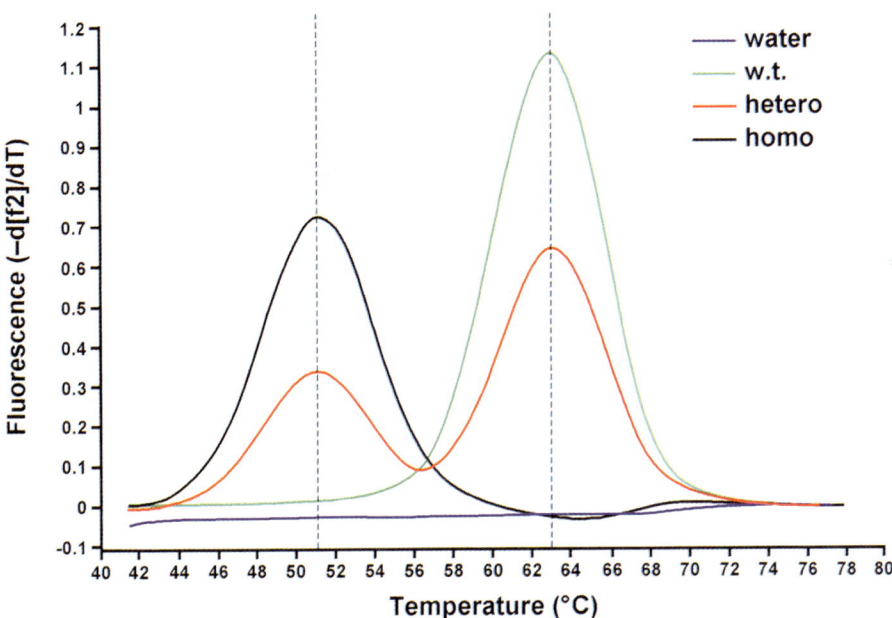

Fig. 2. Melting peaks for Cytochrome P450 2D6*4 Genotyping. Each peak represents the melting temperature of the LCRed640 labeled probes as it "melts off" the template DNA. The melting temperature of the probe complementary to the wildtype template is 63.2°C (*green curve*). The presence of the 2D6*4 point mutation in the homozygote results in a 12°C decrease in the melting temperature (*black curve*). Heterozygote exhibits both peaks representing the normal and mutant alleles (*red curve*). The sample containing only water (*blue curve*) shows no peaks. w.t., wild-type; *hetero*, heterozygote; *homo*, homozygote

Comment

Primer and Probe Design

For the most efficient amplification, primers and the mutant probe should have the T_ms as close to each other as possible. The anchor probe should have a T_m about 5°C above that of mutant probe. The use of computer software, Primer Designer 4, is helpful for designing the primers and probes

Optimization of Assay

The most important factors for optimizing the assay are Mg^{++} concentration, the concentration of each probe, and the temperature profile for final annealing and melting of the probes. Different probe concentrations change the signal intensity generated from the wild-type and mutant alleles (data not shown). Multistep annealing during the melting curve analysis enhances the signal intensity compared to one step annealing (data not shown).

Conclusion

Our results indicate that this assay is a simple, reliable and sensitive method for genotyping of CYP2D6 with excellent turn around time (about 45 min) after DNA preparation. This method should be suitable for routine clinical use and for large series of epidemiological studies.

References

1. Linder MW, Prough RA, Valdes RJ (1997). Pharmacogenetics: a laboratory tool for optimizing therapeutic efficiency. Clin Chem 43:254–266
2. Brosen K, Gran LF (1988). First-pass metabolism of imipramine and desiparmine: impact of the sparteine oxidation phenotype. Clin Pharmacol Ther 43:400–406
3. Linder M, Valdes R (1999). Pharmacogentics: Fundamentals and applications. Therapeutic Drug Monitoring American Association for Clinical Chemistry, Inc. 20:7–23
4. Wong S (2000) Challenges of toxicology for the millennium. Ther Drug Monit 22:52–57
5. Caplin BE Rasmussen RP Bernard PS Wittwer CT (1999) LightCycler Hybridization Probes. Biochemica 1: 5–8

Fluorescent Hybridization Probe Detection of the F508del Cystic Fibrosis Allele on the LightCycler

Cameron N. Gundry*

Introduction

Cystic fibrosis is a prevalent and well-studied autosomal recessive disorder mainly affecting Caucasian populations at a frequency of about 0.05% [1]. The cystic fibrosis transconductance regulator gene that is altered in the disease is on chromosome 7 [2]. Population screening has uncovered nearly 900 variants in the gene to date [3]. Many of these are disease-causing mutations. Approximately 70% of disease-causing alleles have a 3 base pair deletion at codon 508. F508del homozygous individuals comprise approximately 50% of all clinical cases of cystic fibrosis.

The methods used for detecting variants in the cystic fibrosis gene include: direct sequencing [4], heteroduplex analysis [5], denaturing gradient gel electrophoresis [6], nick-translation PCR [7], allele-specific amplification [8], fluorescence-based oligonucleotide ligation [9], template-directed-dye terminator incorporation [10], and a LightCycler-based fluorescent-labeled oligonucleotide melting assay [11].

The assay we describe here uses the fluorimeter of the LightCycler to screen human genomic DNA samples for the F508del mutation, using melting curve analysis. The system is sensitive enough to distinguish the F508del mutation from the F508C single base variant, which may be a factor in male sterility [13, 14]. This assay is currently being used to accelerate throughput of F508del detection at Neogen Screening Inc. (Pittsburgh, PA, USA).

Materials

Equipment

LightCycler Roche Diagnostics, Mannheim (RMB)
LightCycler Capillaries (RMB)

Reagents

Oligonucleotide primers (Idaho Technology Biochem, Salt Lake City, UT)
LCRed640- and fluorescein-labeled probes (Operon Technologies, Alameda, CA)
LightCycler-DNA Master Hybridization probes (RMB)
   The probes were HPLC purified at Idaho Technology Biochem.

---

* Cameron N. Gundry (✉) (e-mail: Cameron.gundry@path.med.utah.edu)
  Department of Pathology, University of Utah, Medical School, Salt Lake City, UT 84132, USA

Procedure

DNA Extraction

Human genomic DNA was extracted from peripheral blood leukocytes according to standard phenol-chloroform DNA extraction and ethanol precipitation techniques [15] and stored at −20°C. Preparation with a Chelex matrix that binds PCR inhibitors from lysed leukocytes [16] also results in equivalent genotyping on the LightCycler.

Primer and Probe Design

The 5'-LCRed640-labeled probe was designed to hybridize over the mutation locus in the middle of the region of the 3 base pair deletion. The fluorescein probe is 29 base pairs long and is 3'-fluorescein-labeled. The LCRed640 probe was made shorter (24 base pairs long) to reduce its thermal stability to a lower value than that of the 3'-fluorescein probe. It is important to design the hybridization probes in such a way that the Tm of the "anchor" probe, the probe which is not overlying the mutation, is significantly higher than that of the probe which overlies the mutation or base change of interest. This ensures that the sequence of interest is being interrogated. The predicted Tm of the LCRed640 probe was 10°C lower than the 3'-fluorescein probe. The 5'-LCRed640 probe was 3' phosphorylated to prevent extension during PCR. Table 1 lists the oligonucleotide sequences used.

Figure 1 shows the relative locations within the product of the forward and reverse primers as well as the adjacent hybridization probes.

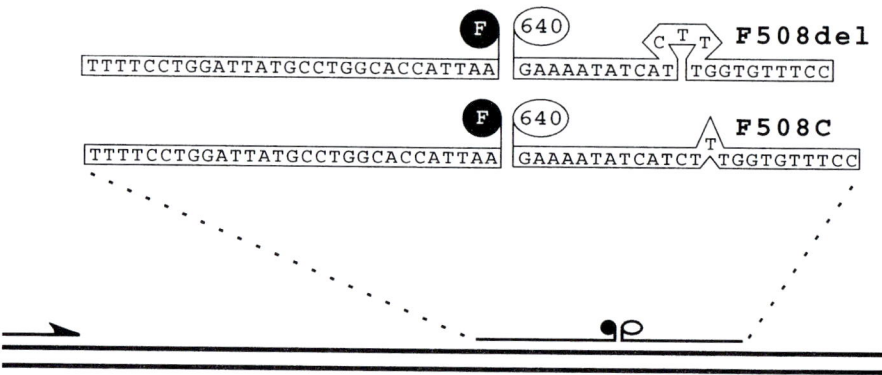

Fig. 1. A schematic showing the orientation of primers and probes. There is a one base pair gap between the adjacent fluorescein- and LCRed640-labeled probes. The probe set consists of a 3'-fluorescein-labeled 29-mer probe that results in fluorescence resonance energy transfer with a 5'-LCRed640-labeled 24-mer probe when both are hybridized. The raised T in the 24-mer probe designates the position of the nucleotide creating a single base pair mismatch when hybridized to the F508C allele. The diagram also shows the position of a 3 base pair bulge loop formed in the probe when hybridized to the F508del allele. The LCRed640-labeled probe was designed to be completely complementary to the wild-type allele

Table 1. Oligonucleotides

Cystic fibrosis gene, exon 10 (Genebank Accession #M55115)				
	Position	Length	GC (%)	T_m (°C)
Primers				
GGAGGCAAGTGAATCCTGAG	245	20	55.0	64.0
CCTCTTCTAG TTGGCATGCT	500 R	20	50.0	63.4
Product	245	256		
Probes				
TTTTCCTGGATTATGCCTGGCACCATTAA-F	395	29	41.4	68.5
LCRed640-GAAAATATCATCTTTGGTGTTTCC-P	425	24	33.3	58.7

Table 2. Master mix

	Volume [µl]	[Final]
Fluorescein probe (1.0 µM)	2.0	0.20 µM
LCRed640 probe (2.5 µM)	1.0	0.25 µM
Primers (5 µM each)	1.0	0.50 µM
LightCycler-DNA Master Hybridization probes	1.0	1x
$MgCl_2$ (25 mM)	1.2	4.0 mM
PCR grade water	2.8	
Master Mix volume	9	
DNA	1.0	
Total volume	10	

Note: This 10× master mix can be stored in the dark at 4°C in the absence of template DNA for several days without negative effects on PCR or subsequent detection.

Reaction Preparation

Table 2 shows the hybridization probe master mix (enough for one 10.0 µl reaction).

The master mix is prepared beforehand and 9 µl are pipetted into the upper chamber of the capillaries. Next, 1.0 µl of extracted genomic DNA (A_{260}=0.1–1.0) was added for a final in-tube volume of 10.0 µl. Capillaries were sealed and centrifuged in a microcentrifuge.

Program the following protocols in the LightCycler software:

- Amplification

Parameter	Value	
Cycles	35	
Type	Quantification	
	Segment 1	Segment 2
Target temperature (°C)	95	63
Incubation time [s]	0	25
Temperature transition rate [°C/s]	20	20
Acquisition mode	None	Single
Display mode	F2	
Fluorimeter gain (F2)	25	

- Melting curve analysis immediately following amplification

Parameter	Value			
Cycles	1			
Type	Melting curves			
	Segment 1	Segment 2	Segment 3	Segment 4
Target temperature (°C)	95	71	37	70
Incubation time [s]	0	0	0	0
Temperature transition rate [°C/s]	20	2.0	0.1	0.1
Acquisition mode	None	None	Cont	Cont.
Display mode	F2			
Fluorimeter gain (F2)	25			

Results

Forty-two samples from cystic fibrosis patients were genotyped for F508del on the LightCycler with perfect correlation to results obtained by allele-specific amplification [11]. The Tm difference obtained during heating was 11.4°C between wild-type and F508del alleles. Two samples had an atypical melting profile and were found to have a T:G transversion within codon 508, known as F508C [12]. The two F508C samples had an extra peak at 56.9 °C, which is 5.7 °C lower than the wild-type peak. Figure 2 shows the derivative melting curves of 4 DNA samples that demonstrate the Tms of wild-type, F508del, and F508C alleles. The amplification protocol took under 17 minutes. The melting and annealing protocols took another 11 minutes to complete. The observed Tms of the different alleles are shown in Table 3.

Fig. 2. Genotyping the F508del, F508C, and wild-type samples. The genotypes are as shown in the upper right-hand corner of the figure. The upper graph plots temperature (T) versus LCRed640 fluorescence (F). The sudden decreases in fluorescence seen in the top graph correspond to the melting peaks in the bottom derivative graph. The heterozygous samples shown two apparent "melting peaks" and identify the T_ms of each allele

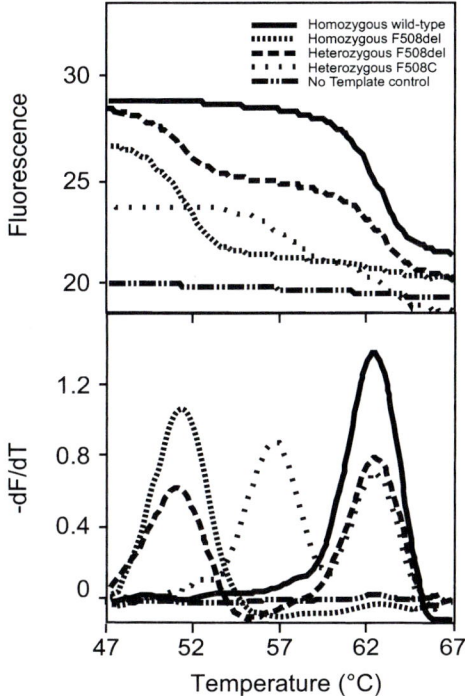

Table 3. Predicted and observed Tms during heating and cooling of different alleles at the cystic fibrosis F508 locus

Locus	Allele	Pairing	Predicted T_m (°C)	Observed T_m (°C) Heating	Cooling
F508	WT	CTT-GAA match	58.7	62.6	62.0
	F508	CT-C mismatch	54.0	56.9	55.2
	F508del	CTT non-pairing	Nd[a]	51.2	51.0

[a] No prediction available.

Comments

The LightCycler can easily amplify and genotype point mutations from genomic DNA [17, 18]. Melting curve analysis identifies each allele by its characteristic Tm. Here we demonstrate that 3-base pair deletions can also be genotyped by the same method. The probe loops over the 3-base pair deletion and is destabilized by an observed 11.4°C, a greater destabilization than seen in most single base mismatches.

Annealing curves during cooling can also be used for genotyping and produce results comparable to traditional melting, or heating analysis [11]. After amplification and a brief denaturation at a high temperature (>90°C) the samples are cooled at a slow rate (0.1°C/s). Probe hybridization is evidenced by an increase of fluorescence and derivative curves can be used to estimate the probe/template hybrid Tm. The Tms estimated by heating are higher than those estimated by cooling, because equilibrium conditions are not achieved [11]. After amplification, both slow cooling and slow heating segments can be performed and the results of both annealing and melting analysis compared (Table 3).

This assay was originally designed to only distinguish F508del from wild-type alleles. However, two samples produced unexplained aberrant hybridization profiles, with melting temperature peaks approximately mid-way between the wild-type and F508del melting temperatures. This raised the possibility of a sequence polymorphism that was different from both the F508del and the wild-type sequences at the locus interrogated. To test this hypothesis, the two aberrant samples were sequenced, revealing the single base alteration F508C in both samples. This demonstrates the utility of fluorescent melting curve analysis to detect multiple mutations under one probe. Other mutations near codon 508 in the CFTR gene under the LCRed640 probe should also give unique Tms, including the benign I506V [19] and the disease-causing I507del and I506del mutations [20]. Even greater multiplexing at different loci can be achieved by different color probes [21].

References

1. NIH Consensus Statement (1997) Genetic testing for cystic fibrosis. 15 (4): 1–37
2. Riordan JR, Rommens JM, Kerem B, Alon N, Rozmahel R, Grzelczak Z, et al (1989) Identification of the cystic fibrosis gene: cloning and characterization of complementary DNA. Science 245: 1066–1073
3. Cystic Fibrosis Genetic Analysis Consortium (1999) Mutation Table. http://www.genet.sickkids.on.ca/cftr-cgi-bin/FullTable.
4. Kobayashi K, Knowles MR, Boucher RC, O'Brien WE, Beaudet AL (1990) Benign missense variations in the cystic fibrosis gene. Am J Hum Genet 47: 611–615
5. Dodson LA, Kant JA (1991) Two-temperature PCR and heteroduplex detection: application to rapid cystic fibrosis screening. Mol Cell Probes 5 (1):21–25
6. Fanen P, Ghanem N, Vidaud M, Besmond C, Martin J, Costes B, Plassa F, Goossens M (1992) Molecular characterization of cystic fibrosis: 16 novel mutations identified by analysis of the whole cystic fibrosis conductance transmembrane regulator (CFTR) coding regions and splice site junctions. Genomics 13 (3): 770–776
7. Lee LG, Connell CR, Bloch W (1993) Allelic discrimination by nick-translation PCR with fluorogenic probes. Nucleic Acids Res 21(16): 3761–3766
8. Wittwer CT, Marshall BC, Reed GH, Cherry JL (1993) Rapid cycle allele-specific amplification: studies with the cystic fibrosis ΔF508 locus. Clin Chem 39 (5): 804–809
9. Eggerding FA, Iovannisci DM, Brinson E, Grossman P, Winn-Deen ES (1995) Fluorescence-based oligonucleotide ligation assay for analysis of cystic fibrosis transmembrane conductance regulator gene mutations. Hum Mutat 5: 153–165
10. Chen X, Zehnbauer B, Gnirke A, Kwok P (1997) Fluorescence resonance energy transfer detection as a homogeneous DNA diagnostic method. Proc Natl Acad Sci U S A 94: 10756–10761
11. Gundry CN, Bernard PS, Herrmann MG, Reed GH, Wittwer CT (1999) Rapid F508del and F508C assay using fluorescent hybridization probes. Genet Test 3(4): 365–370

12. Kobayashi K, Knowles MR, Boucher RC, O'Brien WE, Beaudet AL (1990) Benign missense variations in the cystic fibrosis gene. Am J Hum Genet 47(4): 611–615
13. Meschede D, Eigel A, Horst A, Nieschlag E (1993) Compound heterozygosity for the delta F508 and F508C cystic fibrosis transmembrane conductance regulator (CFTR) mutations in a patient with congenital bilateral aplasia of the vas deferens. Am J Hum Genet 53(1): 292–293
14. Dork T, Dworniczak B, Aulehla-Scholz C, Wieczorek D, Bohm I, Mayerova A et al (1997) distinct spectrum of CFTR gene mutations in congenital absence of vas deferens. Hum Genet 100 (3–4): 365–377
15. Thomas SM, Moreno RF, Tilzer LL (1989) DNA extraction with organic solvents in gel barrier tubes. Nucleic Acids Res 17: 5411
16. Walsh PS, Metzger DA, Higuchi R (1991) Chelex as a medium for simple extraction of DNA for PCR-based typing from forensic material. BioTech 10 (4): 506–513
17. Lay MJ, Wittwer CT (1997) Real-time fluorescence genotyping of factor V Leiden during rapid-cycle PCR. Clin Chem 43 (12): 1–6
18. von Ahsen N, Schutz E, Armstrong VW, Oellerich M (1999) Rapid detection of prothrombotic mutations of prothrombin (G20210A), factor V (G1691A), and methylenetetrahydrofolate reductase (C677T) by real-time fluorescence PCR with the LightCycler. Clin Chem 45(5): 694–696
19. Kobayashi K, Knowles MR, Boucher RC, O'Brien WE, Beaudet AL (1990) Benign missense variations in the cystic fibrosis gene. Am J Hum Genet 47: 611–615
20. Nelson PV, Carey WF, Morris CP (1991) Identification of a cystic fibrosis mutation: deletion of isoleucine506. Hum Genet 86(4): 391–393
21. Bernard PS, Pritham GH, Wittwer CT (1999) 1999 Color multiplexing hybridization probes using the apolipoprotein E locus as a model system for genotyping. Anal Biochem 10;273(2): 221–228

Genotyping β-globin Mutations (Hb S, Hb C, Hb E) by Multiplexing Probe Color and Melting Temperature

Mark G. Herrmann*

Introduction

Human β-globin has over 50 known mutations in exon 1 causing various hemoglobinopathies (1,2). The mutations include base pair substitutions, deletions and splicing defects. Hemoglobin S and C, (codon 6) and E (codon 26) are base pair substitutions that occur often enough to allow for routine screening (Hb S occurs at an allele frequency of 1:400 in African American's)(3). High performance liquid chromatography (4) and isoelectric focusing (5) are routinely used for primary patient screening. However, phenotypic screening does not always necessarily coincide with the genotype, for example Hb S, Hb G$_{Norfolk}$ and Hb$_{Fort-de-France}$ all have the same isoelectrofocusing point (6). Genotyping can be done by allele specific amplification (7), DNA amplification followed by restriction digestion (8) or by fluorescently labeled allele specific primers and gel based detection with color photography (9). These methods require hours to days for completion. Recently, genotyping by determining the melting temperature (T_m) of hybridized fluorescently-labeled oligonucleotide probes was reported (10,11). We have extended the power of this technique by multiplexing different colored probes to simultaneously genotype multiple alleles (12). This procedure has been applied to homogenous genotyping of hemoglobin S, C, and E in a single tube by melting curve analysis by using different colored probes and T_m multiplexing.

Materials

LightCycler Instrument (Roche Diagnostics, Mannheim, Germany)	Equipment
Unlabeled primers (Operon Technology: Alameda, CA)	Reagents
Fluorescein labeled oligonucleotide (Operon Technology: Alameda, CA)	
LC Red 705 labeled oligonucleotide (IT Biochem: Salt Lake City, UT)	
LC Red 640 labeled oligonucleotide (IT Biochem: Salt Lake City, UT)	
LightCycler-DNA Master Hybridization Probes (Roche Diagnostics, Mannheim)	Kits

---

* Mark G. Herrmann (✉) (e-mail: mark.herrmann@path.med.utah.edu)
University of Utah, Department of Pathology, Salt Lake City, UT 84132

Procedure

Primer and Probe Design

Primers and probes were selected from the human β-globin gene sequence. Due to homology between β-globin and δ-globin sequences, primer sites were chosen for β-globin specific amplification. The Hb S and Hb C mutation probe is 3' labeled with LC red 640, and the Hb E mutation probe is 5' labeled with LC Red 705. A single "anchor" probe, labeled on both ends with fluorescein, has a T_m about 15°C higher than the detection probes and is used to excite the detection probe fluorophores through fluorescence resonance energy transfer (FRET). The greater T_m of the fluorescein probe allows it to stay annealed through out the melting of the mutation probes. Figure 1 shows the position of the probes and primers with sequence information specified in Table 1.

The mutation probes are designed to anneal to the antisense strand. The probe detecting the Hb S and Hb C mutations matches the Hb S allele. Destabilizing mismatches occur when the probe is annealed to the wild type sequence (1 mismatch, position 62206-T:T) or to Hb C sequence (2 mismatches, position

Table 1. Oligonucleotides

Human β-globin (Genebank Accession # U01317)					
	Position	Exon	Length	GC (%)	T_m (°C)
Primers					
AGTCAGGGCAGAGCCATCTA	62111	1	20	55 %	64.3
GTTTCTATTGGTCTCCTTAAACCTG	62324R	1	25	40 %	61.7
Product	(62111-62324)		214		
Probes**					
CTCCT<u>G</u>TGGAGAAGTCTGC-LC Red 640	62200	1	19	57.9 %	59.5
LC Red 705-GTTGGTGGT<u>A</u>AGG-CCCTGG-P	62256	1	19	63.2 %	62.6
F-GTTACTGCCCTGTGGGGCAA-GGTGAACGTGGATGA-F	62220	1	35	57.1 %	76.7

** Positions of sequence variation are underlined

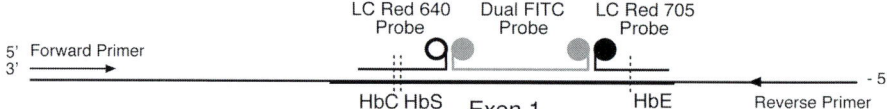

Fig. 1. Hemoglobin Variant Detection Schematic. The anti-sense strand of a 214 bp fragment of the human β-globin gene is shown. The complementary forward primer and fluorescent probes are indicated above the antisense strand. The boundaries of Exon 1 are shown as thick horizontal line with the positions of the hemoglobin variants Hb C, S, E as dashed vertical lines.

62205-G:T, position 62206-T:T) . The probe detecting the Hb E allele is designed to anneal to the antisense strand matching the mutant allele. When the Hb E probe is annealed to wild type sequence a single destabilizing mismatch occurs (position 62265-A:C).

DNA Preparation

Human genomic DNA was extracted by one of several methods, including conventional proteinase K digestion, phenol:chloroform extraction, and ethanol (14) and Chelex100 extraction (15). Between 10 and 100 ng of genomic DNA was added per 10 µl reaction.

PCR

The following master mix was used for n samples:

	Volume [µl]	[Final]
Forward Primer 5 µM	2.0	0.5 µM
Reverse Primer 10 µM	2.0	1.0 µM
DNA Master Hyb. Probes 10x	2.0	1x
$MgCl_2$ Buffer 25 mM	1.6	30 mM
Fluorescein Probe 4 µM	2.0	0.4 µM
LC Red 640 Probe 2 µM	2.0	0.2 µM
LC Red 705 Probe 2 µM	2.0	0.2 µM
PCR Grade H_2O	2.4	
Master Mix Volume	16	
DNA	4.0	
Total Volume	20.0	

16 µl of master mix and 4 µl of genomic DNA template are mixed and then transferred to the capillary tube, sealed, centrifuged and placed in the LightCycler rotor.

- Amplification

Parameter	Value	
Cycles	40	
Type	Quantification	
	Segment 1	Segment 2
Target temperature (°C)	95	63
Incubation time [s]	0	30
Temperature transition rate [°C/s]	20	20
Acquisition mode	None	Single
Gain	F1 = 5, F2 = 25, F3 = 30	

Genotyping Immediately following amplification the probes are melted from the template for genotyping. The following conditions are used:

Parameter	Value		
Cycles	1		
Type	Melting Curve		
	Segment 1	Segment 2	Segment 3
Target temperature (°C)	95	35	80
Incubation time [s]	0	30	0
Temperature transition rate [°C/s]	20	1	0.1
Acquisition mode	None	None	Continuous
Gain	F1 = 5, F2 = 25, F3 = 30		

Color Compensation Due to fluorophore emission overlap, each channel contains an emission composite of all probes. Individual color signals can be obtained by correcting for the amount of overlap in each channel of each fluorophore (16,12). Color compensation was performed according to manufacture instructions (Roche Molecular Biochemicals).

Results

Melting curve analysis of the LC Red 640 detection probe at codon 6 showed clear resolution of each allele. Figure 2A shows the –dF/dT vs. temperature plots of the homozygous genotypes. The melting peak of the matched Hb S probe to the Hb S target has a T_m of 62.7°C. The single mismatch between the probe and wild type target has a T_m of 55.6°C, while the two mismatches between the probe and the Hb C target lowers the T_m further to 50.1°C. Figure 2B shows melting curve analysis of the heterozygous genotypes. The compound heterozygote HbS/C has melting peaks corresponding to both the Hb S and Hb C alleles. The Hb S-trait has melting peaks associated with both the Hb S and wild type alleles. The Hb C-trait is identified by melting peaks associated with both wild type and Hb C.

Melting curve analysis of the LC Red 705 detection probe at codon 26 showed clear resolution of each allele. Figure 2C shows –dF/dT vs. temperature plots for both the homozygous and heterozygous genotypes. The homozygous Hb E genotype has a T_m of 63.8°C. The wild type allele has a T_m of 57.5°C due to the single mismatch between probe and target. The heterozygous Hb E-trait genotype has both the wild type and Hb E peaks. A summary of the observed T_ms of all alleles are shown in Table 2.

Genotyping β-globin Mutations (Hb S, Hb C, Hb E) by Multiplexing Probe Color and Melting Temperature

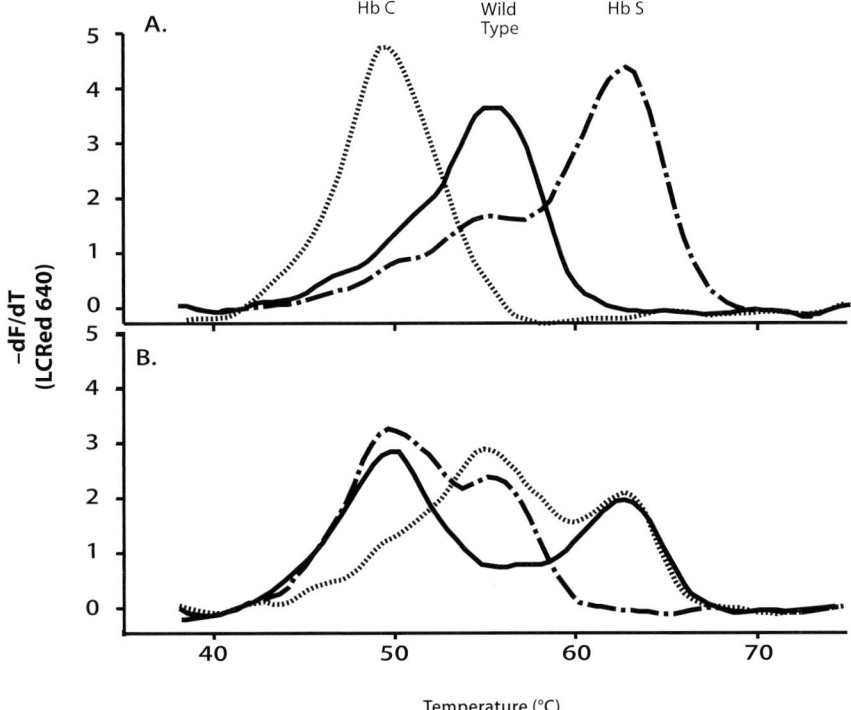

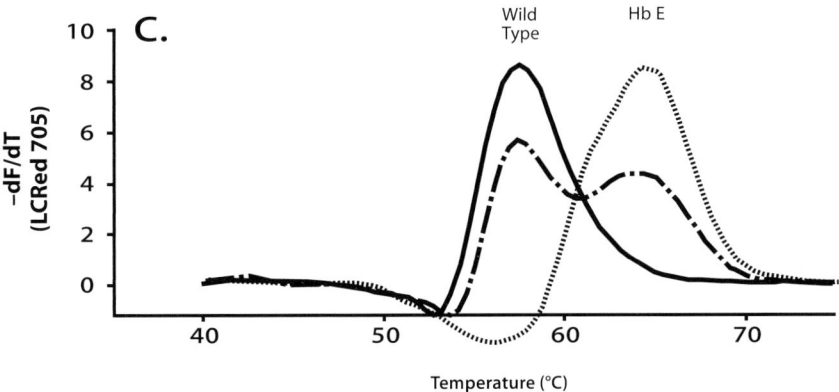

Fig. 2. Multiplex Genotyping by Color and T_m. The –dF/dT melting plots of the LC Red 640 probe for codon 6 are shown in panels A and B. Panel A shows melting peaks of homozygous Hb S (–·–), wild type (——) and Hb C (········) genotypes. Panel B shows melting peaks of heterozygous Hb S-trait (········), Hb C-trait (–·–) and the compound heterozygote Hb S/C (········) genotypes. Panel C shows the –dF/dT melting plots of the LC Red 705 probe for codon 26. Melting peaks for homozygous Hb E (········), wild type (——), and Hb E-trait (–·–)

Table 2. Melting Temperatures of β-globin alleles.

Locus	Allele	Pairing	T_m (°C)
Hb C/Hb S	Wild Type	T-T mismatch	55.6°C
	Hb C	G-T & T-T mismatch	50.1°C
	Hb S	T-A match	62.7°C
Hb E	Wild Type	A-C mismatch	57.5°C
	Hb E	A-T match	63.8°C

Comments

Melting curve analysis with multiple probes is a convenient method for genotyping multiple alleles. This single tube assay correctly genotpyes 3 single nucleotide polymorphisms simultaneously in less than an hour. In contrast, clinical genotyping by PCR amplification and restriction digestion requires several hours and considerable processing after amplification (8).

The thermal stability of a probe is the greatest when it is exactly complementary to the sequence. However, stability is decreased with mismatches and depends upon the type, position and number of mismatches (17,18). The detection probe for codon 6 gave the best separation of alleles when it matched the Hb S sequence. This probe clearly resolved each allele whereas probes complementary to the wild type and Hb C sequences did not (data not shown). Shorter probes exhibited higher mismatch destabilization, allowing for greater discrimination of the alleles. A dual labeled fluorescein probe was used as the FRET donor as a convenient way to excite both acceptor probes.

The ability to multiplex PCR analysis by color and T_m greatly extends the power of monitoring PCR with fluorescence. For example, internal amplification controls are often needed for infectious disease and translocation testing to verify that amplifiable DNA or cDNA is present even if the target amplification is negative. Another common need is for multiplexing a competitor as an internal standard for quantification. Multiplexing also allows multiple amplicons, or multiple mutations within a single amplicon, to be analyzed independently in a single reaction mixture. The β-globin mutations S, C and E can be easily multiplexed by color and T_m on the LightCycler.

References

1. Titus H.J. Huisman, Marianne F.H. Carver, and Georgi D. Efremov. A Syllabus of Human Hemoglobin Variants (1996). The Sickle Cell Anemia Foundation, Augusta, GA, USA.
2. Hardison R, Chui DHK, Riemer C, Miller W, Carver M, Molchanova T, Efremov G, and Huisman THJ (1998). Access to "A Syllabus of Human Hemoglobin Variants (1996)" via the World Wide Web. *Hemoglobin* 22:113-127.
3. Bio-Rad Laboratories. 1994. Instruction manual: VARIANT™ b-thalassemia short program.700:188. Hercules: self published.

4. Newborn Screening for Sickle Cell Anemia Diseases and Other Hemoglobinopathies. NIH Consens Statement Online 1987 Apr 6-8;6(9):1-22
5. Dubart A, Blouquit Y, Goossens M, Chabret C, Testa U, Beuzard Y, Rouyer-Fessard P, Dumez Y, Henrion R, Rosa J (1981). New Techniques for the Prenatal Diagnosis of hemoglobinopathies. *Prog Clin Biol Res* 55:767-78
6. Monte M, Beuzard Y, Rosa J (1976). Mapping of several abnormal hemoglobins by horizontal polyacrylamide gel isoelectric focusing. *Am J Clin Pathol* 66(4):753-9
7. Conner BJ, Reyes AA, Morin C, Itakura K, Teplitz RL, Wallace RB (1983). Detection of sickle cell beta S-globin allele by hybridization with synthetic oligonucleotides. *Proc Natl Acad Sci* Jan;80(1):278-82
8. Zhang YH, McCabe LL, Wilborn M, Therrell BL Jr, McCabe ER (1994). Application of molecular genetics in public health: improved follow-up in a neonatal hemoglobinopathy screening program. *Biochem Med Metab Biol* Jun;52(1):27-35.
9. Embury SH, Kropp GL, Stanton TS, Warren TC, Cornett PA, Chehab FF (1990). Detection of the hemoglobin E mutation using the color complementation assay: application to complex genotyping. *Blood* 76(3):619-23
10. Lay MJ, Wittwer CT (1997). Real-time fluorescence genotyping of factor V Leiden during rapid-cycle PCR. *Clin Chem* 43(12):2262-7
11. Aslanidis C, Schmitz G (1999). High-speed apolipoprotein E genotyping and apolipoprotein B3500 mutation detection using real-time fluorescence PCR and melting curves. *Clin Chem* 45(7):1094-7
12. Bernard PS, Pritham GH, Wittwer CT (1999). Color Multiplexing Hybridization Probes Using the Apolipoprotein E Locus as a Model System for Genotyping. *Anal Biochem.* 273:221-228
13. Wetmur JG (1991).DNA probes: applications of the principles of nucleic acid hybridization. *Crit Rev Biochem Mol Biol* 26(3-4):227-59
14. Thomas SM, Moreno RF, Tilzer LL (1989). DNA extraction with organic solvents in gel barrier tubes. *Nucleic Acids Res* 17(13):5411
15. Walsh PS, Metzger DA, Higuchi R (1991) Chelex 100 as a Medium for Simple Extraction of DNA for PCR-Based Typing from Forensic Material. *BioTechniques* Vol.10, No. 4 : 506-513
16. Bagwell CB, Adams EG (1993). Fluorescence spectral overlap compensation for any number of flow cytometry parameters. *Ann N Y Acad Sci* 677:167-84
17. Doktycz MJ, Morris MD, Dormady SJ, Beattie KL, Jacobson KB (1995). Optical melting of 128 octamer DNA duplexes. Effects of base pair location and nearest neighbors on thermal stability. *J Biol Chem* 270(15):8439-45
18. Guo Z, Liu Q, Smith LM (1997). Enhanced discrimination of single nucleotide polymorphisms by artificial mismatch hybridization. *Nat Biotechnol* 15(4):331-5

Simultaneous Detection of C282Y and H63D Hemochromatosis Mutations Using LCRed 640 and LCRed 705 Labeled Hybridization Probes

CINDY A. MEADOWS, BS CLSP (MB), MAREC PHILLIPS,
MING Y. HUANG, MS and ELAINE LYON*, PH.D

Introduction

Hemochromatosis is the most common genetic illness in people of Northern European descent. This autosomal recessive disorder of iron metabolism occurs with a frequency of approximately 0.5% in Caucasian populations (1). Hemochromatosis leads to organ failure due to iron accumulation and can be misdiagnosed as heart or liver disease. Serious complications include cirrhosis, hepatomas, diabetes, cardiomyopathy, arthritis and hypogonadotrophic hypogonadism (2–4). A cysteine to tyrosine amino acid substitution at codon 282 (C282Y), caused by a G to A transition at nucleotide position 845, is found on 85% to 100% of disease chromosomes from patients of northern European ancestry who meet well defined clinical criteria for iron overload (5–8). Another mutation (H63D) is created by a C to G transversion at nucleotide position 187. This substitution has an estimated penetrance between 0.44 and 1.5% of the homozygous C282Y genotype and is considered pathogenic only when inherited in trans with a C282Y mutation (compound heterozygous; 5–7).

Previously reported methods for hemochromatosis mutation detection include traditional PCR with restriction enzyme digestion (5), and a LightCycler assay with hybridization probes (9). In LightCycler genotyping assays, different alleles are identified by probe melting temperature shifts (9–12). The prior LightCycler hemochromatosis assay distinguished 4 alleles by the melting temperature using a single acceptor color. The T_ms for C282Y were 53°C for the wild-type and 60°C for the mutation while the T_ms for H63D were 63°C for the wild-type and 68.5°C for the mutation (9). Distinguishing all four alleles of two codons by T_m can be difficult. An alternative is to use multiple acceptor probe colors. In this method, C282Y sequences are detected with LCRed640 labeled probes, while H63D sequences are detected with LCRed705 labeled probes.

---

* Elaine Lyon, Ph.D. (✉) (e-mail: e-mail:lyone@arup-lab.com)
  ARUP Laboratories, 500 Chipeta Way, Salt Lake City, UT 84108

Materials

Equipment
LightCycler (Roche Diagnostics, Mannheim, Germany)
LightCycler capillary tubes (Roche Diagnostics, Mannheim, Germany)

Reagents
LightCycler DNA Master Hybridization Probes (Boehringer Mannheim, Germany)
Primers (Idaho Technology, Salt Lake City, UT)
Hybridization probes (Idaho Technology, Salt Lake City, UT)

Procedure

Sample Preparation
DNA was prepared from peripheral blood using phenol/chloroform extraction and ethanol precipitation (13).

Oligonucleotides
Oligonucleotide synthesis and probe labeling were performed as previously described (9). The primers and probes sequences for C282Y and H63D mutations are shown in Table 1.

Table 1. Oligonucleotides

Hemochromatosis Gene (Genebank Accession #Z92910) Codon C282Y					
	Position	Exon	Length	GC (%)	T_m (°C)
Primers					
TGGCAAGGGTAAACAGATCC	6443	intron 3	20	50	60.3
TACCTCCTCAGGCACTCCTC	6838R	intron 4	20	60	62.8
Product	6443–6838		395		
Probes*					
AGATATACGTACCAGGTGGAG-F	6712	4	21	47.6	57.8
LCRed 640-CCCAGGCCTGGATCAG-CCCCTCATTGTGATCTGGG-P	6735	4	35	62.9	76.9
Hemochromatosis Codon H63D					
	Position	Exon	Length	GC (%)	T_m (°C)
Primers					
CACATGGTTAAGGCCTGTTG	4620	2	20	50	61.1
GATCCCACCCTTTCAGACTC	4861R	2	20	55	59.5
Product	4620–4861		241		
Probes*					
ACGGCGACTCTCATCATCATAGA-F	4776R	2	23	47.8	63.3
LCRed 705-CACGAACAGCTGGTCA-TCCACGTAGCCCAAAGCTTCAA-P	4752R	2	38	52.6	75.3

* The position of base alterations in the fluorescein-labeled probes is underlined.

Donor probes were 3'-labeled with fluorescein. Acceptor probes were 5'-labeled with either LCRed640 or LCRed705 and blocked with a 3'-phosphate. The probes were designed to be complementary to the C282Y and H63D mutations.

PCR

The PCR master mix was added to the samples, transferred into capillary tubes, briefly centrifuged to spin the reaction mixture to the bottom of the capillary tube and loaded into the LightCycler. The PCR was performed with the following master mix.

Master Mix	Volume [µl]	[Final]
LightCycler-DNA Master Hybridization Probes (10X)	2	1x
MgCl$_2$ (25mM)	1.6	3.0 mM
C282Y primer (2µM forward primer)	1	0.1µM
C282Y primer (5µM reverse primer)	1	0.25µM
H63D primers (5µM forward primer)	2	0.5µM
H63D primer (2µM reverse primer)	2	0.2µM
C282Y FITC probe (2µM)	2	0.2µM
H63D FITC probe (1µM)	2	0.1µM
C282Y LCRed 640 probe (2µM)	2	0.2µM
H63D LCRed 705 probe (2µM)	2	0.2µM
DNA (50)	2	5
ddH$_2$0	0.4	
Total Volume	20	

PCR was performed on the LightCycler using the following cycling conditions.

Parameter	Value		
Cycles	40		
Type	Quantification		
	Segment 1	Segment 2	Segment 3
Target temperature (°C)	95	58	72
Incubation time [s]	0	10	0
Temperature transition rate [°C/s]	20	20	1
Acquisition mode	None	Single	None
Gains	F1 = 3	F2 = 30	F3 = 30

After 40 cycles, samples were then heated slowly with continuous fluorescence monitoring using the following parameters.

Parameter	Value			
Cycles	1			
Type	Melting Curves			
	Segment 1	Segment 2	Segment 3	Segment 4
Target temperature (°C)	95	65	45	85
Incubation time [s]	5	0	120	0
Temperature transition rate [°C/s]	20	20	1	0.1
Acquisition mode	None	None	Single	Cont.
Gains	F1 = 3	F2 = 30	F3 = 30	

Color Compensation Due to spectral overlap of the fluorophores, each channel was corrected for fluorescent cross-talk as previously described (14), according to the manufacturer's instructions.

Analysis The T_ms at which the probes melt from the templates can be identified by a rapid loss of fluorescence. Data analysis converts these melting curves into derivative curves or "melting peak" plots. Polynomial estimation of the derivative was used with 8°C temperature intervals. C282Y is detected in channel F2 while H63D is detected in channel F3 with the color compensation file enabled.

Results

Melting curve analysis easily distinguishes between the wild-type and mutant alleles at each codon. The melting curves for these two mutations are shown in Figure 1. At the C282Y locus, the wild-type T_m was 57.2°C and the mutant T_m, 63.3°C. At the H63D locus, the wild-type T_m was 57.1°C and the mutant T_m, 67.0°C (Table 2).

In addition to the H63D mutation, the LCRed 705 probe also covers an A-to-T polymorphism at nucleotide 193 that results in a serine-to-cysteine amino acid substitution at codon 65 (S65C) (15). The T_m for this polymorphism was 51.9°C. Figure 2 shows a sample that is heterozygous for both the S65C polymorphism and the H63D mutation. Homozygotes for the wild-type and H63D mutations are also shown.

Table 2.

Locus	Allele	Pairing	Tm (observed)
C282Y	Wild-type	A–C mismatch	57.2°C
	Mutation	A–T match	63.3°C
H63D	Wild-type	C–C mismatch	57.1°C
	Mutation	C–G match	67.0°C
	Polymorphism	C–C mismatch; T–T mismatch	51.9°C

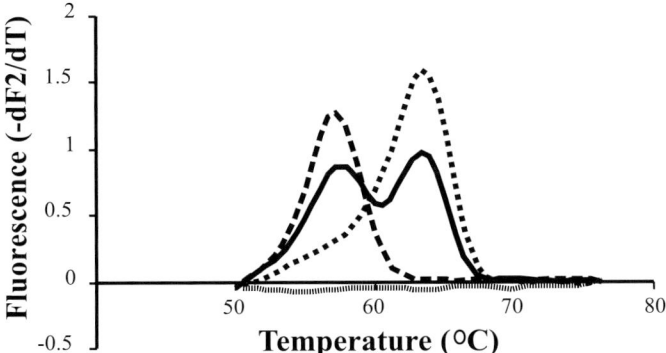

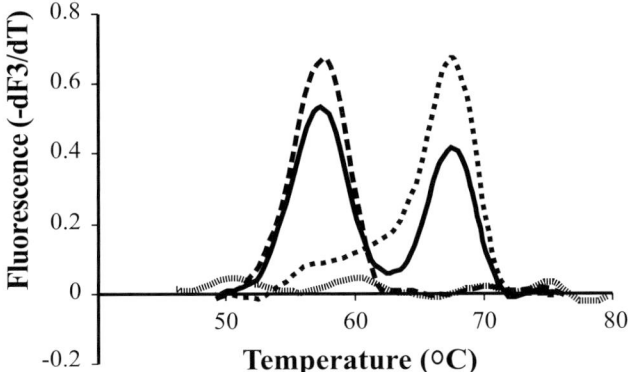

Fig. 1. Melting peaks for hemochromatosis mutations. A. C282Y mutation (F2) with wild-type (– – –)Tm at 57.2°C, mutation (▪▪▪▪▪) Tm at 63.3°C, heterozygote (———) Tm at 57.2°C and 63.3°C, and no template control (═══). B H63D mutation (F3) with wild-type (– – –) Tm at 57.1°C, mutation (▪▪▪▪▪) Tm at 67.0°C, heterozygote (———) Tm at 57.1°C and 67.0°C, and a no template control (═══)

Comments

We describe a color-multiplexed, real-time PCR method for genotyping the two mutations implicated in hemochromatosis. Affected individuals are usually homozygous for the C282Y mutation. Individuals that carry a C282Y mutation and a H63D mutation (compound heterozygotes) are also at risk for the disease, but with a reduced penetrance. Other genotypes have not been significantly correlated with hemochromatosis.

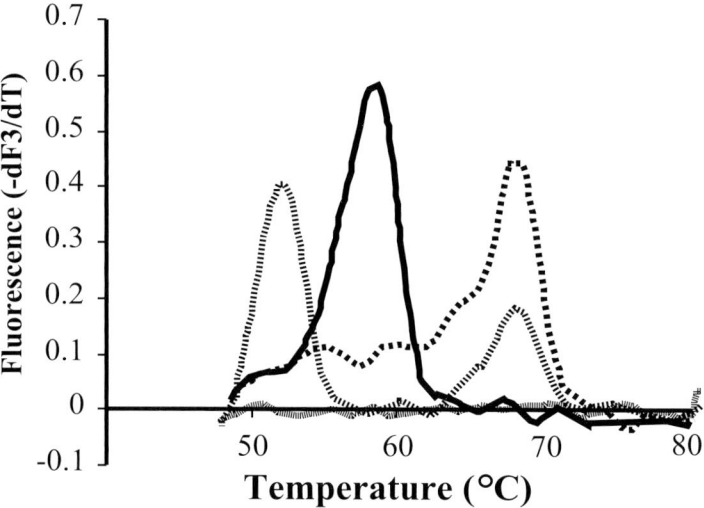

Fig. 2. Melting peaks for hemochromatosis H63D and S65C mutations (F3). H63D homozygous wild-type (———) Tm at 57.1°C, H63D homozygous mutant (- - - -) Tm at 67.0°C, S65C and H63D compound heterozygote (·········) Tm at 51.9°C and 67.0°C, and a no template control (═══════)

LightCycler technology allows multiplex PCR for hemochromatosis either by probe T_m or by color. Stability of a base alteration is affected by the type and position of the mismatch (16, 17), the number of mismatches (18) and the nearest neighbors of the mismatch (19). Due to the many factors that affect probe stability, the actual T_m shift for base alterations can be difficult to predict precisely. By using probes with different fluorophores, designing probes with unique T_ms for each allele is not necessary.

A polymorphism has been described six bases downstream from the H63D mutation with an A to T mutation at the 193 nucleotide position. This polymorphism (S65C) was distinguished from the H63D mutation with our FRET system. The probe was designed complementary to the H63D mutation and is less stable when hybridized to the wild-type allele, thus lowering the T_m. If the DNA sample contains the S65C polymorphism, the probe becomes even less stable upon hybridization, due to a second mismatch. The probe melts from the polymorphic sample earlier than from the wild type creating a peak at a lower temperature than either the wild-type or the mutant. Samples heterozygous for the S65C mutation will have two peaks, one at 51.9°C and the other at 57.1°C (wild-type) or 67.0°C (H63D) depending on the genotype of the other allele (9).

Another polymorphism (G547A) was recently discovered within the reverse primer sequence originally described for C282Y. This polymorphism was found in only C282Y wild-type alleles and could prevent extension from the reverse primer. The resulting alleles specific amplification may result in a heterozygous sample appearing homozygous. Therefore, we used a newly described antisense primer that excludes the site of the polymorphism (20).

PCR with fluorescent monitoring offers several advantages over existing PCR methods. PCR amplification and analysis are performed in a single capillary tube eliminating post-amplification processing. Multiplexing simplifies sample tracking and reduces labor and reagent costs. Rapid cycle PCR shortens the amplification time and increases specificity over traditional PCR (21). Amplification and analysis of 32 samples requires approximately 45 minutes.

References:

1. McLaren CE, Gordeuk VR, Looker AC, Hasselblad V, Edwards CQ, Giffen LM, Kushner JP, Brittenham GM. (1995) Prevalence of heterozygotes for hemochromatosis in the white population of the United States. Blood 86:2021–2027.
2. McKusick VA. Mendelian inheritance in man. (1998) Catalogs of Human Genes and Genetic Disorders. (The Johns Hopkins University Press, Baltimore 12th edition).
3. Bothwell TH, Chartton RW, Motutski AG. (1995) The metabolic and molecular bases of inherited disease (eds. Scriver, CR, Beaudet, AL, Sly, WS, Valle, D) pp. 2237–2269 (McGraw-Hill, New York).
4. Bacon BR, Tavill AS. (1996) Hepatology. A textbook of liver disease (eds. Zakim, D & Boyer, TD) pp. 1439–1472 (W.B. Saunders, Philadelphia).
5. Feder JN, Gnirke A, Thomas W, Tsuchihashi Z, Ruddy DA, Basava A, Dormishian F, Domingo R Jr, Ellis MC, Fullan A, Hinton LM, Jones NL, Kimmel BE, Kronmal GS, Lauer P, Lee VK, Loeb DB, Mapa FA, McClelland E, Meyer NC, Mintier GA, Moeller N, Moore T, Morikang E, Prass CE, Quintana L, Starnes SM, Schatzman RC, Brunke KJ, Drayna DT, Risch NJ, Bacon BR, Wolff RK. (1996) A novel MHC class I-like gene is mutated in patients with hereditary heamochromatosis. Nature Genet 13:399–408.
6. Beutler E, Gelbart T, West C, Lee P, Adams M, Blackstone R, Pockros P, Kosty M, Venditti CP, Phatak PD, Seese NK, Chorney KA, Ten Elshof AE, Gerhard GS, Chorney M. (1996) Mutation analysis in hereditary hemochromatosis. Blood Cells, Molecules, and Disease 22:187–194.
7. Jouanolle AM, Fergelot P, Gandon G, Yaouanq J, Le Gall JY, David V. (1997) A candidate gene for hemochromatosis: frequency of the C282Y and H63D mutations. Hum Genet 100:544–547.
8. Jazwinska EC, Cullen LM, Busfield F, Pyper WR, Webb SI, Powell LW, Morris CP, Walsh TP. (1996) Haemochromatosis and HLA-H. Nature Genet 14:249–251.
9. Bernard PS, Ajoika RS, Kushner JP, Wittwer CT. (1998) Homogenous multiplex genotyping of hemochromatosis mutations with fluorescent hybridization probes. Am J Pathol 153:1055–1061.
10. Lay MJ, Wittwer CT. (1997) Real-time fluorescence genotyping of factor V Leiden during rapid-cycle PCR. Clin Chem 43:2262–2267.
11. Bernard PS, Lay MJ, Wittwer CT. (1998) Integrated amplification and detection of the C677T point mutation in the methylenetetrahydrofolate reductase gene by fluorescence resonance energy transfer and probe melting curves. Anal Biochem 255:101–107.
12. Lyon E, Millson A, Phan T, Wittwer CT. (1998) Detection and identification of base alterations within the region of factor V Leiden by fluorescent melting curves. Molec Diag 3:203–210.
13. Thomas SM, Moreno RF, Tilzer LL. (1989) DNA extraction with organic solvents in gel barrier tubes. Nucleic Acids Res 17:5411.
14. Bernard PS, Pritham GH, Witwer CT.(1999) Color multiplexing hybridization probes using the apolipoprotein E locus as a model system for genotyping. Anal Biochem 273: 221–228.
15. Douabin V, Deugnier Y, Jouanolle AM, Moirand R, Macqueron G, Gireau A, Le Gall JY, David V. (1997) Polymorphisms in the haemochromatosis gene. International Symposium on Iron in Biology and Medicine. Saint-Malo France, p 267.
16. Ikuta S, Takagi K, Wallace RB, Itakura K. (1987) Dissociation kinetics of 19 base paired oligonucleotide-DNA duplexes containing different single mismatched base pairs. Nucleic Acids Res 15:797–811.

17. Aboul-ela F, Koh D. Tinoco Jr I. (1985) Base-base mismatches. Thermodynamics of double helix formation for $dCA_3XA_3G + dCT_3YT_3G$ (X,Y=A,C,G,T). Nucleic Acids Res 13:4811–4824.
18. Guo Z, Lui Q, Smith LM. (1997) Enhanced discrimination of single nucleotide polymorphisms by artificial mismatch hybridization. Nature Biotechnol 15:331–335.
19. Ke S-H, Wartell RM. (1993) Influence of nearest neighbor sequence on the stability of base pair mismatches in long DNA: determination by temperature-gradient gel electrophoresis. Nucleic Acids Res 21:5137–5143.
20. Jeffrey GP, Chakrabarti S, Hegele RA, Adams PC. (1999) Polymorphism in intron 4 of HFE may cause overestimation of C282Y homozygote prevelence in haemochromatosis. Nature Genetics 22: 325–326.
21. Wittwer CT, Reed GB, Ririe KM. (1994) Rapid cycle DNA amplification. (eds. Mullis KB, Ferré F, Gibbs RA) pp174–181. The polymerase chain reaction. (Birkhauser, Boston).

Genotyping of Angiotensin-Converting Enzyme and Angiotensinogen Polymorphisms with the LightCycler System

Eiichi Sakai, Minori Tajima, Mitsuko Mori, Reiko Inage, Manabu Fukumoto, Kan-ichi Nakagawara*

Introduction

The renin-angiotensin system regulates blood pressure, and maintains electrolyte homeostasis in humans [1]. In this system, angiotensinogen (AGT) is catalyzed by renin to form angiotensin I, which is then cleaved by angiotensin-converting enzyme (ACE) to yield angiotensin II, a potent vasopressor and effector on renal function [1]. Recently, several studies have focused on the correlation between physiological disorders and the genetic variation of peptides in the renin-angiotensin system. A specific mutation in the angiotensinogen gene was reported to be associated with essential hypertension [2,3]. Individuals with homozygous deletion alleles of ACE were reported to have a higher level of serum ACE and an increased risk of ischemic heart disease, sudden death, left ventricular hypertrophy, increased blood glucose levels, diabetic nephropathy, and premature death [4–7]. Detection of mutations in genes that constitute the renin-angiotensin system may be important in the prevention and control of disorders in the cardiovascular system, glucose metabolism, and urinary function.

In the ACE gene, insertion/deletion polymorphisms are observed in intron 16 [4–7]. For ACE genotyping, PCR with electrophoresis of PCR products has been the standard method. In the AGT gene, a point mutation in codon 235 (methionine to threonine, ATG to ATC) is observed, and for its genotyping, PCR and digestion of PCR products by a restriction enzyme at 3' end of codon 235 have been reported [3,8]. These conventional methods for genotyping take several hours and are limited by laboratory equipment, restricting the highly parallel analysis needed to establish relationships between genetic mutations and diseases.

In this study, genotyping was performed on the newly available LightCycler system, which shortens the time required for PCR and allows genotyping by melting curve analysis. ACE polymorphisms were determined by melting curve analysis with SYBR Green I. AGT polymorphisms were investigated using specifically designed probes with fluorescent dyes. This work presents a strategy for genotyping these polymorphisms quickly and accurately.

---

* Kan-Ichi Nakagawara (✉) (e-mail: nakagawara@ngrl.co.jp)
  2-3-36 Ohgi-machi, Miyagino-ku, Sendai 983-0034, Japan

Materials

Equipment LightCycler Instrument
Oligo primer analysis software

Reagents PCR Primers (Nihon Gene Research Laboratory)
Hybridization Probes (Nihon Gene Research Laboratory)
LightCycler-DNA Master SYBR Green I (Roche Diagnostics, Mannheim, Germany)
LightCycler-DNA Master Hybridization Probes (Roche Diagnostics, Mannheim, Germany)

Procedure

Sample Preparation Whole blood was sampled, and immediately mixed with sodium EDTA. Genomic DNA in blood cells was isolated following the phenol-chloroform extraction method [9]. DNA samples were stored at 4°C until use.

Primer and Probe Design For ACE genotyping, we used a primer set previously reported for detection of the insertion and deletion alleles [8]. We used SYBR Green I as the fluorescence marker for PCR products.

For AGT genotyping, we designed unique PCR primers and hybridization probes. The primer sequences were based on the published angiotensinogen sequence [10]. Hybridization probes were designed with the aid of the Oligo primer analysis software, following published guidelines [11]. A probe surrounding codon 235 was selected as a mutation probe. The primers and probes used are shown in detail in Table 1.

LightCycler PCR Angiotensin-Converting Enzyme Before PCR, boiling of DNA template, primers and $MgCl_2$ without LightCycler-DNA Master SYBR Green I was performed.
Reaction mix for each 20 µl reaction:

	Volume [µl]	[Final]
DNA template (30 ng/µl)	1.0	1.5 ng/µl
$MgCl_2$ (25 mM)	3.2	5 mM
Primers (10 µM each)	1.0 + 1.0	0.5µM
H_2O (PCR grade)	11.8	
Total volume	18.0	

- The reaction mix was boiled in a microcentrifuge tube for 10 min
- 2 µl LightCycler-DNA Master SYBR Green I was added
- 20 µl were loaded into each capillary tube

Table 1. Oligonucleotides

Angiotensin-converting enzyme (Genebank Accession #X62855 and #J04144)					
	Position	Ex/Int	Length	GC(%)	T_m (°C)
Primers					
CTGGAGACCACTCCCATCCTTTT	1402(X62855)	Int 16	24	54	71.3
GATGTGGCCATCACATTCGTCAGAT	2352R(J04144)	Ex 16	25	48	70.9
Angiotensinogen (Genebank Accession #M24686)					
Primers GATGCGCACAAGGTCCTG-TCTG	921	2	22	59	68.8
CCAGAGCCAGCAGAGAGGTT	1279R	2	20	60	67.0
Product	921–1279	2	359		
Probes CTGTCCACACTGGCTCCC-ATCA-F (mutation probe)	1233R	2	22	59	68.8
LCRed640GAGCAGCCAGTCTTCCATC-CTGTCAC-P (anchor probe)	1192R	2	26	58	71.9

Variable nucleotides in the probes are underlined.

The following PCR protocol was used for amplification and SYBR Green I detection:

- Denaturation for 2 min at 95°C
- Amplification

Parameter	Value		
Cycles	40		
Type	Quantification		
	Segment 1	Segment 2	Segment 3
Target temperature (°C)	95	58	72
Incubation time (s)	30	30	45
Temperature transition rate (°C/s)	20	20	20
Acquisition mode	None	None	Single
Gains	F1=5		

- Melting curve analysis

Parameter	Value		
Cycles	1		
Type	Melting curves		
	Segment 1	Segment 2	Segment 3
Target temperature (°C)	95	60	95
Incubation time (s)	0	20	0
Temperature transition rate (°C/s)	20	20	0.2
Acquisition mode	None	None	Cont
Gains	F1=5		

- Cooling at 40°C for 1 min

Angiotensinogen The following PCR protocol was used for amplification and detection of fluorescence from hybridization probes:

- Hybridization probe master mix for each 20 µl reaction:

	Volume [µl]	[Final]
LightCycler DNA Master Hybridization Probes	2.0	1×
MgCl$_2$ (25 mM)	1.6	3.0 mM
Primers (10 µM each)	1.0 + 1.0	0.5 µM
Probe 1 (8 µM)	1.0	0.4 µM
Probe 2 (16 µM)	1.0	0.8 µM
H$_2$O (PCR grade)	11.4	
Total volume	19.0	

A total of 19 µl of master mix, and 1 µl (30 ng) of DNA template were loaded in each capillary tube.
- Denaturation for 2 min at 95°C
- Amplification

Parameter	Value		
Cycles	40		
Type	Quantification		
	Segment 1	Segment 2	Segment 3
Target temperature (°C)	95	58	72
Incubation time (s)	5	15	25
Temperature transition rate (°C/s)	20	20	20
Acquisition mode	None	Single	None
Gains	F1=1, F2=10		

- Melting Curve Analysis

Parameter	Value		
Cycles	1		
Type	Melting curves		
	Segment 1	Segment 2	Segment 3
Target temperature (°C)	95	40	80
Incubation time (s)	0	20	00
Temperature transition rate (°C/s)	20	20	0.2
Acquisition mode	None	None	Cont
Gains	F1=1, F2=10		

- Cooling at 40°C for 1 min

Results

A representative time course of melting curve analysis is shown in Fig. 1 (panels A and B, ACE; panels C and D, AGT). In panel A, the fluorescence of sample A and B decreases at around 85°C, and 92°C, respectively. In panel B, the fluorescence of sample C decreases biphasically, at around 85°C and 92°C. The rapid decrease of fluorescence indicates a conformational change of DNA from double strand to

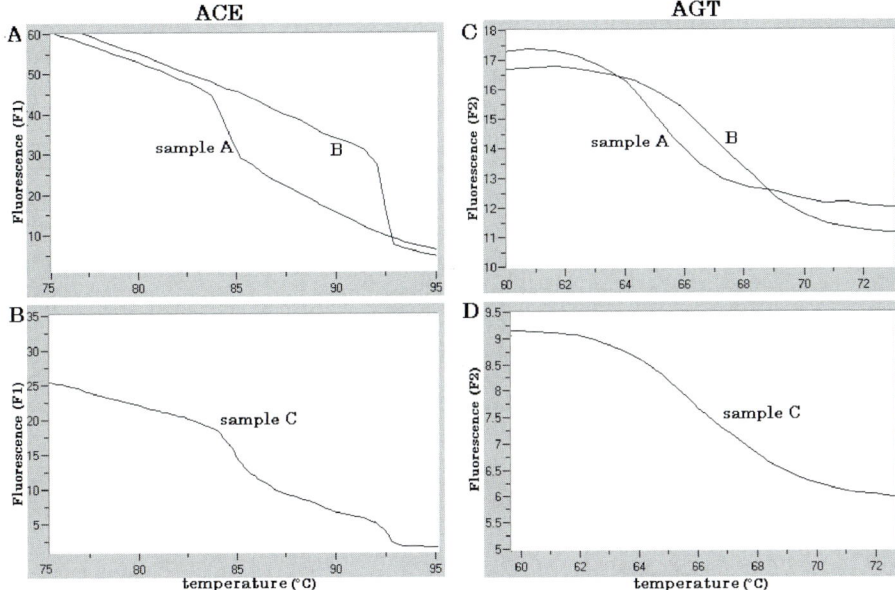

Fig. 1A–D. Relationship between the fluorescence intensity of PCR products and temperature in melting curve analysis. **A,B,** ACE analysis; **C, D,** AGT analysis

single strand, and different melting temperatures indicate different DNA sequences. The fluorescence curves of samples A and B indicate the templates are homozygous, and the curve of sample C indicates that the template as heterozygotes. Similarly, in panel C, fluorescence curves indicate two different genotypes.

Figure 2 shows the relationship between the first negative derivative of fluorescence and temperature. This transformation shows the melting temperature as a peak and helps to distinguish the genotype easily. In panel A, a peak at 85°C in the trace of sample A, and at 92°C in the trace of sample B are shown, indicating melting temperatures of deletion (D/D) and insertion type (I/I) homozygotes, respectively. Therefore, three types of curves A, B, and C in panel B indicate genotypes of D/D, I/I and I/D, respectively. Similarly, in panel C, two peaks at 65°C and 68°C indicate melting temperatures of mutation (M/M) and wild type (T/T) homozygotes. The biphasic curve of sample C indicates a heterozygote (M/T). The allele melting temperatures are summarized in Table 3.

Using the LightCycler system, we genotyped ACE and AGT on DNA samples from a cohort whose ACE and AGT genotypes had been studied by Kawazu et al. with conventional PCR and restriction enzyme digestion [8]. Our genotyping of both molecules (52 samples for ACE, and 48 samples for AGN) completely agreed with the results from conventional methods.

Comments

In ACE analysis, boiling of DNA mixture for 10 min. before PCR was necessary for successful amplification, presumably to insure complete denaturation.

The observed T_ms of the II and DD alleles were 92°C and 85°C, respectively. As shown in Table 2, the observed T_m of II is comparable with that calculated by an empirical equation [12]. The observed T_m of DD was lower than the calculated T_m, possibly because of atypical nearest-neighbor frequencies, local sequence, or conformational state [13].

In this study, we analyzed two different types of mutations: deletion/insertion and point mutations. With the LightCycler system, the experimental time for PCR and melting curve analysis was 90 min for ACE and 70 min for AGT. Compared with conventional methods, the experimental time per sample was drastically reduced. Also, the amount of template DNA was reduced from 250 [3,8] to 30 ng, because of the high detection sensitivity of the LightCycler system. The LightCy-

Table 2. Melting temperature of ACE alleles

Genotype	Base pairs	GC (%)	Calculated T_m (°C)	Observed T_m (°C)
II	479	54	89	85
DD	191	49	93	92

We calculated T_m following an empirical equation [12], where (Na$^+$) and (K$^+$) concentration was taken as 0 and 50 mM, respectively.

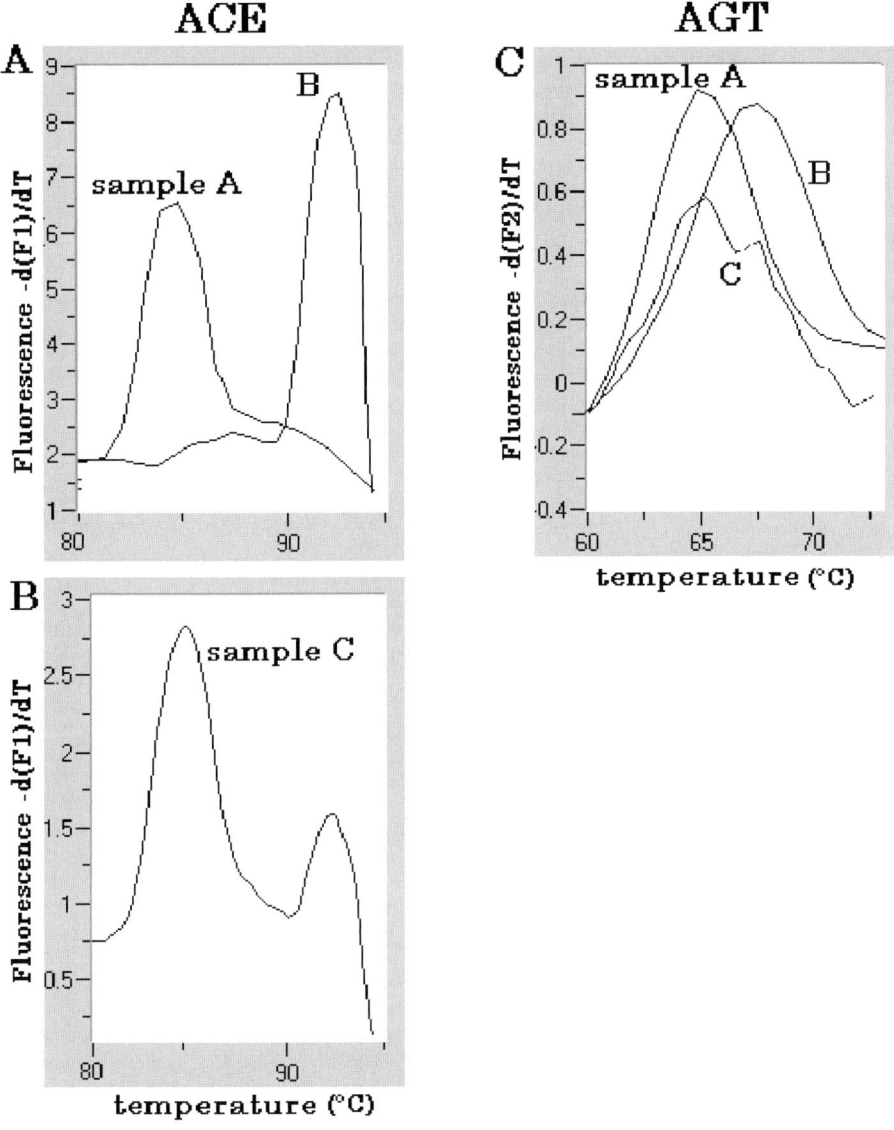

Fig. 2A–C. Relationship between the first negative derivative of fluorescence and temperature in melting curve analysis. **A, B**, ACE analysis; **C**, AGT analysis

cler system allows analyses of many samples in a limited time and of many genes from a limited amount of template. Though the balance of cost and performance of analysis should be considered, the LightCycler system will advance mass analyses for many genes responsible for diseases and clinical disorders.

Table 3. Melting temperatures of AGT alleles

Locus	Allele	Pairing	T_m (observed)
AGT	WT	CG match	68
	M235T	CC mismatch	65

References

1. Hardman JG, Limbird LE, Molinof PB, Ruddon RW, Gilman AG (1996) Renin and angiotensin. In: The pharmacological basis of therapeutics, (9th edn. McGraw-Hill, New York, pp 733–758
2. Jeunemaitre X, Soubrier F, Kotelevtsev YV, Lifton RP, Williams CS, Charru A, Hunt SC, Hopkins PN, Williams RR, Lalouel JM, Corvol P (1992) Molecular basis of human hypertension: role of angiotensinogen. Cell 71: 169–180
3. Caulfield M, Lavender P, Farrall M, Munroe P, Lawson M, Turner P, Clark AJ (1994) Linkage of the angiotensinogen gene to essential hypertension. N Engl J Med 330: 1629–1633
4. Cambien F, Poirier O, Lecerf L, Evans A, Cambou JP, Arveiler D, Luc G, Bird JM, Bara L, Ricard S, Tiret L, Amouyel P, Alhenc-Gelas F, Soubrier F (1992) Deletion polymorphysm in the gene for angiotensin-converting enzyme is a potent risk factor for myocardial infarction. Nature 359: 641–644
5. Morris BJ, Zee RY, Schrader AP (1994) Different frequencies of angiotensin-converting enzyme genotypes in older hypertensive individuals. J Clin Invest 94:1085–1089
6. Iwai N, Ohmichi N, Nakamura Y, Kinoshita M (1994) DD genotype of the angiotensin-converting enzyme gene is a risk factor for left ventricular hypertrophy. Circulation 90: 2622–2628
7. Schunkert H, Hense HW, Holmer SR, Stender M, Perz S, Keil U, Lorell BH, Riegger GA (1994) Association between a deletion polymorphism of the angiotensin-converting enzyme gene and left ventricular hypertrophy. N Engl J Med 330: 1634–1638
8. Ohno T, Kawazu S, Tomono S (1996) Association analysis of the polymorphisms of angiotensin-converting enzyme and angiotensinogen genes with diabetic nephropathy in Japanese non-insulin-dependent diabetics. Metabolism 45: 218–222
9. Sambrook J, Fritsch EF, Maniatis T (1989) Appendix E: Commonly used techniques in molecular cloning. In: Molecular cloning: a laboratory manual 2nd edn. Cold Spring Harbor Laboratory, Cold Spring Harbor, NY
10. Gaillard I, Clauser E, Corvol P (1989) Structure of human angiotensinogen gene. DNA 8:87–99
11. Caplin BE, Rasmussen RP, Bernard PS, Wittwer CT (1999) Light CyclerTM Hybridization probes. Biochemica 1: 5–8
12. Wetmur JG (1995) Formation and structure of nucleic acid hybrids. In: Molecular biology and biotechnology: a comprehensive desk reference. Wiley-VCH, New York, pp 605–608
13. Blake RD (1995) Denaturation of DNA. In: Molecular biology and biotechnology: a comprehensive desk reference. Wiley-VCH, New York, pp 207–210

Genotyping of the Most Common Thiopurine Methyltransferase Mutations with the LightCycler

Optimization of Hybridization Probe Assays for the Detection of Mutations Causing Stable Mismatches in DNA Duplexes

Ekkehard Schütz*, Nicolas von Ahsen

Introduction

Thiopurine derivatives are commonly used for immunosuppressive therapy after organ transplantation (azathioprine) and in the therapy of leukemia (mercaptopurine and thioguanine). Their pharmacological efficacy is based on an in vivo toxification pathway that ultimately leads to 6-thioguanine nucleotides, which act as anti-metabolites and interfere with nucleic acid synthesis [1]. Two detoxification pathways are known, one via xanthine oxidase, a stably abundant enzyme in the Caucasian population, and the second via thiopurine methyltransferase (TPMT, EC.2.1.1.67). The latter enzyme is subject to a genetic polymorphism, that in 11% of the Caucasian population leads to a heretozygous deficiency and in 0.3% to a homozygous deficiency of this enzyme [3, 4, 6]. If patients with a homozygous deficiency of TPMT are given thiopurines at the usual dosage, this will lead to a severe myelosuppression [9, 10] often with life-threatening pancytopenia [5]. Phenotyping of this enzyme is possible [2], but the method is laborious and technically demanding; it is therefore only performed in specialized laboratories. Genotyping of this defect is, on the other hand, hampered by the fact that to date eight mutation are known [11–13] that lead to a deficient phenotype (Fig. 1). Around 90% of TPMT deficiencies can be attributed to one of these known mutations.

The TPMT gene is located on chromosome 6p22.3 and consists of 10 exons, with a cDNA of about 3000 bp and an ORF of 735 bp that encodes for a 245AA peptide with a molecular mass of 32.5 kDa. Since the expression of a pseudogene has been described [14], genotyping on the basis of genomic DNA is mandatory. It has been shown that among the inactivating mutations, ca. 75% are represented by the TPMT*3 group, which is categorized into the subtypes A to D. The most frequent (55%) of these is the TPMT*3 A variant, which incorporates two mutations. The first of these mutations, G460A, leads to an Ala154Thr amino acid (AA) exchange and is located on exon 7, whereas the second, A719G, leading to a Thr240Cys AA exchange, is found on exon 10. TPMT*3B is defined by the isolat-

---

* Ekkehard Schütz (✉) (e-mail: eschuetz@med.uni-goettingen.de)
  Department Clinical Chemistry, Georg-August-University, Robert-Koch-Str. 40, 37075 Göttingen, Germany

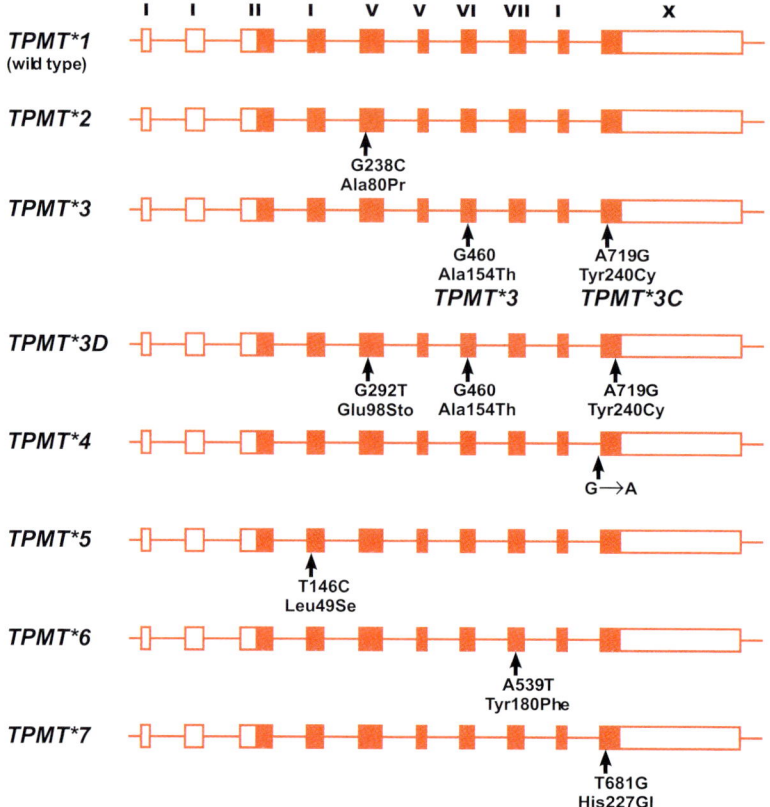

Fig. 1. Thiopurine methyltransferase genetic polymorphism. Boxes represent the exons, closed boxes show ORF (Reprint with permission from E. Schütz et al., CLIN CHEM 2000; 46:1728–37)

ed presence of G460A (7%) and TPMT*3 C by the isolated A719G mutation (13%). One case has been described with G460A and A719G that showed an additional mutation (G292T) leading to a premature stop codon known asTPMT*3D. On the basis of this knowledge, detection of both the G460A and the A719G mutations enables the diagnosis of 75% of all mutations that occur in a Caucasian population.

Despite the superiority of phenotyping for most clinical situations where a homozygous deficiency must be excluded – e.g., prior to an intravenous loading dose with azathioprine in accelerated regimens used in chronic inflammatory diseases [15, 17] – genotyping has relevance. Firstly, it is of high scientific interest since there are still unknown mutations (ca. 10% of deficiencies are not TPMT*2 – TPMT*7) that need to be clarified. Secondly, patients are usually first subject to investigation of thiopurine pharmacogenetics after myelosuppression has already been seen under therapy. In such cases, patients will often need transfusions due to a severe anemia. If so, phenotyping is no longer reliable, since activity of TPMT, a cytosolic enzyme, is measured in erythrocyte preparations from patients. After blood transfusion, the sample will contain a combination of patient and donor red blood cells, leading to false high TPMT activity. In such cases genotyping will be more reliable.

Outline

In this chapter, the following subjects are presented: (1) a procedure will be shown for genotyping TPMT*3 A, *3B and *3 C; (2) using TPMT*3 C, the optimization of probe design to achieve a sensitive and specific assay will be illustrated.

Materials

LightCycler Instrument **Equipment**
MeltCalc analysis software

Amplification Primers **Reagents**
Hybridization Probes
Colour-Compensation-Kit (Roche Diagnostics, Mannheim, Germany)
Native Taq polymerase and PCR buffer (Life Technologies)
Bovine serum albumin (BSA, New England BioLabs)
Dimethylsulfoxide (DMSO, Sigma)
dNTPs (Boehringer Mannheim)

Oligonucleotides were custom synthesized by MWG-Biotech, Ebersberg, Germany, except for probe set 1 and 2 for TPMT3c, which were designed and synthesized by TibMolbiol, Berlin, Germany.

Procedure

Genomic DNA was prepared by standard methods or with a rapid alkaline lysis **LC-PCR**
method [13], where 10 µl EDTA-anticoagulated whole blood lysed and 1 µl of the genomic DNA solution was used for analysis. All PCR reactions were carried out in 10 µl in LightCycler capillaries. Capillaries were sealed, centrifuged in a microcentrifuge and placed into the LightCycler rotor.
   The reaction mixture consisted of:
0.5 U Taq DNA polymerase
1 µl 10×PCR buffer
0.2 mM each dNTP
2.5 mM $MgCl_2$
500 mg/l bovine serum albumin
50 ml/l dimethylsulfoxide

0.5 μM amplification primers
0.3 μM labeled anchor probe
0.1 μM labeled detection probe

Alternatively the ready made LightCycler-DNA Master Hybridization Probes mix might be used, but this may not function for every assay described herein (data not shown).

Amplification primers, detection, and oligonucleotides used are presented in Table 1.

Table 1. Oligonucleotides

Thiopurine methyltransferase (Genebank Accession U30515 and U30518)				
	Position	Length	GC(%)	T_m(°C)
TPMT*3B primers				
CTCCACACCCAGGTCCACACATT	746	23	56.5	66.1
GTATAGTATACTAAAAAATTAAGACAG	1035 R	27	22.2	51.4
Product	746-1035	290		
TPMT*3B probes				
GGCAACTAATGCTCCTCTATCC-F	958 R	22	50.0	58.7
Cy5.5- ATCATGTCAAATTTGCCAATA-TTTGTCCTACCA	933 R	33	33.3	63.7
TPMT*3C primers				
AATCCCTGATGTCATTCTT	1	19	36.8	53.4
CACATCATAATCTCCTCTCC	401 R	20	45.0	54.3
Product	1-401	402		
TPMT*3C probes				
Set 1 (wild type)				
TTTACTTTTCTGTAAGTAGATATAACTT-F	169 R	28	21.4	53.0
LCRed640- TCAAAAAGACAGTCAATTCCCCAA	140 R	24	37.5	59.5
Set 2 (mutant)				
LCRed640- TCTGTAAGTAGACATAACTTTTCAA	161 R	25	28.0	54.2
AGTGTGATTTTATTTTATCTATGTCTCATTTACTT-F	197 R	35	22.9	58.7
Set 3 (wild type)				
Cy5.5-TTATATCTACTTACAGAAAAGTAAA	145	25	20.0	49.7
TTGGGGAATTGACTGCCTTTTTGAA-F	117	25	40.0	62.2

The following settings for the LightCycler were used for all genotyping assays described herein:
- Denaturation: 45 s, 95°C
- Amplification

Parameter	Value		
Cycles	55		
Type	Quantification		
	Segment 1	Segment 2	Segment 3
Target temperature [°C]	95	50	72
Incubation time [s]	0	5	10
Temperature transition rate [°C/s]	20	20	20
Acquisition mode	None	Single	None
Gains	F1=5	F2=20	F3=40

- Melting Curve

Parameter	Value		
Cycles	40		
Type	Quantification		
	Segment 1	Segment 2	Segment 3
Target temperature [°C]	95	40	80
Incubation time [s]	5	45	0
Temperature transition rate [°C/s]	20	20	0.1
Acquisition mode	None	None	Cont.

Color Compensation

The color compensation feature eliminates the cross-talk of fluorescence emission into channels other than the color specific detection channels. This is due to the broad emission spectra of the fluorescence dyes and the bandwidth of the photometer filters.

The procedure and the Color-Compensation-Kit provided by the manufacturer was used as recommended. The compensation for Channel 3, works equally well for LCRed705 and Cy5.5, an alternative dye with essentially the same characteristics as LCRed705. Alternatively a specific Color Compensation-file for Cy5.5 can be generated by using an appropriate amount of Cy5.5 instead of LCRed705 during color calibration run. The advantage of Cy5.5 is the price, since the costs of an equal amount of a labelled probe-set is substantially lower that that of a LCRed probe set. The method of linking the dye to the oligonucleotide does not significantly influence its hybridization or fluorescence characteristics. A more detailed discussion of color compensation in LightCycler assays has been published recently [16, 18].

Results

The first two hybridization-probe sets for the detection of the TPMT*3 C mutation were arbitrarily designed and placed on the antisense strand, one as wild-type compatible and one as a mutation compatible probes set. As shown in Fig. 2, only the mutation compatible probe discriminated in such a way that in the case of heterozygousity a typical double peak was present. The wild-type compatible probe did not discriminate; the resulting melting profile showed one broad peak. Using the MeltCalc analysis software, the predicted differences in T_m between mutation

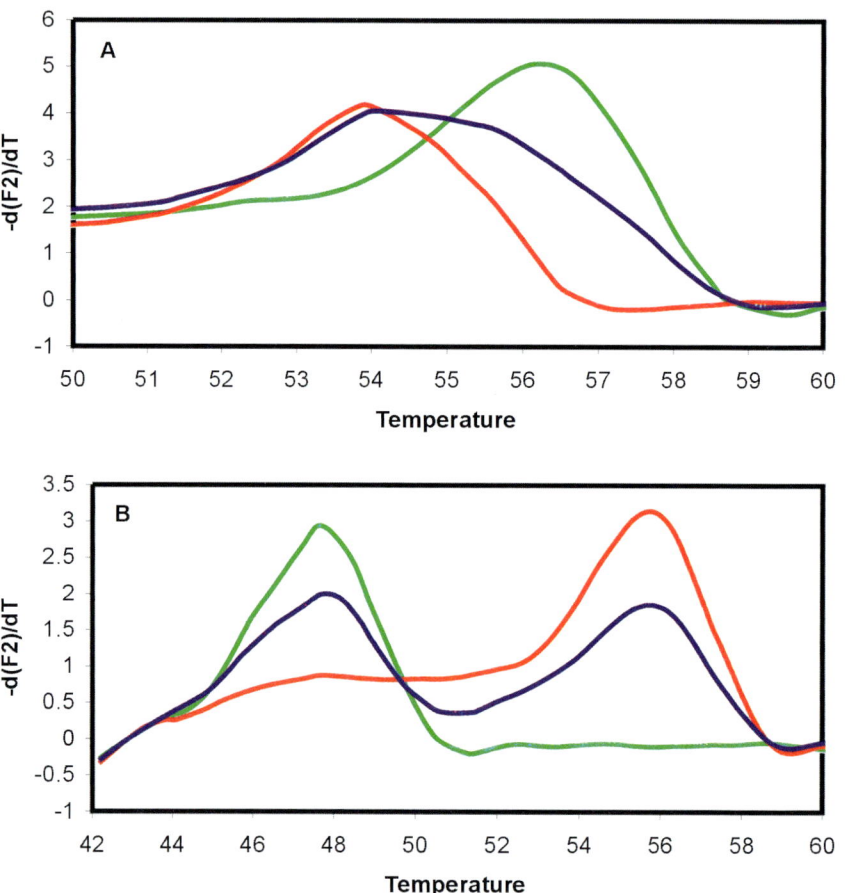

Fig. 2. Melting curves from TPMT*3 C hybridization-probe assays (probe set 1). Both the wild-type (panel A) and the mutation compatible (panel B) probes were designed to hybridize with the antisense strand. Here the mutation compatible probe shows a reliable discrimination between the two alleles, whereas the wild-type compatible probe does not. (*Green line*, wild-type; *red line,* TPMT*3 C; *blue line,* TPMT*1/TPMT*3 C)

and wild-type were 1.9°C for the wild-type compatible probe, and 6.4°C for the mutation compatible probe, which correspond well to the experimental results.

The remainder of the herein used probe-sets were designed to achieve a maximum T_m difference, calculated on the basis of the thermodynamic nearest neighbor model. We have performed thermodynamic calculations with MeltCalc, a Microsoft Excel add-in by Ekkehard Schütz and Nicolas von Ahsen. A version with basic functionality is available free for non-commercial use at URL http://www.meltcalc.de. Using this approach, it was possible to design a probe set for TPMT*3 C that is both wild-type compatible and discriminated within the sensitivity limits of the LightCycler between the alleles. The probe was designed in the complementary orientation compared to the first probe set from Tib-Molbiol. The best discrimination could be achieved for the detection of the TPMT*3B mutation which, on the other hand, is much less problematic in the case of a wild-type compatible probe, since the resulting C/A mismatch (NN: gc/ca; ct/aa) is substantially less stable (increase in free energy: 4.9 kcal/mol) than the A/C mismatch (NN: ta/ac; at/ca) that is present in the TPMT*3 C mutation (increase in free energy: 3.15 kcal/mol). The results of the melting analyses for both TPMT*3B and TPMT*3 C are shown in Fig. 3.

Comments

After 5 decades of using thiopurines in immunosupression and anti cancer therapy, our understanding of the pharmacogenetics has increased dramatically during the last few years. Now it is possible to attribute the well known side effects of these drugs in most cases to a biochemical basis, the inherited deficiency of thiopurine methyltransferase, the key enzyme in the detoxification of these drugs. The LightCycler facilitates a rapid screening for the most common mutations of this enzyme, which are responsible for 75% of deficient phenotypes. Methods for genotyping of all eight known TPMT mutations with the LightCycler were recently published [19]. The TPMT*3 C mutation can serve as an example for the role of the mismatch type and their base neighbors for melting curve analyses. In this case, a wild type compatible detection probe shows only a small T_m difference between wild-type and the specific mutation. A mutation compatible probe allows a much better discriminaton to exclude the specific mutation, but is hampered by the fact that eight other possible mutations (mismatches) under this specific probe were calculated to give T_m values within ±0.5°C of the wild-type T_m and 22 possible mismatches were within the range of ±1.0°C. Keeping the limitations of routine genotyping with the LightCycler in mind, exclusion of TMPT*3 C with the mutation specific probe is not a definite evidence for wild-type. Therefore a wild-type compatible probe was designed, based on thermodynamic calculations that discriminates TPMT*3 C with reliable performance. We have recently shown the validity of a thermodynamic nearest neighbor model for LightCycler hybridization-probe assays [17]. Such calculation of T_m differences of all possible mismatches under a used detection probe compared to the perfect match are therefore strongly recommended for the interpretation of genotyping. We therefore

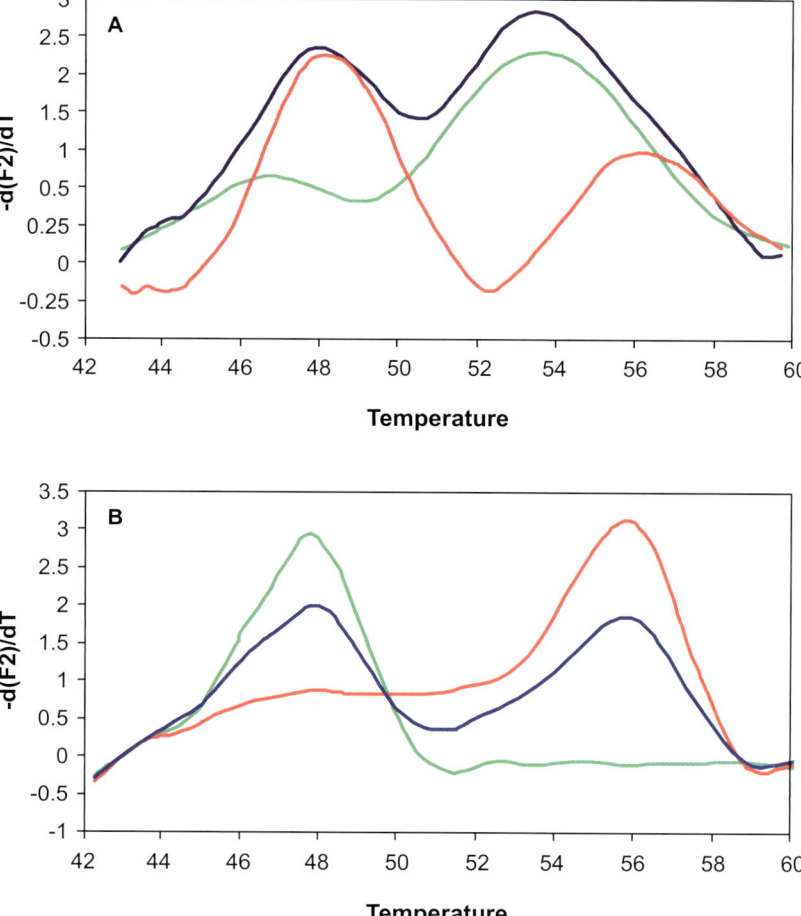

Fig. 3. Melting curves from TPMT*3 C (panel A) and final TPMT*3 C (panel B) hybridization-probe assay. Both hybridization-probe sets were designed using a thermodynamic model to yield the highest possible melting point difference between perfect match (wild-type) and specific mutation. (*Green line,* wild-type; *red line,* TPMT*3B or C rsp.; *blue line,* heterozygous)

integrated a feature that gives a list of all possible mismatches under a probe with their respective T_m values in the MeltCalc software [8], that is available for non-commercial use or at http://www.meltcalc.de. Scientists working on genotyping with hybridization-probes or other hybridization-based techniques are encouraged to use this software to get a deeper insight into the power but also the shortcomings of this mutation detection principle. In our experience, it is possible for all genes that we have investigated so far, to set up a hybridization-probe assay that reliably detects the wild-type and/or mutation of interest by careful selection of a detection/anchor probe set, based on thermodynamic n-n modelling (see upper cases). This is due to the fact, that for a single point mutation, four basically different probes are possible (sense and antisense, and for each a wild type and mutation compatible probe). It can be demonstrated that there is a dramatically large difference in T_m discrimination (e.g., deltaT_m 9.2°C vs 2.4°C) between the best probe and worst probe that can be chosen, even if the probes have the same T_m for the perfect match (in this case 55°C). For insertion/deletion polymorphisms we could show, that thermodynamic calculations may also be very helpful in designing reliable hybridization probe assays for the LightCycler [20]. Therefore, such a strategy will increase the likelihood of synthesizing hybridization probe sets with optimal allele discrimination.

References

1. Ellion GB (1989) The purine path to chemotherapy. Science 244: 41–53
2. Weinshilboum RM, Raymond FA, Pazmino PA (1978) Human erythrocyte thiopurine methyltransferase: radiochemical microassay and biochemical properties. Clin Chim Acta 85:323–333
3. Weinshilboum RM, Sladeck SL (1980) Mercaptopurine pharmacogenetics: monogenetic inheritance of erythrocyte thiopurine methyltransferase activity. Am J Hum Genet 32: 651–662
4. McLeod HL, Lin JS, Scott EP, Pui CH, Evans WE (1994) Thiopurine methyltransferase activity in American white and black subjects. Clin Pharmacol Ther 55:15–20
5. Schütz E, Gummert J, Mohr FW, Oellerich M (1993) Azathioprine induced myelosuppression in thiopurine methyltransferase deficient heart transplant recipient. Lancet 341:426
6. Schütz E, Gummert J, Armstrong VW, Mohr FW, Oellerich M (1996) Azathiorpine Pharmacogenetics: the relationship between 6-thioguanine nucleotides and thiopurine methyltransferase in patients after heart and kidney transplantation. Eur J Clin Chem Clin Biochem 34: 199–205
7. Schütz E, Pickenpack A, Lang B, Oellerich M (1999) Pharmocokinetics of 6-thioguanine nucleotides under i.v. therapy with azathioprine. Ther Drug Monit 21:476 [Abstract]
8. Schütz E, von Ahsen N (1999) A simple and versatile standard spreadsheet software for the thermodynamic melting point prediction for oligonucleotide hybridization with and without mismatches. Biotechniques 27: 1218–1224
9. Lennard L, Gibson BES, Nicole T, Lilleyman JS (1993) Congenital thiopurine methyltransferase deficiency and 6-mercaptopurine toxicity during treatment for acute lymphoblastic leukaemia. Arch Dis Child 69: 577–579
10. Leipold G, Schütz E, Haas JP, Oellerich M (1997) Azathioprine-induced severe pancytopenia due to a homozygous two-point mutation of the thiopurine methyltransferase gene in a patient with juvenile HLA-B27-associated spondylarthritis. Athritis Rheum 40: 1896–1898
11. Krynetski EY, Schuetz JD, Galpin AJ, Pui CH, Relling MV, Evans WE (1995) A single point mutation leading to loss of catalytic activity in human thiopurine methyltransferase. PNAS 92: 949–953
12. Tai HL, Krynetski EY, Yates CR, Loennechen T, Fessing MY, Krynetskaia NF et al (1996) Thiopurine-S-methyltransferase deficiency: two nucleotide transitions define the most prevalent mutant allele associated with loss of catalytic activity in Caucasians. Am J Hum Genet 58: 694–702
13. Otterness C, Szumlanski C, Lennard L, Klemetsdal B, Aarbakke J, Park-Hah JO et al (1997) Human thiopurine methyltransferase pharmacogenetics: gene sequence polymorphisms. Clin Pharmacol Ther 62: 60–73

14. Lee D, Szumlanski C, Houtman J, Honchel R, Rojas K, Overhauser J et al (1995) Thiopurine methyltransferase pharmacogenetics: Cloning of liver cDNA and a processed pseudogene on human chromosome 18q21.1. Drug Metab Dispos 23; 398–405
15. Sandborn WJ, V. Os EC, Zins BJ, Tremaine J, Mays DC, Lipsky JJ (1995) An intravenous loading dose of azathioprine decreases the time to response in patients with Crohn's disease. Gastroenterology 109: 1808–1817
16. Bernard PS, Pritham GH, Wittwer CT (1999) Color multiplexing hybridization probes Using the apolipoprotein E locus as a model system for genotyping. Anal Biochem 273: 221–228
17. von Ahsen N, Oellerich M, Armstrong VW, Schütz E (1999) Application of a thermodynamic nearest-neighbor model to estimate nucleic acid stability and optimize probe design: prediction of melting points of multiple mutations of apolipoprotein B-3500 and factor V with a hybridization probe genotyping assay on the LightCycler. Clin Chem 45: 2094–2101
18. von Ahsen N, Oellerich M, Schütz E (2000) Using two reporter dyes without interference in a single tube rapid cycler PCR: alpha1-antitrypsin genotyping by multiplex real-time fuorescence PCR with the LightCycler. Clin Chem 46: 156–161
19. Schütz E, von Ahsen N, Oellerich M (2000) Genotyping of eight Thiopurine Methyltransferase mutations: Three-Color multiplexing, "Two-Color/shared" anchor and fluorescence-quenching hybridization probe assays based on thermodynamic nearest-neighbor probe design Clin Chem 46:1728–37)
20. von Ahsen N, Oellerich M, Schütz E (2000) DNA base bulge vs unmatched end formation in probe-based diagnostic insertion/deletion genotyping: Genotyping the UGT 1A1 $(TA)_n$ polymorphism by real-time fluorescence PCR Clin Chem 46 (12: in press)

Detection of the Mitochondrial DNA Mutation MELAS3243 Using Hybridization Probes

Stephanie Kleinle*, Sabina Gallati

Introduction

A number of human diseases have been associated with mitochondrial (mt)DNA mutations, such as large-scale rearrangements or point mutations within either protein-coding-genes or, more commonly, tRNA-genes. Mitochondrial disorders present with a spectrum of neuromuscular and non-neuromuscular symptoms, and several clinical entities have been defined and associated with specific mtDNA mutations, such as Kearns-Sayre syndrome (KSS), mitochondrial myopathy, encephalopathy, lactic acidosis, and stroke-like episodes (MELAS) [1], and myoclonus epilepsy with ragged red fibers (MERRF).

The mutation at position nt 3243 of the mt genome is one of the most common mtDNA mutations causing the MELAS syndrome with a prevalence of 1:6000 adults[2].

Pathogenic mtDNA mutations are usually heteroplasmic, presenting as a mixture of mutated and wild-type mtDNA. The severity of the disease correlates with the amount of mutated mtDNA found in affected tissues.

The LightCycler System provides a simple and elegant tool for real-time detection of point mutations using hybridization probes.

We established a protocol for detection of mtDNA point mutations on the example of the MELAS3243 mutation. In addition, the sensitivity of melting curves was studied.

Materials

Equipment

LightCycler Instrument
Oligo Primer analysis software Vs.6.0 (MedProbe, Oslo)

Reagents

Amplification primers (Microsynth, Balgach)
Hybridization probes (TIB MOLBIOL, Berlin)
LightCycler-DNA Master Hybridization Probes (Roche Diagnostics, Mannheim, Germany)
DNA Extraction Spin columns (Qiagen, Basel)

---

* Stephanie Kleinle (✉) (e-mail: skleinle@dkf2.unibe.ch)
  Molekulare Humangenetik, Medizinische Universitäts-Kinderklinik, Inselspital, 3010 Bern, Switzerland

Procedure

Sample material and preparation

Human total DNA was taken from control individuals and patients carrying different levels of heteroplasmy of the MELAS3243 mutation. Heteroplasmy was quantified by restriction-fragment-length-polymorphism analysis (RFLP), using fluorescence labeled primers on an ABI377 [3]. Total DNA was extracted from peripheral blood cells and tissue via spin columns[3].

Primer Design

Primers (Table 1) were designed and tested for primer dimers and false priming sites in the mt genome using the Oligo Primer analysis software.

Table 1. Primers and hybridization probes

Primers	Nt (Acc.no. J01415)	Length	GC (%)	T_m (°C)
CTACTTCACAAAGCGCCTT	3149–3167	19	47.4	56
AAGCATTAGGAATGCCATTG	3372–3353	20	40.0	56
Hybridization Probes				
GATTACCGGGCTCTGCCATCTT-F	3254–3233	22	54.5	61.8
LCRed640-ACAAACCCTGTTCTTG-GGTGGGTGT-P	3231–3207	25	52.0	66.9

Probe Design

The hybridization probe corresponded to wild-type mtDNA and covered nt 3243. It was designed with a lower annealing temperature than the anchor probe. One base separated the probes (Table 1).

Hybridization Probe Master Mix:

	Volume [µl]	[Final]
LightCycler-DNA Master Hybridization Probes	2	1×
$MgCl_2$ (25 mM)	1.6	3 mM
Primers (10 µM each)	1 + 1	0.5 µM
Hybridization probe-F (4 µM)	1	0.2
Anchor probe-LC640 (8 µM)	1	0.4
H_2O	10.4	
Total DNA (10 ng/µl)	2	

- Denaturation for 30 s at 95°C
- Amplification

Parameter	Value		
Cycles	40		
Type	Quantification		
	Segment 1	Segment 2	Segment 3
Target temperature [°C]	95	58	72
Incubation time (h:min:s)	0	10	10
Temperature transition rate [°C/s]	20	20	20
Acquisition mode	None	Single	None
Gains	F1=1	F2=20	

- Melting Curve Analysis

Parameter	Value		
Cycles	1		
Type	Melting Curves		
	Segment 1	Segment 2	Segment 3
Target temperature [°C]	95	45	85
Incubation time [h:min:s]	0	10	0
Temperature transition rate [°C/s]	20	20	0.2
Acquisition mode	None	None	Cont.
Gains	F1=1	F2=20	

Results

Analysis of mtDNA from two patients with different levels of heteroplasmy is shown in the figure.

The MELAS3243 mutation introduces a single base mismatch G::T under the hybridization probe and decreases the melting temperature (T_m) by 3.5°C from 68°C to 64.5°C. Two different peaks can be identified, originating from wild-type (T_m=68°C) and mutated mtDNA (T_m=64.5°C).

Analysis of patients with heteroplasmy levels of 70% MELAS3243 to 30% wild-type and 15% MELAS3243 to 85% wild-type revealed that heteroplasmy proportions could be detected, but fluorescence signals were not separated clearly into two peaks.

Comments

Template DNA

A total of 10 ng of human total DNA was used for LC PCR. Due to the high copy number of mtDNA, less total DNA is required for amplification than with nuclear genes [4]. The average crossing point was 20.

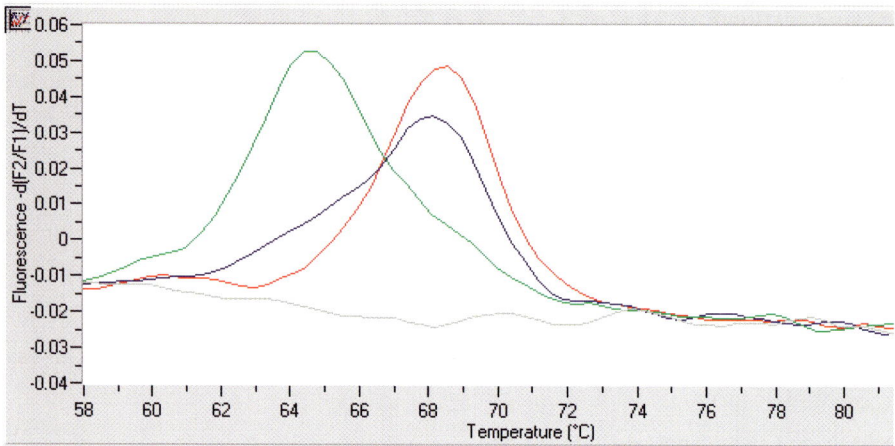

Fig. 1. Derivative melting curve ($-dF/dT$) showing detection of the MELAS3243 mutation. Samples are wild-type (blood) (*red*), patient with 70% mutated mtDNA in skeletal muscle (*green*), patient with 15% mutated mtDNA in blood (*blue*), and negative control (*grey*).

Sensitivity

We were not able to detect low levels of heteroplasmy. Our current aim is to increase the sensitivity of melting curve analysis for quantification of mtDNA point mutations.

Primers presented in this protocol formed no primer dimers and therefore ensured optimal amplification and high sensitivity.

Preliminary data show that sensitivity of the hybridization probes may be increased by decreasing the transition rate of the renaturation step in order to obtain optimal probe annealing.

Hybridization probes may also be relocated to create an A::C mismatch at nt 3243.

References

1. Hammans SR, Sweeney MG, Hanna MG, Brockington M, Morgan-Hughes JA, Harding AE (1995) The mitochondrial DNA transfer RNALeu(UUR) A->G(3243) mutation. A clinical and genetic study. Brain 118, 721–734
2. Majamaa K, Moilanen JS, Uimonen S, Remes AM, Salmela PI, Karppa M, Majamaa-Voltti KA, Rusanen H, Sorri M, Peuhkurinen KJ, Hassinen IE (1998) Epidemiology of A3243G, the mutation for mitochondrial encephalomyopathy, lactic acidosis, and strokelike episodes: prevalence of the mutation in an adult population. Am J Hum Genet 63:447–54
3. Kleinle S, Schneider V, Moosmann P, Brandner S, Krähenbühl S, Liechti-Gallati S (1998) A novel mitochondrial tRNA^{Phe} mutation inhibiting anticodon stem formation associated with a muscle disease. Biochem Biophys Res Commun 247:112–115
4. Wallace DC (1992) Diseases of the mitochondrial DNA. Annu Rev Biochem 61:1175–1212

Acquired Genetic Alterations in Human Diseases

Detection of *p53* Allele Deletions in Human Cancer by Quantification of Genomic Copy Number 159
Jochen Wilhelm, Alfred Pingoud, Meinhard Hahn

Monitoring of Residual Disease in Patients with Chronic Myelogenous Leukemia Using Specific Fluorescent Hybridization Probes for Real-Time Quantitative RT-PCR 171
Andreas Hochhaus, Michael Emig, Andreas Weisser, Susanne Saussele, Martin Müller, Paul La Rosée, Christian Kuhn, Peter Paschka, Rüdiger Hehlmann

Development of Quantitative RT-PCR for the Expression of Wilms' Tumor WT1 Suppressor Gene in Leukemia on the LightCycler 187
Yoji Ishida, Kazunori Murai

Real-Time Detection of Minimal Residual Disease by Amplifying Immunoglobulin Genes in Acute Lymphoblastic Leukemia on the LightCycler ... 197
Makoto Nakao, Bart Janssen, Claus R. Bartram

HER2/*neu* Gene Amplification Quantified by PCR and Melting Peak Analysis Using a Single Base Alteration Competitor as an Internal Standard 207
Elaine Lyon, Ph.D, Alison Millson, MT(ASCP), Arminda Suli, B.Sc.

Quantification of Residual Tumor Cells in Monoclonal B-cell Lymphoma 219
Thomas Pfitzner, Andreas Engert, Stefan Barth

Development of PCR-Based Assays for the Detection of Chromosomal Translocations Using SYBR Green I 231
Sandra D. Bohling, Kojo S. J. Elenitoba-Johnson

Relative Quantification of the HER2/*neu* Oncogene Using SYBR Green I 241
Rachel Woods

Detection of *p53* Allele Deletions in Human Cancer by Quantification of Genomic Copy Number

JOCHEN WILHELM, ALFRED PINGOUD, MEINHARD HAHN*

Introduction

In most cases cancer is a genetic disease, characterized by a multiple step process [1], which involves the inactivation of tumor suppressor genes (e.g., by deletion of alleles) and the activation of oncogenes (e.g., by amplification of alleles). In healthy cells these misregulated genes are important regulators of cellular growth and cellular division, DNA repair, and apoptosis. *p53* is the most prominent and most important tumor suppressor gene which becomes inactivated in nearly all cancer types, roughly 50% of all human tumors being affected by *p53* inactivation [2]. The deletion of an allele or the inactivation by point mutations are the most prominent mechanisms for loss of function of *p53*. Tumors containing inactivated *p53* are highly aggressive and often resistant to chemo- as well as radiotherapy. As the functional status of *p53* has many clinical and therapeutical implications [3, 4, 5], rapid and sensitive tests for the presence or absence of functional *p53* genes are needed. Several techniques have been described for the detection of *p53* allele deletions: (a) analysis of polymorphic DNA markers (RFLPs, minisatellites, microsatellites) for loss of heterozygosity [6, 7]; (b) Southern blotting techniques [2]; (c) fluorescence in situ hybridizations [5, 8]; (d) comparative genomic hybridization [8, 9]; and (e) quantitative competitive PCR [10] which all have several disadvantages, e.g., that they require large amounts of high molecular DNA (b); that they involve cell culture techniques and fluorescence microscopy (c, d); that they depend on the presence of heterozygous allelotypes of DNA markers, application of radioactivity, or fluorescent labels plus DNA sequencers (a); that they require internal control templates (e); and that they only detect allele imbalance but in most cases do not allow discrimination between deletions and amplifications (a).

Here we present a quantitative rapid-cycle real-time PCR with external standardization which quantifies both the mean cellular copy number of *p53* and *IGF-1*. The later target is a well suited internal reference or housekeeping gene for the analysis of tumor DNAs, e.g., bladder cancer [11]. This assay circumvents most of the limitations mentioned above.

---

* Meinhard Hahn (✉) (e-mail: Meinhard.U.Hahn@chemie.bio.uni-giessen.de)
 Institut für Biochemie, FB 08, Justus-Liebig-Universität Giessen, Heinrich-Buff-Ring 58, 35392 Giessen, Germany

Materials

Equipment and Software

LightCycler instrument and software 3 (Roche Diagnostics, Mannheim, Germany)
LightCycler capillaries, centrifuge adapters and cooling blocks (Roche Diagnostics, Mannheim)
OLIGO primer analysis software 5.0 (National Biosciences Inc., Plymouth, USA)
Human blood and tumor tissue samples
Bladder tumor tissue, shock-frozen in liquid nitrogen and stored at –70°C
K-EDTA stabilized blood both of tumor patients and of healthy probands

Reagents

Deoxynucleotides (dATP, dCTP, dGTP, dTTP), PCR-grade (Roche Diagnostics, Mannheim)
Taq DNA polymerase 10× reaction buffer (100 mM Tris-HCl, 15 mM $MgCl_2$, 500 mM KCl, pH 8.3 at 20°C) (Roche Diagnostics, Mannheim)
Bovine serum albumin (BSA), 20 mg/ml, special quality for molecular biology (Roche Diagnostics, Mannheim)
$MgCl_2$, 25 mM (Roche Diagnostics, Mannheim)
Taq DNA polymerase, recombinant, 5 U/µl (Roche Diagnostics, Mannheim)
Oligodeoxynucleotides/PCR primers, HPSF-grade (MWG-Biotech AG, Ebersberg, Germany)
(for sequences: see Table 1)
Hybridization Probes (TIB MOLBIOL, Berlin, Germany)
Pure water (Merck Eurolab GmbH, Darmstadt, Germany)
Dilution buffer: 10 mM Tris-HCl, pH 8.5
RNase A, DNase-free, 7000 U/ml (QIAGEN GmbH, Hilden, Germany)
QIAamp DNA Blood Mini Kit (QIAGEN)
QIAamp DNA Mini Kit (QIAGEN)

Procedure

Genomic DNA Preparation

Human genomic DNA was isolated from K-EDTA treated blood, using the QIAamp DNA Blood Mini Kit, or from tumor tissue samples by the QIAamp DNA Mini Kit according to the manufacturer's protocol, including an RNase A incubation step for complete degradation and removal of RNA. After elution of the genomic DNA in kit buffer AE, the uv-absorbance of DNA solutions was recorded in the range of 220–320 nm. DNA concentrations were calculated using the dsDNA-specific relation 1 $A^{260\ nm} \cong 50$ µg/ml. Standard DNA stock solutions of 12 ng DNA/µl were prepared from the extracted blood DNA of healthy individuals, by diluting the DNA samples in 10 mM Tris-HCl, pH 8.5. This corresponds to 4000 copies of template DNA per microliter for a human single copy gene, assuming a value of 3 pg DNA per haploid human genome. A 1:2 dilution series in the range of 4000–125 copies/µl of this DNA is used for external standardization of quantitative LightCycler experiments. Tumor tissue DNA and blood DNA extracts of patients were adjusted to a concentration of 3 ng DNA/µl.

Hybridization probes reaction master mix for each 10 µl reaction:

LightCycler PCR

	Volume [µl]	[Final]
10× *Taq* reaction buffer	1.0	1×
MgCl$_2$ (25 mM)	1.8	6 mM
BSA (5 mg/ml) [12]	1.0	0.5 mg/ml
dNTP mix (2 mM each)	1.0	0.2 mM each
Primer mix (5 µM each)	1.0	500 nM each
Hybridization probes (1 µM each)	2.0	200 nM each
Taq DNA polymerase (5 U/µl)	0.1	0.5 U
H$_2$O	0.1	
Total volume	8.0	

For a series of n PCRs, depending on the number of reactions, the (n+1) or (n+2)-fold amount of the reaction master mix was prepared, first of all by thoroughly mixing all components without *Taq* DNA polymerase. This enzyme was added to the mix immediately before its use, when 8 µl of master mix and 2 µl of DNA template were given into each LightCycler capillary. For analyzing unknown samples in five parallel reactions, the corresponding amounts of master mix and sample DNA were mixed and dispensed into five capillaries. Capillaries were sealed, centrifuged in a microcentrifuge at low speed (3000 rpm) using centrifuge adapters and transfered into the LightCycler sample carousel. The following PCR protocol was used for amplification and acceptor fluorescence detection:

- Denaturation at 95°C for 30 s
- Amplification

Parameter	Value		
Cycles	50		
Type	Quantification		
	Segment 1	Segment 2	Segment 3
Target temperature [°C]	95	55	72
Incubation time [s]	0	5	10
Temperature transition rate [°C/s]	20	20	20
Acquisition mode	None	Single	None
Gains	F1=5, F2=15		

Analysis of LightCycler Data

The quantitative analysis of the resulting LightCycler fluorescence raw data was carried out, using LightCycler Software 3. The background readings were subtracted from the raw data and were analyzed using the fit points method. The threshold value was adjusted to minimize the error of the calibration curve.

Simulation of LOH

Before starting to analyze clinical samples in order to detect *p53* allele deletions, the precision and reproducibility of the assay has to be checked for each LightCycler system by quantification of simulated partial *p53* allele deletions. This is done by titrating template DNA, isolated from the blood of a healthy individual, in a narrow range of concentrations (2000; 1750; 1500; 1250; 1000 template copies/capillary). These "unknown" samples are amplified in the LightCycler in five parallel experiments. The PCR products generated are detected on-line using a *p53* specific system of primers/hybridization probes (Table 1). Simultaneously, a dilution series of the same blood DNA (8000–250 copies/capillary; one PCR for each concentration), which is defined as the external standard, is amplified in the same manner (see Fig. 1). The corresponding quantification experiment is carried out using primers/hybridization probes specific for the internal reference gene *IGF-1* (see Fig. 2).

Both *p53* and *IGF-1* copy numbers are calculated for the "unknown" samples by the LightCycler software as described above. In the next step the LOH values of the simulated samples are calculated using the definition for LOH (see below), considering "unknown" sample of 2000 copies/capillary to be the "healthy tissue" sample containing two *p53* and two *IGF-1* alleles in each of 1000 cells and the "unknown" sample containing 1000 *p53* copies/capillary to be the "tumor sam-

Table 1. Oligonucleotides

	Position	Exon/Intron	Length	GC (%)	T_m (°C)
p53 (Genebank Accession #X54156)					
PCR primers:					
TTC CTA GCA CTG CCC AAC A	14,673	I8/E9	19	52.6	68.4
GAC TGG AAA CTT TCC ACT TG	14,797	I9	20	45.0	63.4
PCR product			125	48.0	86.2
Hybridization probes					
CCC CAG CCA AAG AAG AAA CCA CTG GAT GGA GAA T-F	14,707	E9	34	50.0	77.2
LCRed640-TTT CAC CCT TCA GGT ACT AAG TCT TGG GAC CTC TT-P	14,741	E9	35	45.7	75.4
IGF-1 (Genebank Accession #M12659)					
	Position	Exon/Intron	Length	GC (%)	T_m (°C)
PCR primers:					
ACA GCT CGG CAT AGT CTT	1,133	Promoter	18	50.0	65.2
CCA AGT GAG GGG TGT GA	1,257	Promoter	17	58.8	65.4
PCR product			125	50.8	94.5
Hybridization probes					
ATG AGA CAG TGC CCT AAA GGG ACC AAT CCA ATG-F	1,174	Promoter	33	48.5	75.7
LCRed640-CTG CCT GCC CCT CCA TAG GTT CTA GGA AAT GAG-P	1,207	Promoter	33	54.5	76.7

P, 3' terminal phosphate; F, 3' terminal fluorescein; LCRed640, 5' terminal fluorophor LightCycler RED640

ple", with only one *p53* allele in each of 1000 cells. The other "unknown" samples represent mixtures of cells with one or two *p53* copies per cell. The assay is validated when all data points in a plot of theoretical versus calculated LOH (see Fig. 3) fall on a curve described by the equation y=x.

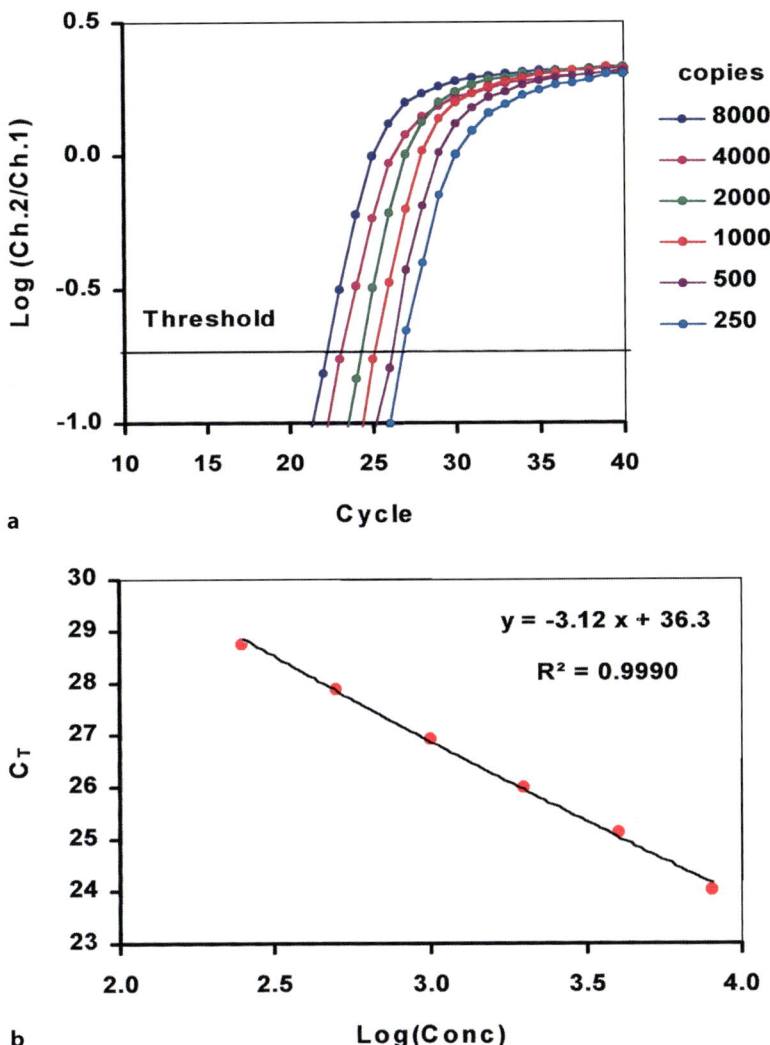

Fig. 1 a,b. Copy number quantification of the target gene p53. **a** A dilution series of human genomic DNA from blood was used as standard DNA template for the amplification of a *p53*-specific 125-bp fragment. The logarithm of the ratio of fluorescence signal intensities (channel 2/channel 1) is shown for the amplification process. The exponential phase of PCR corresponds to the linear segment of the curves. **b** The calibration curve shows the logarithmic concentration of the *p53* template copy number per reaction versus the threshold cycle (C_T) of each PCR

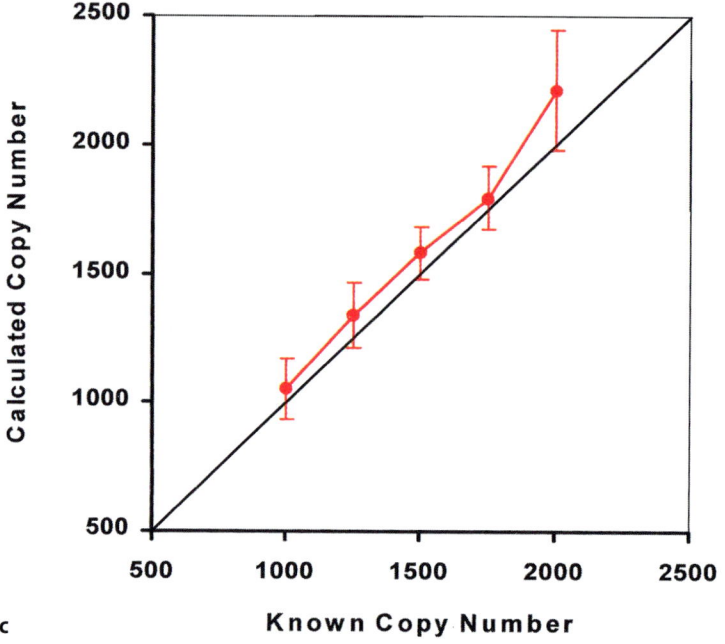

Fig. 1 c. Using the calibration curve of **b**, the *p53* copy number was calculated for several "unknown" samples, e.g., blood DNA dilutions. For each sample the mean value for the copy number determined in five independent experiments is shown together with its standard deviation. The diagonal black line represents the theoretical curve

LOH Quantification for Unknown Tumor Samples

For LOH analysis of clinical samples, both the DNA of the tumor and the DNA of a control sample of the same patient (healthy tissue or blood sample) are amplified, using *p53* as well as *IGF-1* specific primers. Each test is carried out in five independent experiments. In the same LightCyler-run, the gene specific amplifications of a standard DNA series are included (8000; 4000; 2000; 1000; 500 copies/capillary; blank). The LOH is calculated using the mean values for each sample and the LOH-formula given below.

Results

The system of primers/hybridization probes specific for the human tumor suppressor gene *p53* and the internal reference gene *IGF-1* described here, yields optimal amplification signal courses (e.g., high signal intensities, clear exponential phases, proper sigmoidal shapes) and allows, therefore, exact copy number quantifications with low standard deviations. Even standard titrations with small dilution factors (<2) result in distinguishable equidistant amplification curves (Fig. 1a, 2b) resulting in standard calibration curves with low error values (Fig. 1b, 2b). For both genes, the quantified copy numbers match the theoretical values (Fig. 1c, 2c).

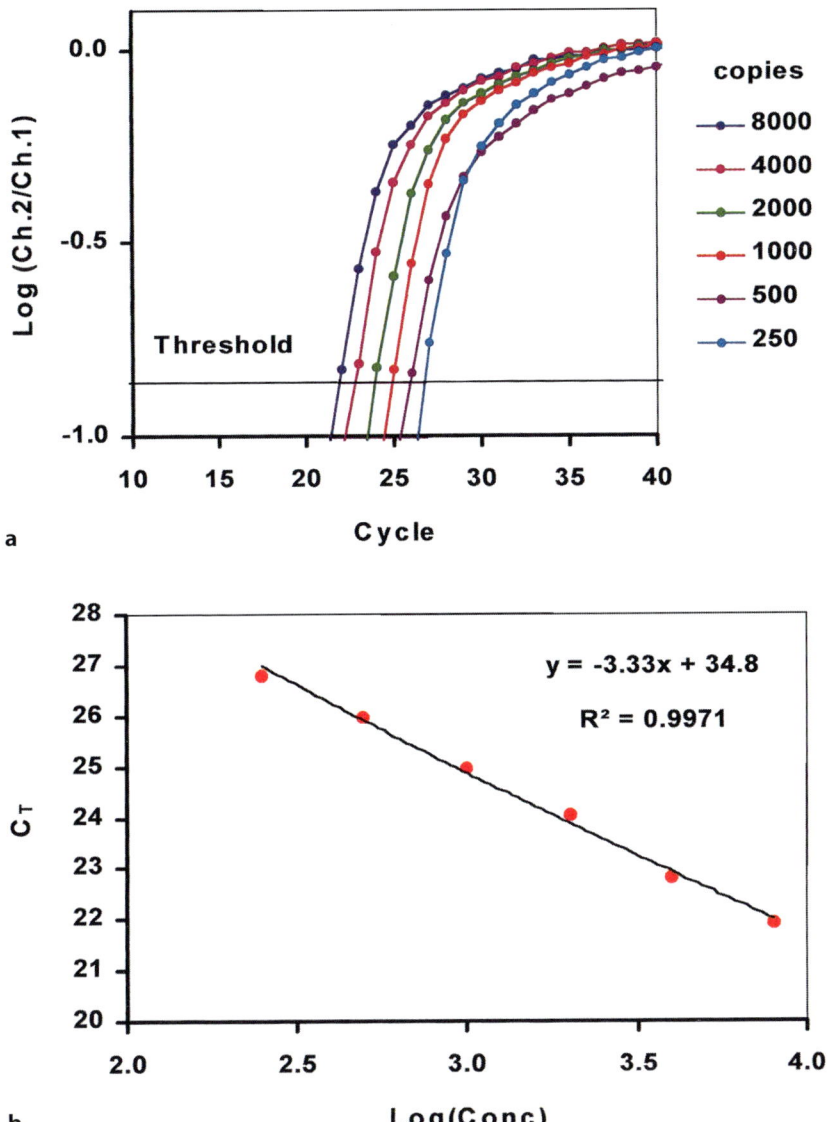

Fig. 2 a–c. Copy number quantification of the internal control gene *IGF-1*. The quantification of *IGF-1* copy number was performed as described in the legend of *Fig. 1*

By LOH simulation (Fig. 3) it is shown that precision as well as reproducibility of this method are sufficient to measure LOH values of less than 0.80. This would correspond to tumor tissue samples containing only 20% of cells with a *p53* allele deletion or the discrimation of *p53* concentrations differing by a factor as small

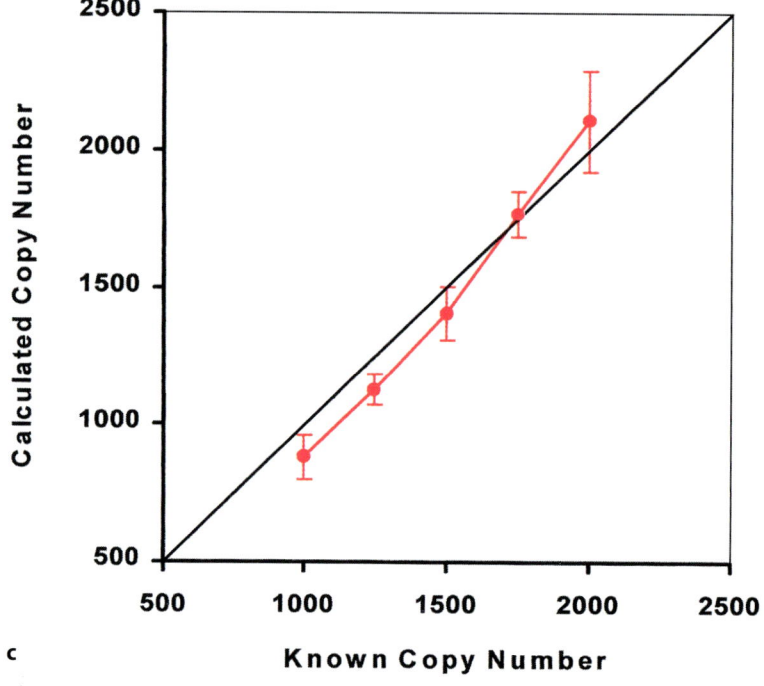

Fig. 2 c.

as 1.15, respectively. The reproducibility of the LOH assay was determined for four independent quantifications. The interassay variation was approximately 15% of the mean values.

The suitability of the assay for clinical tumor samples was demonstrated for human bladder tumor samples. These samples were previously characterized for *p53* LOH by the well established analysis of microsatellite markers [7]. Figure 4 presents the data of a tumor which resulted in a LOH of 0.39. This corresponds well with the LOH of 0.53 determined by the quantitative LightCycler PCR. In comparison to the microsatellite analysis, the method described here is very rapid, avoids problems of carry over contaminations of PCR products (closed capillary format), and allows for direct quantification of the particular gene of interest instead of indirect quantifications, as typically obtained when determining allelic imbalances. This is even possible with patients showing non-informative allelotypes of established polymorphic markers.

Troubleshooting

When developing a new amplification system of PCR primers/hybridization probes specific for a target gene, one has to check very carefully for the absence

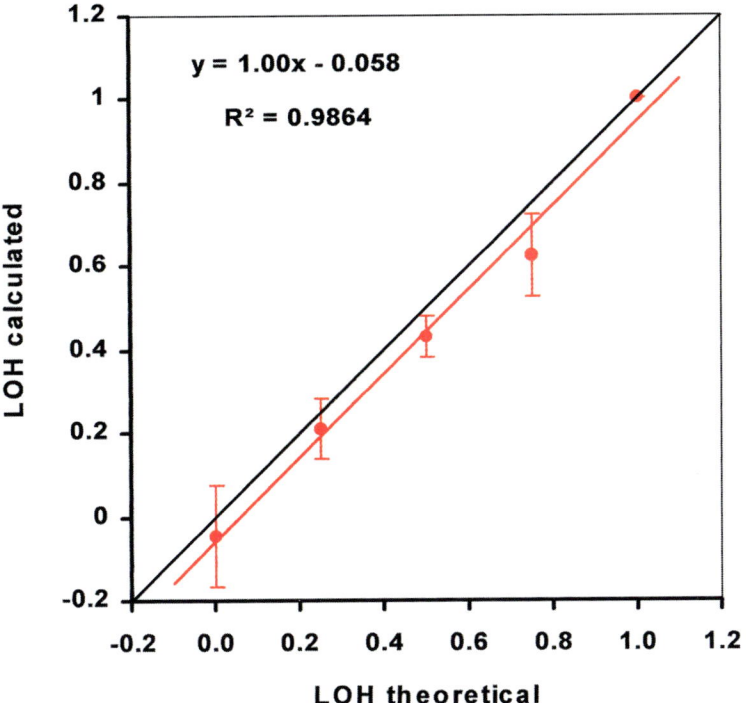

Fig. 3. Quantification of simulated partial deletions (LOH) of a *p53* allele. Using different amounts of "sample" template (e.g., genomic blood DNA), varying in the range of 1000–2000 copies of *p53* per reaction, the initial *p53* copy numbers were calculated according to *Fig. 1*. These data were normalized (see comments) with the quantification results which were obtained for a PCR with an initial amount of blood DNA template corresponding to 2000 copies of the *IGF-1* reference gene. *Red points* in the diagram represent the measured mean values of *p53* LOH for $n=4$ independent amplification series, *red bars* depict the standard deviations. The *diagonal black line* represents the theoretical curve. LOH definition: LOH=1, 100% cells within the analyzed sample have 2 *p53* alleles; LOH=0, all cells have lost one *p53* allele

of non-specific PCR products as well as primer dimers by native polyacrylamide gelelectrophoresis. This has to be done despite sequence specific amplicon detection by hybridization probes; although these undesirable products are not detected, they can interfere with this highly precise kind of quantitative analysis. If such minor products are amplified, one can try to avoid this by the application of hot start PCR techniques (TaqStart Antibody, Clontech, or HotStarTaq DNA polymerase, QIAGEN, which is activated by an initial heating step) or touch down amplification protocols which make use of a high stringency in the initial PCR cycles. A further critical factor is the selection of a human single copy gene which is used for the normalization of tumor suppressor gene data. This internal reference gene must not be affected by amplifications or deletions in the tumor type examined. A critical factor in the precise quantification of subtle differences in

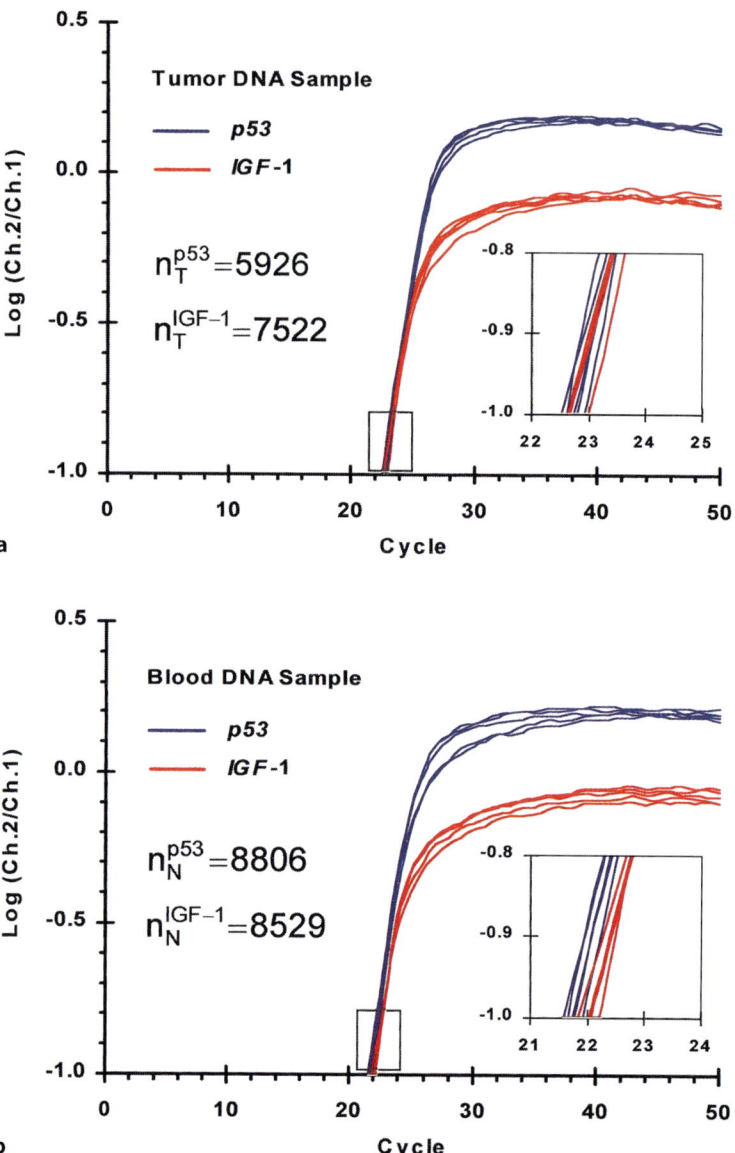

Fig. 4. Quantification of a *p53* allele deletion within a human bladder tumor. **a** The semi-logarithmic plot of fluorescence signal intensities shows the data for the copy number quantification of *p53* and *IGF-1*. In each capillary 6 ng template DNA were used which was extracted from a patient's human bladder tumor sample. Each gene was quantified in a fivefold set-up. The *insert* shows a blow-up of the boxed segment to demonstrate the high degree of reproducibility. **b** This diagram shows the data for blood DNA of the same tumor patient according to **a**. Using the *p53* and *IGF-I* copy number data of this patient's blood and tumor DNA, a LOH-value of 0.53 is calculated. This could mean that 53% of the cells have two *p53* alleles, while 47% have only one *p53* allele

sample concentrations, is the total volume error of all pipetting steps. Therefore, one has to check for proper calibration of all pipettes but also for a constant quality of pipette tips to minimize this factor. We regularly use Gilson pipettes and original Gilson tips for this work.

Comments

Usually, polymorphic DNA markers are used to detect allele imbalances or deletions of tumor suppressor genes. For reasons of comparison, we use the previously introduced "LOH value" [7] as a measure for the proportion within a sample of "normal" cells carrying two copies of the specific suppressor gene. The LOH is defined as given below when using the data of quantitative PCR experiments for its calculation:

$$\text{LOH} = 2 \times \frac{n_T^{p53} / n_T^{IGF-1}}{n_N^{p53} / n_N^{IGF-1}} - 1$$

n_T^{p53}: initial number of *p53* template copies in the tumor sample
n_N^{p53}: initial number of *p53* template copies in the healthy or "normal" sample
n_T^{IGF-1}: initial number of *IGF-1* template copies in the tumor sample
n_N^{IGF-1}: initial number of *IGF-1* template copies in the healthy or "normal" sample
LOH=1 no difference in allele status between tumor and healthy control
0≤LOH<1 one *p53* allele is affected by a deletion
LOH>1 one *p53* allele can be amplified (unlikely) or *IGF-1* is affected by a deletion

To achieve high precision, the DNA standards should enclose the expected range of sample DNA concentration with narrow dilution intervals. In addition, because of tumor microheterogeneity, it is recommended to use microdissected tumor tissue samples, ideally containing only tumor cells, to allow for sensitive detection of LOH-positive cells. It should be noticed that LOH-analysis by quantitative LightCycler PCR is especially recommended for the detection of amplified oncogenes because of a higher dynamic range of copy number variation.

Applications

The application of LightCycler PCR for the purpose of quantitative PCR with external standardization allows for highly accurate quantification of a gene's copy number. Therefore, by this approach it is possible to directly detect minor losses of tumor suppressor gene alleles (e.g., *p53*, *p16*) or amplifications of oncogenes (*c-myc*, *HER-2*) in tumor DNA samples, even in cases where only a fraction of 20% of sample cells has an allelic imbalance. This approach makes the LOH-analysis of heterozygous polymorphic markers (RFLPs, VNTRs, STRs) unnecessary. The technique can also be applied to quantify the nuclear copy number of transfected gene sequences in recombinant organisms.

References

1. Fearon ER, Vogelstein B (1990) A genetic model for colorectal tumorigenesis. Cell 61: 759–767
2. Nigro JM, Baker SJ, Preisinger AC, Jessup JM, Hostetter R, Cleary K, Bigner SH, Davidson N, Baylin S, Devilee P, Glover T, Collins FS, Weston A, Modali R, Harris CC, Vogelstein B (1989) Mutations in the *p53* gene occur in diverse human tumour types. Nature 342; 705–708
3. Harris CC, Hollstein M (1993) Clinical implications of the p53 tumor-suppressor gene. N Engl J Med 329: 1318–1327
4. Gallagher WM, Brown R (1999) p53-oriented cancer therapies: current progress. Ann Oncol 10: 139–150
5. Döhner H, Fischer K, Bentz M, Hansen K, Benner A, Cabot G, Diehl D, Schlenk R, Coy J, Stilgenbauer S, Volkmann M, Galle PR, Poustka A, Hunstein W, Lichter P (1995) p53 gene deletion predicts for poor survival and non-response to therapy with purine analogs in chronic B-cell leukemias. Blood 85: 1580–1589
6. Hahn M, Fislage R, Pingoud A (1995) Polymorphism of the pentanucleotide repeat d(AAAAT) within intron 1 on the human tumor suppressor gene p53 (17p13.1). Hum Genet 95: 471–472
7. Hahn M, Matzen SE, Serth J, Pingoud A (1995): Semiautomated quantitative detection of loss of heterozygosity in the tumor suppressor gene p53. Biotechniques 18: 1040–1047
8. Mahdy E, Yoshihiro S, Zech L, Wester K, Pan Y, Busch C, Dohner H, Kallioniemi O, Bergerheim U, Malmstrom PU (1999) Comparision of comparative genomic hybridization, fluorescence in situ hybridization and flow cytometry in urinary bladder cancer. Anticancer Res 19: 7–12
9. Moch H, Presti JC Jr, Sauter G, Buchholz N, Jordan P, Mihatsch MJ, Waldman FM (1996): Genetic aberrations detected by comparative genomic hybridization are associated with clinical outcome in renal cell carcinoma. Cancer Res 56, 27–30
10. Hahn M, Pingoud A (1999) Quantitative PCR with internal standardization and OLA-PCR product analysis for the *p53* tumor suppressor gene. In: Kochanowski B, Reischl U. (eds) Methods in Molecular Medicine, Vol. XXVI, Quantitative PCR Protocols. Humana Press Inc., Totowa, NJ
11. Cairns JP, Chiang P-W, Ramamoorthy S, Kurnit DM, Sidransky D (1998) A comparison between microsatellite and quantitative PCR analyses to detect frequent *p16* copy number changes in primary bladder tumors. Clin Cancer Res 4: 441–444
12. Wittwer CT, Herrmann MG, Moss AA, Rasmussen RP (1997) Continuous fluorescence monitoring of rapid cycle DNA amplification. Biotechniques 22: 130–138

Monitoring of Residual Disease in Patients with Chronic Myelogenous Leukemia Using Specific Fluorescent Hybridization Probes for Real-Time Quantitative RT-PCR

Andreas Hochhaus*, Michael Emig, Andreas Weisser, Susanne Saussele, Martin Müller, Paul La Rosée, Christian Kuhn, Peter Paschka, Rüdiger Hehlmann

Introduction

In many ways, chronic myelogenous leukemia (CML) serves as a paradigm for the utility of molecular methods to diagnose malignancy or to monitor patient response to therapy [1]. CML constitutes a clinical model for molecular detection and therapy surveillance since this entity was the first leukemia known to be associated with a specific chromosomal rearrangement, the Philadelphia (Ph) translocation t(9;22)(q34;q11), and the presence of two chimeric genes, BCR-ABL on chromosome 22 and ABL-BCR on chromosome 9. BCR-ABL is transcribed to a specific BCR-ABL mRNA and encodes in most patients a 210-kDa chimeric protein with increased tyrosine kinase activity. The central role of BCR-ABL in several pathways which lead to uncontrolled proliferation has been shown in vitro and in vivo. Several approaches have been introduced that can specifically detect the Ph translocation or its products, such as fluorescent in situ hybridization, Southern blotting, western blotting, and reverse transcriptase polymerase chain reaction (RT-PCR) [2–4]. Of these, RT-PCR for BCR-ABL mRNA is by far the most sensitive and consequently has received the most attention in the context of minimal residual disease.

Since non-quantitative RT-PCR analysis after therapy gives only limited information, quantitative or semi-quantitative RT-PCR assays have been developed that enable the kinetics of residual BCR-ABL transcripts to be monitored over time in patients after allogeneic stem cell transplantation or interferon-α (IFN-α) therapy. Currently, the most common method to quantify BCR-ABL transcripts is by competitive PCR. Although this method has been established in a number of laboratories, it is cumbersome and requires large numbers of individual amplification reactions to be performed for each datapoint.

Novel real-time PCR procedures promise to simplify existing protocols [5,6]. We have established an alternative real-time RT-PCR approach for detection and quantification of BCR-ABL fusion transcripts, using the new LightCycler technology [7,8]. Fluorescein is used as a donor fluorophore, LC Red640 (Roche Diagnostics, Mannheim, Germany) as the acceptor fluorophore. We demon-

---

* PD Dr. Andreas Hochhaus (✉) (e-mail: hochhaus@uni-hd.de)
   III. Medizinische Universitätsklinik, Fakultät für Klinische Medizin Mannheim der Universität Heidelberg, Wiesbadener Str. 7–11, 68305 Mannheim, Germany

strate that hybridization probes in combination with the LightCycler enables rapid and robust quantification of BCR-ABL and control transcripts in clinical specimens [9].

Patients and Methods

Patients and Samples

A total of 254 (222 peripheral blood, PB, and 32 bone marrow, BM) samples from 120 CML patients (76 male; 44 female) were analyzed. Between one and nine samples were investigated from each patient during the course of their disease. At diagnosis, 117 patients were Ph-positive and three Ph-negative/BCR-ABL-positive. In all, 219 samples were obtained on or after therapy with IFN-α, 17 after allogeneic BMT, 11 after chemotherapy, and seven at diagnosis.

Cytogenetic Analysis

Cytogenetic studies were performed on BM aspirates according to standard protocols. Metaphases from direct and/or short-term (24-h or 48-h) cultures were examined after Giemsa banding [10]. The cytogenetic response was evaluated as follows: complete, 0% Ph+ metaphases; partial, 1–34% Ph+ metaphases; minor, 35–94% Ph+ metaphases; and none, >94% Ph+ metaphases [11]. A minimum of ten metaphases were analyzed. A total of 81 RT-PCR results obtained by the LightCycler were available for comparison with contemporaneous cytogenetic analyses.

Southern Blot Analysis

Southern blotting was performed as reported [12,13]. DNA obtained from PB or BM was digested with *Bgl*II, *Eco*RI, *Bam*HI, *Xba*I, and/or *Hin*dIII, electrophoresed on a 0.7% agarose gel, blotted, and hybridized to the 2 kb *Bgl*II/*Hin*dIII (5'M-bcr) and 1.2 kb *Hin*dIII/*Bgl*II (3'M-bcr) probes. Response was expressed as complete (BCR ratio 0%), partial (1–24%), minor (25–49%) and none (50%) [13].

RNA Extraction, cDNA Synthesis

Total leukocyte RNA was extracted from 10–20 ml of PB and/or from 1–5 ml of BM aspirate after lysis of red blood cells [14]. Samples were processed as soon as possible after aspiration, although some spent 1–3 days in transit. RNA extraction was performed by CsCl gradient centrifugation [14] or by commercially available extraction kits (RNeasy Qiagen, Hilden, Germany, or HighPure, Roche Diagnostics, Mannheim, Germany). RNA was reverse transcribed using random hexamer priming and MMLV reverse transcriptase as described [14]. cDNA samples were stored at –20°C. A cDNA equivalent of up to 0.5 µg RNA was used for one PCR reaction.

Competitive PCR

The type of BCR-ABL transcript was determined by multiplex PCR [14,15] or nested PCR as described. An estimate of the number of BCR-ABL transcripts and total ABL (i.e., BCR-ABL plus ABL) transcripts as an internal standard was made by a competitive PCR titration assay [16–18]. The final result of quantification by competitive PCR was expressed as the ratio between BCR-ABL and ABL transcripts in percent.

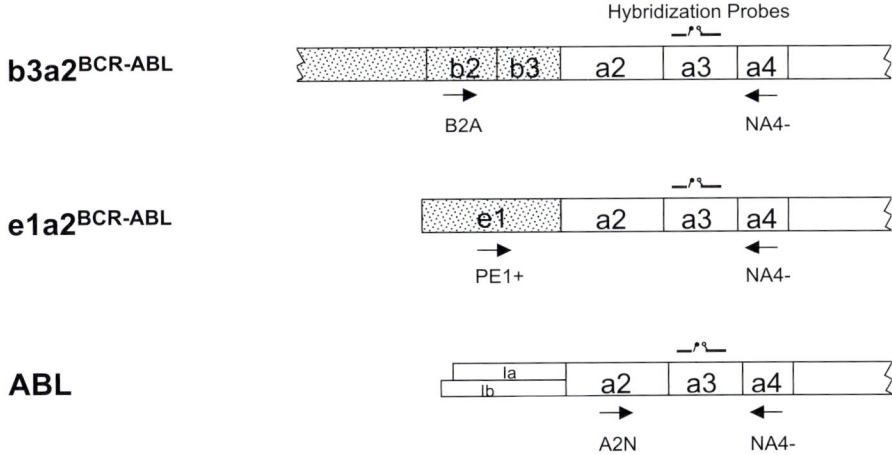

Fig. 1. Schematic map of the amplification and detection of BCR-ABL and ABL transcripts. Specific primer pairs were used to amplify various types of BCR-ABL and ABL transcripts. The detection format was uniform, using a pair of adjacent fluorescently labeled hybridization probes to ABL exon 3

Real-Time Quantitative PCR

LightCycler instrument (Roche Diagnostics, Mannheim, Germany)

LightCycler-DNA Master Hybridization Probes (Roche Diagnostics, Mannheim)
dNTPplus (dATP, dUTP, dGTP, dCTP) (Roche Diagnostics, Mannheim)
Heat labile uracil DNA glycosylase (UDG) (Roche Diagnostics, Mannheim)
Amplification primers (MWG, Ebersberg, Germany)
Hybridization probes (TIB MOLBIOL, Berlin, Germany)

**Materials
Equipment
Reagents**

Procedure

PCR was performed using 2 µl mastermix (LightCycler-DNA Master Hybridization Probes; Roche Diagnostics, Mannheim, Germany, containing buffer, dATP, dCTP, dGTP, dUTP, and Taq polymerase); 3 mM $MgCl_2$, 0.25 µM of each 3' and 5' fluorescent hybridization probes (TIB MolBiol, Berlin, Germany), 0.5 µM of each 3' and 5' oligonucleotide primer (highly purified salt free, HPSF, grade, MWG, Ebersberg, Germany) (Table 1), 1 U heat labile uracil DNA glycosylase (UDG, Roche Diagnostics, Mannheim, Germany), 2 µl cDNA, and water to a final volume of 20 µl. Prior to amplification, mixes were incubated for 5 min at room temperature to allow degradation of specific contaminating PCR products from previous amplifications by UDG. To improve reliability of the assay, mastermixes for 35 samples were prepared. The reaction mix for BCR-ABL, ABL, and G6PD transcript quantification was:

	Volume [μl]	[Final]
LightCycler-DNA Master Hybridization Probes	2	1×
MgCl$_2$ (25 mM)	2.4	4 mM
Probe mix (10 μM each)	0.5	0.25 μM
Primers (100 μM each)	0.1 + 0.1	0.5 μM
Uracil DNA glycosylase (1 U/μl)	1	1 U
cDNA	2	
dH$_2$O	11.9	
Total volume	20	

Table 1. Oligonucleotides

BCR (GeneBank Accession # U07000)					
	Position	Exon	Length	GC (%)	T$_m$ (°C)
Primers					
TTCAGAAGCTTCTCCCTGACAT	123597	BCR b2 (e13)	22	45	65.8
CAGATCTGGCCCAACGATGG	16272	BCR e1	20	60	67.8
ABL (GeneBank Accession # U07563)					
Primers					
CCCAACCTTTTCGTTGCACTGT	49996	ABL a2	22	50	68.7
CGGCTCTCGGAGGAGACGTAGA	58610R	ABL a4	22	64	71.5
Probes					
LCRed640-AATGGGGAATGGTGT-GAAGCCCAAA-P	50658	ABL a3	25	48	70.4
TGAAAAGCTCCGGGTCTTAGGC-TATAATCA-F	50627	ABL a3	30	43	70.0
G6PD (GeneBank Accession # X55448)					
Primers					
CCGGATCGACCACTACCTGGGCAAG	15116	G6PD 6	25	64	74.7
GTTCCCCACGTACTGGCCCAGGACCA	16443R	G6PD 9	26	65	77.7
Probes					
LCRed640-CAAATCTCAGCACCATGA-GGTTCTGCAC-P	15165R	G6PD 7/6	28	50	70.9
GTTCCAGATGGGGCCGAAGATCC-TGTTG-F	15376R	G6PD 7	28	57	73.5

Heat labile UDG was deactivated by an initial denaturation step of 1 min at 95°C. The PCR protocol was:

Parameter	Value		
Cycles	45		
Type Quantification	Segment 1	Segment 2	Segment 3
Target temperature (°C)	95	64	72
Incubation time (s)	1	10	26
Temperature transition rate (°C/s)	20	20	2
Acquisition mode	None	None	Single
Gains	F1=1, F2=15		

According to the primer pair used b2a2, b3a2, b2a3, b3a3, or e1a2 BCR-ABL, total ABL, or G6PD transcripts were amplified. Amplification, fluorescence detection, and post-processing calculations were performed using the LightCycler apparatus (Roche Diagnostics, Mannheim, Germany).

The hybridization probes were labeled 3' with fluorescein or 5' with LCRed640. Fluorescence was measured after each annealing step and expressed as the ratio between fluorescence at 640 nm (designated F2) and at 530 nm (designated F1).

The fluorescence signal was plotted against the cycle number for all samples and external standards. These standards consisted of serial dilutions (10^1–10^7 molecules per reaction) of plasmids pGD210 ($b3a2^{BCR-ABL}$), pB190 ($e1a2^{BCR-ABL}$), and pGdBBX (G6PD). The crossing point was determined for each standard dilution, i.e., the point at which the signal rose above the background level. The higher the initial number of starting molecules, the earlier the signal appears above the background (Fig. 2). A standard curve for each run was constructed by plotting the crossing point against the log of the number of molecules in the standard). The number of target molecules in each sample was then calculated automatically by reference to this curve. Results were expressed as the number of target molecules per 2 µl cDNA. Normalized levels of disease were calculated as the ratio of BCR-ABL and BCR-ABL to G6PD transcripts in 2 µl of cDNA.

Statistical Analysis

The ratios BCR-ABL/ABL and BCR-ABL/G6PD were compared with the proportion of Ph+ metaphases for all samples that had contemporaneous cytogenetic analyses. The comparison of the BCR-ABL/ABL and BCR-ABL/G6PD ratios between the response groups determined by cytogenetics or Southern blot analysis was made using the Kruskal-Wallis test. The correlation between ABL and G6PD transcript numbers as normalization controls, between the ratios BCR-ABL/ABL derived from competitive PCR versus real-time PCR, and between paired results from PB and BM was evaluated using Spearman's rank correlation coefficient. The equation of the regression line of ABL and G6PD transcript numbers and results from PB and BM were calculated by least-squares analysis. Results from contemporaneous PB and BM samples were compared by the paired Student's *t*-test. All statistical tests were performed using the GraphPad Prism software package.

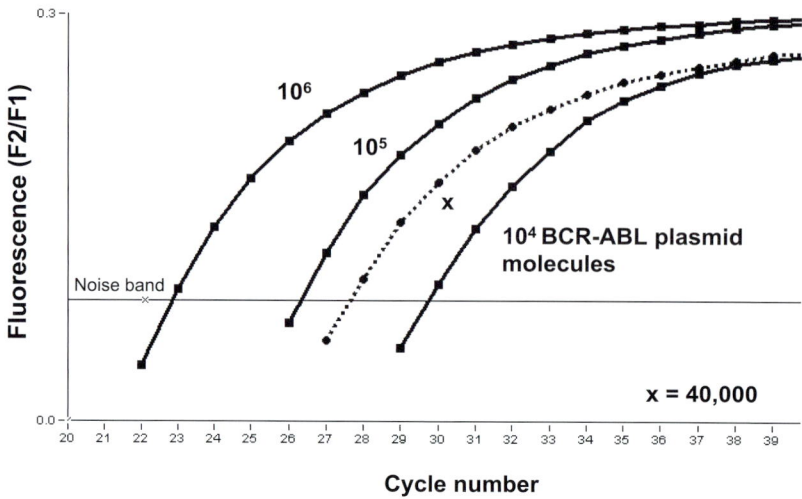

Fig. 2. Example of real-time PCR. Three standard dilutions of plasmid pGD210 (b3a2$^{BCR-ABL}$) were compared with a patient's sample of unknown BCR-ABL concentration. Each *point* represents the fluorescence intensity (F2/F1) measured after each PCR cycle. Plotting the cycle threshold of the unknown sample on the standard curve revealed that 40,000 BCR-ABL transcripts were present at the start of the reaction

Results

Sensitivity of Real-Time PCR

In repeated tests, we could reliably amplify 10 b3a2 or e1a2 BCR-ABL, 10 ABL, and 10 G6PD molecules per reaction. To calculate the maximum sensitivity, serial dilutions of 5, 50, 500, and 5000 myeloid PB cells from a b3a2$^{BCR-ABL}$ positive CML patient at diagnosis in 5×10^7 leukocytes from a healthy donor were processed and analyzed. We routinely achieved a sensitivity of 10^{-5}, i.e., we could detect one CML cell in 10^5 normal white blood cells. The same approach was performed with a dilution of BV173 (b2a2$^{BCR-ABL}$ positive cell line) and K562 (b3a2$^{BCR-ABL}$ positive cell line) cells in HL60 (BCR-ABL negative cell line) cells. In this regard, a sensitivity of 10^{-6} for BV173 and 10^{-7} for K562 was reached.

Reliability of Real-Time PCR

Quantitative analysis of BCR-ABL in 20 identical samples of 500 molecules of plasmid pGD210 (b3a2$^{BCR-ABL}$) in one run (intraassay comparison) resulted in a coefficient of variation (CV, calculated for the determined concentrations) of 18%; the respective cycle threshold crossing points resulted in a CV of 2.9%. The analysis of 20 identical samples in 20 runs on 20 days (interassay comparison, day-to-day variation with new mixtures of reagents) resulted in a CV of 17%, and a crossing point CV of 2.8%.

Patient Samples: Qualitative Nested RT-PCR

All 254 cDNA samples were BCR-ABL positive by conventional nested PCR. A total of 64 patients expressed b3a2, 40 b2a2, 13 both b3a2 and b2a2, and one each b3a3, b2a3, and e1a2 BCR-ABL transcripts.

Of the 254 samples that were BCR-ABL positive by nested PCR, 245 were also BCR-ABL positive by the LightCycler. Between 10 and 3,700,000 (median 10,000) BCR-ABL transcripts were detected. All 254 samples were positive for total ABL transcripts; between 150 and 11,600,000 (median 540,000) ABL transcripts were detected (Table 2). Nine BCR-ABL negative samples showed ABL transcript levels of 150 to 1,390,000 (median 119,000).

Patient Samples: Real-Time PCR

In 249 samples, G6PD transcripts were quantified as a second normalization standard. All samples were positive, between 300 and 7,000,000 (median 630,000) G6PD transcripts were detected (Table 2). The number of G6PD transcripts was at a median of 1.5 times higher than the number of ABL transcripts. G6PD and ABL transcript numbers correlated with $r=0.74$ ($p<0.0001$; Fig. 3).

Comparison of ABL and G6PD Transcript Numbers as Normalization Controls

In 201 cDNA samples the ratio BCR-ABL/ABL was determined by nested competitive PCR which compared well with the same ratio determined by real-time PCR ($r=0.90$, $p<0.0001$, Fig. 4). Five of nine samples that were negative on the LightCy-

Comparison of Real-Time PCR and Competitive PCR

Table 2. Quantification of BCR-ABL, ABL, and glucose-6-phosphate dyhydrogenase (G6PD) transcripts using real-time PCR as compared to the ratio BCR-ABL/ABL determined by competitive PCR

	Real-time PCR					Competitive PCR
	BCR-ABL	ABL	G6PD	Ratio BCR-ABL/ABL	Ratio BCR-ABL/G6PD	Ratio BCR-ABL/ABL
n	254	254	249	254	249	201
Median	10,000	540,000	630,000	4.6%	3.7%	6.8%
Minimum	10 (9 negative)	150	300	0%	0%	0.0002%
Maximum	3,700,000	11,600,000	7,000,000	145%	183%	133%

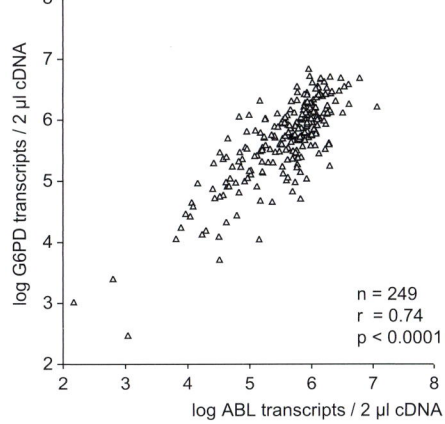

Fig. 3. Comparison of the quantification of total ABL and G6PD transcripts as normalization controls in 249 consecutive samples. The transcript numbers of both genes correlate with $r=0.74$, $p<0.0001$

cler but positive by nested RT-PCR were tested by competitive PCR and found to be positive at a median BCR-ABL transcript level of 0.036% (range 0.0034–0.9%).

Comparison of Real-Time PCR and Cytogenetics

Contemporaneous metaphase cytogenetic results were available for 81 samples investigated by real-time PCR. They showed a complete response ($n=17$), partial response ($n=20$), minor response ($n=15$), and nonresponse ($n=29$) and correlated well with the ratios BCR-ABL/ABL ($p<0.0001$) and BCR-ABL/G6PD ($p<0.0001$), respectively (Fig. 5).

Comparison of Real-Time PCR and Southern Blot Analysis

The BCR ratio as determined by quantitative Southern blot analysis in 122 DNA samples extracted from the same PB or BM samples from which RNA was extracted for PCR analysis revealed complete response ($n=50$), partial response ($n=30$), minor response ($n=14$), or nonresponse ($n=28$) and correlated with the BCR-ABL/ABL ratio as determined by real-time PCR ($p<0.0001$).

BCR-ABL Levels in PB and BM

In eight cases, PB and BM samples obtained on the same day were available for comparison. The ratios BCR-ABL/ABL and BCR-ABL/G6PD in the two tissues were not significantly different. The ratios BCR-ABL/ABL ($r=0.90$, $p=0.0046$) and BCR-ABL/G6PD ($r=0.98$, $p=0.0004$) correlated well between PB and BM samples (Fig. 6).

Gel Analysis of PCR Products

Exemplary amplification products performed in the LightCycler were recovered from the capillaries. Samples were diluted in sucrose/bromphenol blue loading buffer, heated to 65°C for 5 min, loaded on 1.5% ethidium bromide stained agarose gels in 0.5% TBE buffer and electrophoresed. The estimated size of the

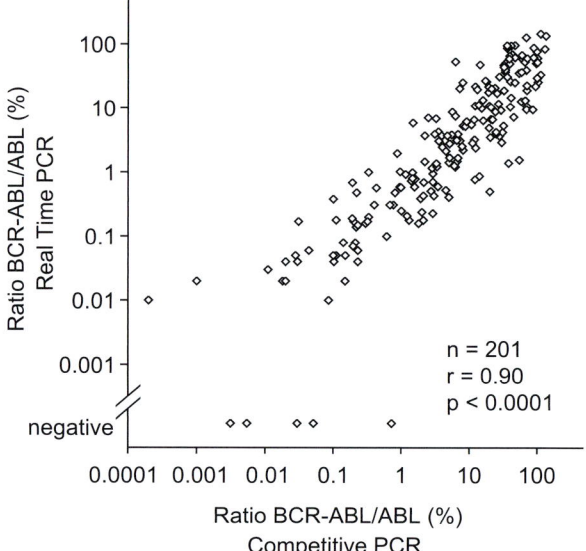

Fig. 4. Comparison of BCR-ABL/ABL ratios determined by real-time and competitive PCR. In five samples positive with competitive PCR, no BCR-ABL transcripts were detected by real-time PCR. The ratios derived by both methods correlate with $r=0.90$, $p<0.0001$

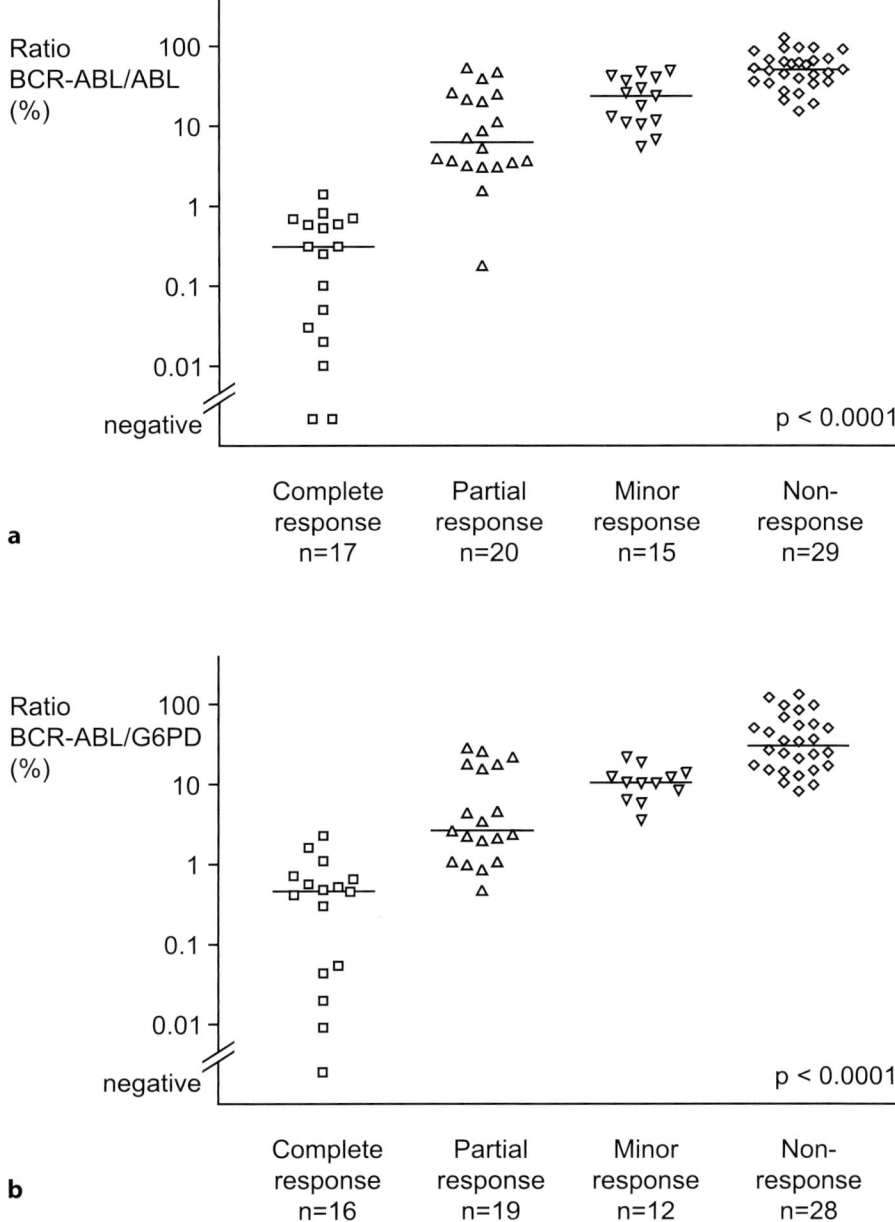

Fig. 5. a, b. Ratios BCR-ABL/ABL (**a**) and BCR-ABL/G6PD (**b**) derived from real-time PCR according to contemporaneous cytogenetic response. The ratios BCR-ABL/ABL and BCR-ABL/G6PD are significantly different between complete, partial, minor, and non-responders ($p<0.0001$)

amplified fragments matched the calculated size [19,20] for b3a2 BCR-ABL (671 bp), b2a2 BCR-ABL (596 bp), b3a3 BCR-ABL (497 bp), b2a3 BCR-ABL (422 bp), e1a2 BCR-ABL (540 bp), ABL (386 bp), and G6PD (343 bp) in all cases.

Discussion

The assessment of the proportion of malignant BM or PB cells is an important prognostic indicator for CML patients after therapy. Cytogenetic analysis is still the standard technique for defining the response of patients to treatment, but this technique suffers from serious limitations. Bone marrow metaphases are required and aspiration and cultivation of proliferating cells is not always sufficient. Furthermore, this technique is relatively insensitive since typically a maximum number of about 30 metaphases are analyzed. The advantages of quantitative PCR techniques for monitoring CML patients during or after treatment have been published previously [2]. Competitive PCR results correlate well with the cytogenetic response in individuals treated with IFN-α [18] and, moreover, the actual levels of detectable residual disease in complete cytogenetic responders may vary by as much as four orders of magnitude [21,22]. After BMT, rising or persistently high levels of BCR-ABL mRNA can be detected in patients prior to cytogenetic or hematologic relapse [17,23–26]. Quantitative PCR data has been used to initiate donor lymphocyte transfusions for treatment of relapse [27] and to monitor the response [28].

The advent of real-time PCR enables the direct measurement of the amount of PCR product during the amplification process and therefore reflects the dynamics of the reaction. Although the final amount of PCR product at the plateau phase usually does not correlate well with the number of starting molecules in the reaction, the time point (cycle number) at which the fluorescence for a particular

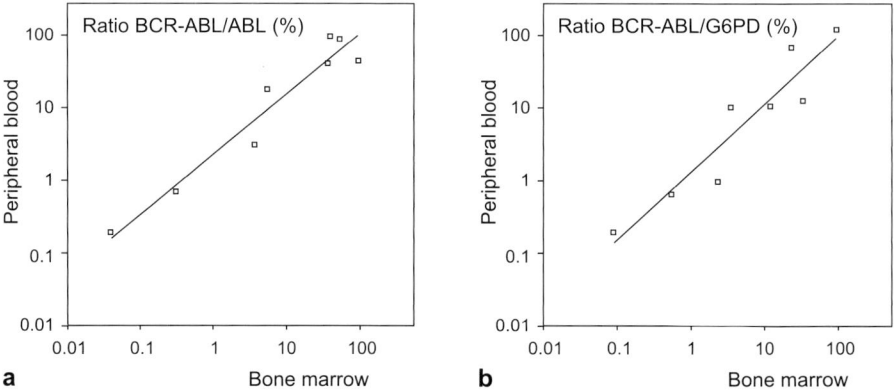

Fig. 6. a, b. Correlation of the $\log_{10}$ ratio BCR-ABL/ABL (**a**) and BCR-ABL/G6PD (**b**) obtained from eight pairs of contemporaneous PB and BM samples. The coefficient of regression is $r=0.90$ ($p=0.0046$) for the ratios BCR-ABL/ABL and $r=0.98$ ($p=0.0004$) for the ratios BCR-ABL/G6PD

sample rises above the background (crossing point or threshold cycle) is a much more accurate indicator of the initial number of PCR targets. Amplification and *specific detection* of fusion transcripts by real-time PCR is possible by simultaneous amplification and fluorescence melting curve analysis using non-specific labeling by the double strand DNA specific dye SYBR Green I [29]. However, if specific detection and *accurate quantification* is required, it is necessary to use adjacent fluorescently labeled hybridization probes. Real-time PCR does not require post-PCR sample handling and it therefore considerably reduces the risk of contamination. In addition, the possibility of contamination is further reduced by the use of heat labile UDG prior to amplification.

Real-time PCR for the detection and quantification of BCR-ABL transcripts using the TaqMan system with a fluorescent probe [30,31] has been described [5,6]. We have developed a related system using fluorescent hybridization probes in the LightCycler [8,32]. One advantage of this system is the rapid cycling time (<40 s/cycle) due to amplification being performed in glass capillaries and being driven by forced air heating. This system enables the temperature in each reaction to be changed very rapidly. By minimizing denaturation and annealing times, specificity, speed, and yield of the reaction are significantly improved [8].

We have optimized and validated this technique in samples from 120 CML patients. The use of probes that match ABL exon 3 sequences in combination with appropriate primers allows the detection of all known BCR-ABL variants [33,34], including e1a2 [35], e6a2 [34], c3a2 (e19a2) [36], and BCR-ABL transcripts lacking ABL exon 2 (b2a3 [37] or b3a3 [38,39]). The sensitivity of the single-step real-time PCR with hybridization of fluorescent probes is almost as high as standard nested PCR. The reason why the real-time assay is slightly less sensitive is almost certainly due to the fact that we used 2 µl of cDNA in the LightCycler but 5 µl of cDNA for standard nested PCR.

To standardize BCR-ABL mRNA levels for variability in RNA and cDNA quality, we amplified transcripts of housekeeping genes as normalization controls. Appropriate control sequences should be expressed at similar levels as the target genes in non-responders and should be cDNA specific, i.e., they should span long introns and pseudogenes in genomic DNA must not exist [40–43]. Therefore, we decided to use the amplification of total ABL cDNA sequences between exons a2 and a4. The advantage of using ABL is that the same pair of hybridization probes used to quantify BCR-ABL may be applied. We tested the variability of the ABL quantification by comparing ABL and G6PD transcript levels in samples of different qualities and different levels of BCR-ABL. The levels of ABL and G6PD were significantly correlated and similar results were obtained when either BCR-ABL/ABL or BCR-ABL/G6PD ratios were compared with cytogenetic results. These data suggest that either gene may be used as an internal standard. For individual samples, however, significant differences in the G6PD/ABL ratio are seen occasionally and it remains to be determined which is the most appropriate control gene for use in quantitative RT-PCR analysis.

The reliability of this system has been demonstrated by comparison with an existing method of PCR quantification: nested competitive PCR. Virtually identical results were determined with both methods, with a coefficient of correlation of $r=0.90$ ($p<0.0001$). The degree of concordance with metaphase cytogenetics and Southern blot analysis was also high. However, some overlap between the conventional cytogenetic response groups may be explained at least in part by the statistical significance of sampling small numbers of metaphases [44]. Results with PB and BM were concordant in all cases, indicating that either tissue may be used for residual disease studies in CML. This result confirms similar data that have previously been described using competitive PCR [18,45].

We conclude that quantitative real-time PCR with hybridization probes is a rapid, sensitive and reliable method for monitoring CML patients after therapy. The method offers the opportunity to standardize the assay and to develop rigorous standards and controls [46]. It is likely that real-time PCR will enable the quantitative analysis of residual disease to become more widely available, not only for patients with CML but also for other malignancies that are characterized by specific molecular markers.

The study was supported by the Deutsche José Carreras Leukämiestiftung e.V. and the Forschungsfonds der Fakultät für Klinische Medizin Mannheim der Universität Heidelberg, Germany. We thank Drs. H. Wittor, H. Leying, and G. Betzl (Roche Diagnostics, Mannheim/Penzberg, Germany) for continuous technical advice and support. The M-bcr probes for Southern blot analysis were provided by Prof. C.R. Bartram (University of Heidelberg, Germany). Standard plasmid pGD210 (b3a2$^{BCR-ABL}$) was provided by Dr. R. van Etten (Harvard Medical School, Boston, MA, USA), plasmid pB190 (e1a2$^{BCR-ABL}$, GenBank accession #AF113911) by Dr. T. Burmeister (Free University Berlin, Germany), and plasmid pGdBBX by Dr. Ph. Mason (Imperial College School of Medicine, London, UK). We wish to thank the hematologists and cytogeneticists participating in the German CML trials and in the UK Medical Research CML trials for sending samples and providing clinical and cytogenetic data. The study has been published by Emig et al. in *Leukemia*, 1999 [9].

References

1. Sawyers CL (1999) Chronic myeloid leukemia. N Engl J Med 340:1330–1340
2. Cross NCP (1997) Assessing residual leukaemia. Baillieres Clin Haematol 10:389–403
3. Lion T (1996) Monitoring of residual disease in chronic myelogenous leukemia: methodological approaches and clinical aspects. Leukemia 10:896–906
4. Morley A (1998) Quantifying leukemia. N Engl J Med 339:627–629
5. Mensink E, van de Locht A, Schattenberg A, Linders E, Schaap N, Guerts van Kessel A, de Witte T (1998) Quantitation of minimal residual disease in Philadelphia chromosome positive chronic myeloid leukaemia patients using real-time quantitative RT-PCR. Br J Haematol 102:768–774
6. Preudhomme C, Révillion F, Merlat A, Hornez L, Roumier C, Duflos-Grardel N, Jouet JP, Cosson A, Peyrat JP, Fenaux P (1999) Detection of BCR-ABL transcripts in chronic myeloid leukemia (CML) using a 'real-time' quantitative RT-PCR assay. Leukemia 13: 957–964
7. Wittwer CT, Herrmann MG, Moss AA, Rasmussen RP (1997) Continuous fluorescence monitoring of rapid cycle DNA amplification. Biotechniques 22:130–138

8. Wittwer CT, Ririe KM, Andrew RV, David DA, Gundry RA, Balis UJ (1997) The LightCycler: A microvolume multisample fluorimeter with rapid temperature control. Biotechniques 22: 176–181
9. Emig M, Saußele S, Wittor H, Weißer A, Reiter A, Willer A, Berger U, Hehlmann R, Cross NCP, Hochhaus A (1999) Accurate and rapid analysis of residual disease in patients with CML using specific fluorescent hybridization probes for real-time quantitative RT-PCR. Leukemia 13: 1825–1832
10. Harnden DG, Klinger HP (eds) (1985) ISCN. An international system for human cytogenetic nomenclature. Karger, Basel
11. Kantarjian HM, Smith TL, O'Brien S, Beran M, Pierce S, Talpaz M, and the Leukemia Service (1995) Prolonged survival in chronic myelogenous leukemia after cytogenetic response to interferon-α therapy. Ann Intern Med 122:254–261
12. Grossman A, Silver RT, Szatrowski TP, Gutfriend A, Verma RS, Benn PA (1991) Densitometric analysis of Southern blot autoradiographs and its application to monitoring patients with chronic myeloid leukemia. Leukemia 5:540–547
13. Reiter A, Skladny H, Hochhaus A, Seifarth W, Heimpel H, Bartram CR, Cross NCP, Hehlmann R (1997) Molecular response of CML patients treated with interferon-α monitored by quantitative Southern blot analysis. Br J Haematol 97:86–93
14. Cross NCP, Feng L, Bungey J, Goldman JM (1993) Minimal residual disease after bone marrow transplant for chronic myeloid leukaemia detected by the polymerase chain reaction. Leuk Lymphoma 11 [Suppl 1]:39–43
15. Cross NCP, Melo JV, Feng L, Goldman JM (1994) An optimized multiplex polymerase chain reaction (PCR) for detection of BCR-ABL fusion mRNAs in haematological disorders. Leukemia 8:186–189
16. van Rhee F, Marks DI, Lin F, Szydlo RM, Hochhaus A, Treleaven J, Delord C, Cross NCP, Goldman JM (1995) Quantification of residual disease in Philadelphia-positive acute lymphoblastic leukemia: comparison of blood and bone marrow. Leukemia 9:329–335
17. Cross NCP, Feng L, Chase A, Bungey J, Hughes TP, Goldman JM (1993) Competitive polymerase chain reaction to estimate the number of BCR-ABL transcripts in chronic myeloid leukemia patients after bone marrow transplantation. Blood 82:1929–1936
18. Hochhaus A, Lin F, Reiter A, Skladny H, Mason PJ, van Rhee F, Shepherd PCA, Allan NC, Hehlmann R, Goldman JM, Cross NCP (1996) Quantification of residual disease in chronic myelogenous leukemia patients on interferon-α therapy by competitive polymerase chain reaction. Blood 87:1549–1555
19. Chissoe SL, Bodenteich A, Wang YF, Wang YP, Burian D, Clifton SW, Crabtree J, Freeman A, Iyer K, Jian L, Ma Y, McLaury HJ, Pan HQ, Sarhan OH, Toth S, Wang Z, Zhang G, Heisterkamp N, Groffen J, Roe BA (1995) Sequence and analysis of the human ABL gene, the BCR gene, and regions involved in the Philadelphia chromosomal translocation. Genomics 27:67–82
20. Chen EY, Cheng A, Lee A, Kuang WJ, Hillier L, Green P, Schlessinger D, Ciccodicola A, D'Urso M (1991) Sequence of human glucose-6-phosphate dehydrogenase cloned in plasmids and a yeast artificial chromosome. Genomics 10:792–800
21. Hochhaus A, Lin F, Reiter A, Skladny H, van Rhee F, Shepherd PCA, Allan NC, Hehlmann R, Goldman JM, Cross NCP (1995) Variable numbers of BCR-ABL transcripts persist in CML patients who achieve complete cytogenetic remission with interferon-α. Br J Haematol 91:126–131
22. Hochhaus A, Reiter A, Saußele S, Reichert A, Emig M, Kaeda J, Schultheis B, Berger U, Shepherd PCA, Allan NC, Hehlmann R, Goldman JM, Cross NCP for the German CML Study Group and the UK MRC CML Study Group) (2000) Molecular heterogeneity in complete cytogenetic responders after interferon-α therapy for chronic myeloid leukemia: low levels of minimal residual disease are associated with continuing remission. Blood 95:62–66
23. Lin F, Kirkland MA, van Rhee F, Chase A, Coulthard S, Bungey J, Goldman JM, Cross NCP (1996) Molecular analysis of transient cytogenetic relapse after allogeneic bone marrow transplantation for chronic myeloid leukaemia. Bone Marrow Transplant 18:1147–1152

24. Lin F, van Rhee F, Goldman JM, Cross NCP (1996) Kinetics of increasing BCR-ABL transcript numbers in chronic myeloid leukemia patients who relapse after bone marrow transplantation. Blood 87:4473–4478
25. Gaiger A, Lion T, Kalhs P, Mitterbauer G, Henn T, Haas O, Fodinger M, Kier P, Forstinger C, Quehenberger P, Hinterberger W, Jäger U, Linkesch W, Mannhalter C, Lechner K (1993) Frequent detection of BCR-ABL specific mRNA in patients with chronic myeloid leukemia (CML) following allogeneic and syngeneic bone marrow transplantation (BMT). Leukemia 7:1766–1772
26. Lion T, Henn T, Gaiger A, Kalhs P, Gadner H (1993) Early detection of relapse after bone marrow transplantation in patients with chronic myelogenous leukaemia. Lancet 341:275–276
27. van Rhee F, Lin F, Cullis JO, Spencer A, Cross NCP, Chase A, Garicochea B, Bungey J, Barrett J, Goldman JM (1994) Relapse of chronic myeloid leukemia after allogeneic bone marrow transplant: the case for giving donor leukocyte transfusions before the onset of hematologic relapse. Blood 83:3377–3383
28. Raanani P, Dazzi F, Sohal J, Szydlo R, van Rhee F, Reiter A, Lin F, Goldman JM, Cross NCP (1997) The rate and kinetics of molecular response to donor leucocyte transfusions in chronic myeloid leukemia patients treated for relapse after allogeneic bone marrow transplantation. Br J Haematol 99:945–950
29. Bohling SD, King TC, Wittwer CT, Elenitoba-Johnson KSJ (1999) Rapid simultaneous amplification and detection of the MBR/JH chromosomal translocation by fluorescence melting curve analysis. Am J Pathol 154:97–103
30. Gibson UEM, Heid CA, Williams PM (1996) A novel method for real time quantitative RT-PCR. Genome Res 6:995–1001
31. Heid CA, Stevens J, Livak KJ, Williams PM (1996) Real time quantitative PCR. Genome Res 6:986–994
32. Higuchi R, Dollinger G, Walsh PS, Griffith R (1992) Simultaneous amplification and detection of specific DNA sequences. Biotechnology 10:413–417
33. Melo JV (1996) The diversity of BCR-ABL fusion proteins and their relationship to leukemia phenotype. Blood 88:2375–2384
34. Hochhaus A, Reiter A, Skladny H, Melo JV, Sick C, Berger U, Guo JQ, Arlinghaus RB, Hehlmann R, Goldman JM, Cross NCP (1996) A novel BCR-ABL fusion gene (e6a2) in a patient with Philadelphia chromosome negative chronic myelogenous leukemia. Blood 88:2236–2240
35. Melo JV, Myint H, Galton DA, Goldman JM (1994) P190$^{BCR-ABL}$ chronic myeloid leukaemia: the missing link with chronic myelomonocytic leukaemia? Leukemia 8:208–211
36. Pane F, Frigeri F, Sindona M, Luciano L, Ferrara F, Cimino R, Meloni G, Saglio G, Salvatore F, Rotoli B (1996) Neutrophilic-chronic myelogenous leukemia (CML-N): a distinct disease with a specific marker (BCR-ABL with c3a2 junction). Blood 88:2410–2414
37. van der Plas DC, Soekarman D, van Gent AM, Grosveld G, Hagemeijer A (1991) bcr-abl mRNA lacking abl exon a2 detected by polymerase chain reaction in a chronic myelogeneous leukemia patient. Leukemia 5:457–461
38. Paldi Haris P, Barta A, Lengyel L, Batai A, Masszi T, Remenyi P, Denes R, Paloczi K, Kelemen E, Foldi J (1994) Molecular background of a new case of chronic myelogenous leukemia with bcr-abl chimera mRNA lacking the A2 exon [letter]. Leukemia 8:1791
39. Iwata S, Mizutani S, Nakazawa S, Yata J (1994) Heterogeneity of the breakpoint in the ABL gene in cases with BCR/ABL transcript lacking ABL exon a2. Leukemia 8:1696–1702
40. Lion T (1996) Control genes in reverse transcriptase-polymerase chain reaction assays. Leukemia 10:1527–1528
41. Soutar RL, Dillon J, Ralston SH (1997) Control genes for reverse-transcription-polymerase chain reaction: A comparison of beta actin and glyceraldehyde phosphate dehydrogenase. Br J Haematol 97:247–248
42. Cross NCP, Lin F, Goldman JM (1994) Appropriate controls for reverse transcription polymerase chain reaction (RT-PCR). Br J Haematol 87:218
43. Taylor JJ, Haesman PA (1994) Control genes for reverse transcriptase/polymerase chain reaction (RT-PCR). Br J Haematol 86:444–447

44. Hook EB (1977) Exclusion of chromosomal mosaicism: tables of 90%, 95%, and 99% confidence limits and comments on use. Am J Hum Genet 29:94–97
45. Lin F, Goldman JM, Cross NCP (1994) A comparison of the sensitivity of blood and bone marrow for the detection of minimal residual disease in chronic myeloid leukaemia. Br J Haematol 86:683–685
46. Hochhaus A, Weisser A, La Rosée P, Emig M, Müller MC, Reiter A, Kuhn C, Berger U, Hehlmann R, Cross NCP (2000) Detectron and quantification of residual disease in chronic myelogenous leukemia. Leukemia 14:998–1005

Development of Quantitative RT-PCR for the Expression of Wilms' Tumor WT1 Suppressor Gene in Leukemia on the LightCycler

Yoji Ishida*, Kazunori Murai

Introduction

Wilms' tumor (WT) is a childhood neoplasm which arises as a result of inactivation of the WT1 gene. The WT1 suppressor gene encodes a transcription factor of the zinc finger family. Recently, Miwa et al. [1] reported that WT1 was expressed in human leukemia. In addition, Inoue et al. [2] reported that WT1 was a new prognostic factor and a new marker for the detection of minimal residual disease (MRD) in acute leukemia. However, their methods were qualitative, or at best, semi-quantitative. Since then, several quantitative RT-PCR methods using external standards and competitors have appeared but are labor intensive and time-comsuming. We have developed a new technique to quantify WT1 expression in bone marrow cells in acute leukemia and myelodysplastic syndrome (MDS) using LightCycler technology.

Materials

Equipment

LightCycler Instrument
Oligo Primer analysis software

Reagents

Leukemic cell line K562 (ATCC)
Amplification Primers (Nihon Gene Research Laboratories)
Hybridization Probes (Nihon Gene Research Laboratories)

The following reagents were purchased from Roche Diagnostics, Mannheim, Germany:
RNA/DNA Stabilization Reagent for Blood/Bone Marrow
mRNA Isolation Kit for Blood/Bone Marrow
LightCycler-RNA Amplification Kit SYBR Green I
LightCycler-DNA Master SYBR Green I
LightCycler-RNA Amplification Kit Hybridization Probes
LightCycler-DNA Master Hybridization Probes

---

* Yoji Ishida (✉) (e-mail: yishida@iwate-med.ac.jp)
  Division of Hematology, IIIrd Department of Internal Medicine, Iwate Medical University School of Medicine, 19-1, Uchimaru Morioka, 020-8505, Japan

Procedure

Sample Preparation

Ten volumes of RNA/DNA Stabilization Reagent for Blood/Bone Marrow were added to one volume of bone marrow aspirate. Resulting lysates were stored at –20 C. After thawing at room temperature for 30 min, RNA was extracted using the mRNA Isolation Kit for Blood/Bone Marrow. RNA was eluted from the spin column in a volume of 30 μl.

Primer Design

Published primers were used for the outer set [2]. The inner primer selection was performed with the aid of the Oligo computer program. Hybridization probes were designed with a moderate stringency according to published guidelines [3]. Oligonucleotide characteristics are shown in Table 1.

Magnesium Concentration

The concentration of $MgCl_2$ is one of the most important factors that influence the efficiency as well as the specificity of the amplification reaction. The melting temperatures of PCR products at $MgCl_2$ concentrations from 3–7 mM were analyzed using the SYBR Green detection method.

Table 1. Oligonucleotide primers and hybridization probes for WT1

(Genbank Accession # X 51630)	Position	Exon	Length	GC (%)	T_m (°C)
Outer Primers					
GGCATCTGAGACCAGTGAGAA	1319	7	21	52.4	67.4
GAGAGTCAGACTTGAAAGCAGT	1801R	10	22	45.5	65.6
Inner Primers					
CCCTTCATGTGTGCTTAC	1344	7	18	50	60.1
TTTGGTCATGTTTCTCTG	1712R	10	18	38.9	56.8
Hybridization Probes					
CTTGTCAGCGAAAGTTCTCCCGG-F	1540	9	23	56.5	69.4
LCRed640-CGACCACCTGAAGACCCACACCA	1565	9	23	60.9	73.6

RT-PCR for WT1 (values for actin are in parentheses when different)

● One-step RT-PCR

	Volume [μl]	[Final]
LightCycler RT-PCR Reaction Mix Hybridization Probes	2	1 ×
LightCycler RT-PCR Enzyme Mix	0.4	1 ×
$MgCl_2$ (25 mM)	1.6 (0.8)	5 mM (4 mM)
Outer Primers (5 μM)	2+2 (2.4+2.4)	0.5 μM (0.6 μM)
Fluorescein Probe (2 μM)	2	0.2 μM
LCRed 640 Probe (4 μM)	2	0.4 μM
H_2O (PCR grade)	6	
Total Volume	18	

Development of Quantitative RT-PCR for the Expression of Wilms' Tumor WT1 Suppressor Gene in Leukemia

- Nested PCR

Two µl of RNA were added and the samples centrifuged into LightCycler cuvettes. The effect of Mg^{++} on amplification specificity was studied by melting curves, using the Reaction Mix from the LightCycler Amplification Kit SYBR Green I, removing the fluorescent probes, and varying the Mg^{++} concentration from 3 to 7 mM.

The following RT-PCR protocol was used for reverse transcription and amplification of WT 1 transcripts (values for actin are in parentheses when different)

– Reverse transcription for 10 min at 55°C
– Denaturation for 2 min (30 sec) at 95°C

- Amplification

Parameter	Value		
Cycles	27 (35)		
Type	Quantification		
	Segment 1	Segment 2	Segment 3
Target Temperature	95	60 (57)	72
Incubation Time	1	15 (10)	19 (18)
Temperature Transition Rate	20	20	2
Acquisition Mode	None	Single	None
Gains (SYBR Green I)	F1=5		
Gains (Hybridization Probes)	F1=1	F2=10	

- Denaturation

Melting Curve Protocol (SYBR Green I)

Parameter	Value		
Cycles	1		
Type	Melting Curve		
	Segment 1	Segment 2	Segment 3
Target Temperature	95	65	95
Incubation Time	0	10	0
Temperature Transition Rate	20	20	0.2
Acquisition Mode	None	None	Cont
Gains	F1=5		

- Cooling to 40°C

Table 2. Oligonucleotide primers and hybridization probes for β actin

(Genbank Accession # NM 001101)				
	Position	Length	GC (%)	T_m (°C)
Primers				
CCAACCGCGAGAAGATGACA	414	20	57.9	64.9
GGAAGGAAGGCTGGAAGAGT	873R	20	55.0	64.2
Hybridization Probes				
CCTCCCCCATGCCATCCTGCGTC-F	593	23	69.6	82.3
LCRed640-GGACCTGGCTGGCCGGGACCTGA	607	23	73.9	84.0

- Nested PCR for WT 1

	Volume [μl]	[Final]
LightCycler-DNA Master Hybridization Probes	2	1 X
MgCl$_2$ (25 mM)	3.2	5 mM
Inner primers (20 μM)	0.5+0.5	0.5 μM
Fluorescein Probe (2 μM)	2	0.2 μM
LCRed640 Probe (4 μM)	2	0.4 μM
H$_2$O (PCR grade)	7.8	
Total Volume	18	

Two μl of the undiluted product from the outer amplification were added and the samples centrifuged into LightCycler cuvettes. The following protocol was used for nested amplification of WT1 transcripts with inner primers.

- Denaturation at 95°C for 30 sec

- Amplification

Parameter	Value		
Cycles	35		
Type	Quantification		
Temperature target	Segment 1	Segment 2	Segment 3
Target Temperature	95	51	72
Incubation Time	0	15	15
Temperature Transition Rate	20	20	20
Acquisition Mode	None	Single	None
Gains	F1=1	F2=10	

- Cooling to 40°C

The nested PCR products were diluted in sucrose from capillaries containing 0.5 µg/ml ethidium bromide. Samples were diluted in sucrose/bromphenol blue loading buffer and loaded on 1.3% agarose gels in TAE buffer (40 mM Tris, 20 mM Acetic acid, and 1 mM EDTA pH 8.3) containing 0.05 µg/ml of ethidium bromide.

Gel Analysis of PCR Products

Results

Figure 1 shows SYBR Green I melting curves of amplified WT 1 RT-PCR products at Mg^{++} concentrations from 3 to 7 mM. Five µM Mg^{++} minimized the amount of low melting products. At higher or lower Mg^{++} concentrations, significant amounts of low melting products can decrease the sensitivity of SYBR Green I quantification, although hot start techniques (such as an anti-Taq antibody) and/or high temperature fluorescence acquisition can minimize the effects. Quantification sensitivity can also be improved by using specific hybridization probes.

Magnesium Optimization and Low Melting Products

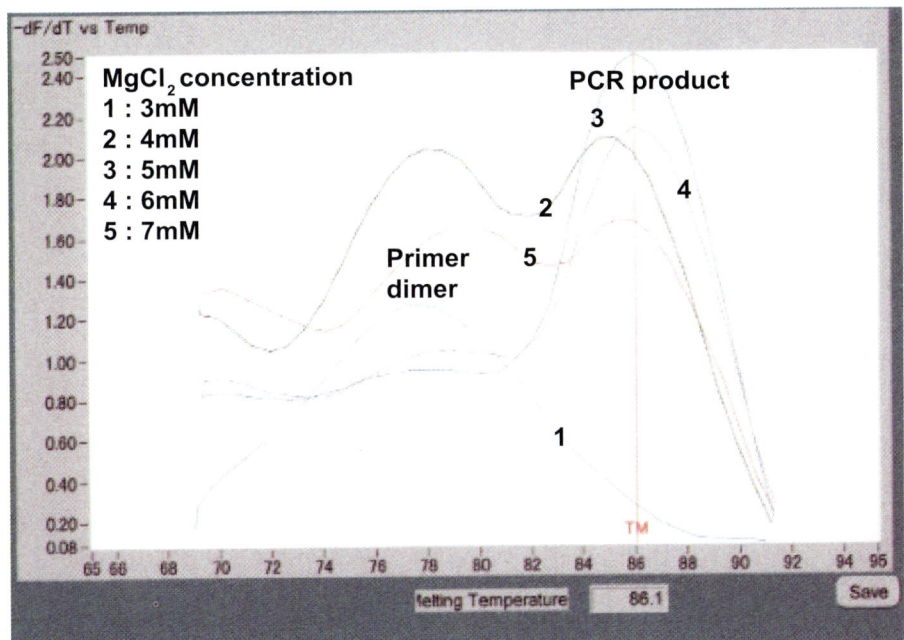

Fig. 1. $MgCl_2$ effects on amplified products of RT-PCR. The effect of Mg^{++} on product specificity after amplification in the presence of SYBR Green I and 3–7 mM Mg^{++} with the outer primer set. A derivative melting curve plot is shown

Quantitative RT-PCR of WT1 Expression with Hybridization Probes

Standard curves of diluted mRNA from K562 cells resulting from RT-PCR and RT-nested-PCR are shown in Figures 2 and 3, respectively. All amplification curves were approximately parallel, indicating similar efficiency of amplification. Linear plots of log concentration vs crossing point were obtained with errors of less than 10E-3.

Gel Analysis of PCR Products

As expected, the major nested PCR product was a 369 bp band on a 1.3 % agarose gels (Figure 4).

Primer Dimer Formation A significant amount of primer dimer accumulated, as shown in the melting curve (Fig. 1) in samples with low target copy number. The formation of these dimers can result in false positive values using the SYBR Green detection method and often reduce the sensitivity of the PCR for low copytargets. Much attention should be paid to primer design. Fortunately, the primer dimer was not formed at 5 mM $MgCl_2$ in this study.

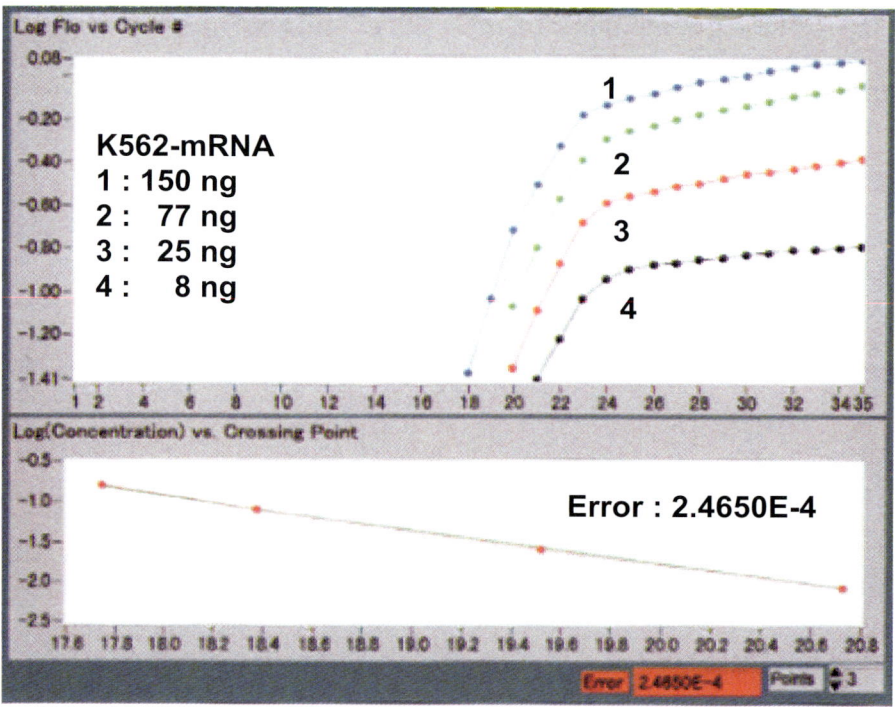

Fig. 2. Amplification (fluorescence vs cycle number) and linearity (log concentration vs crossing point) plots of RT-PCR with the outer primer pair

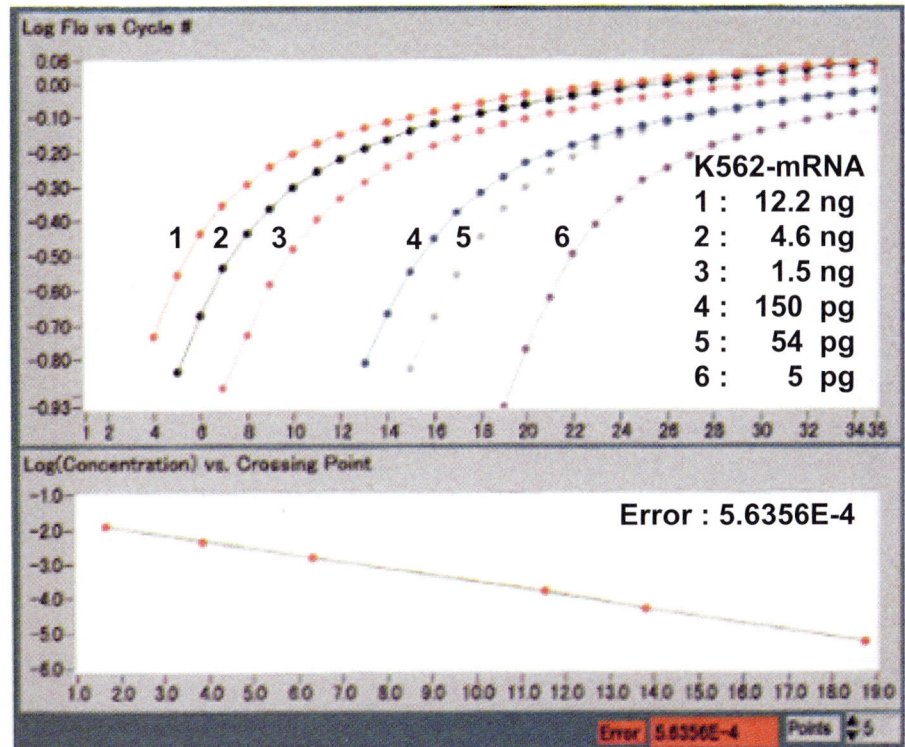

Fig. 3. Amplification (fluorescence vs cycle number) and linearity (log concentration vs crossing point) plots of nested RT-PCR. Nested PCR was performed using 2 µl of undiluted amplicon from the first RT-PCR and detected with the hybridization probe method

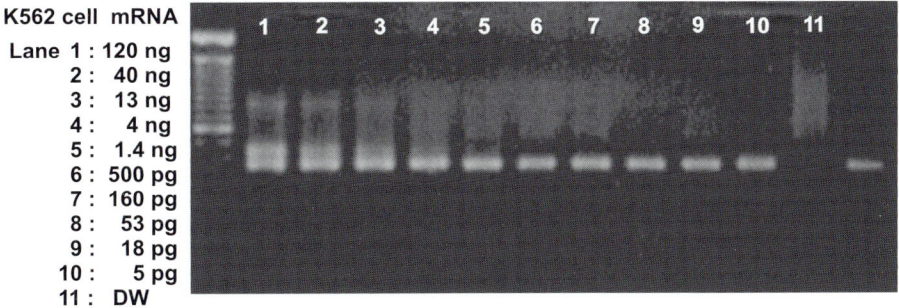

Fig. 4. Gel analysis of PCR products. First round PCR product was serially diluted and amplified with nested primers. The PCR products were analyzed on a 1.3 % agarose gel in TAE buffer containing 0.5 µg/ml ethidium bromide. The predominate PCR product is 369 bp

Titration Performance of the LC-PCR

Figures 2 and 3 showed the amplification of titrated cDNA from K562 in the first and the nested PCRs. All amplification curves had approximately the same slope and the measured values were close to the theoretically expected values.

Gel Analysis of PCR Products

All PCR products amplified in the nested PCR showed bands with 369 bp on 1.3% agarose gel electrophoresis (Fig. 4).

Comments

Sample Preparation

Sample preparation is one of the most critical aspects of quantitative PCR. Isolation of high quality mRNA is an important step for the quantification of gene expression. mRNA in bone marrow or blood is stabilized from RNAse by simple mixing with a stabilization reagent. Lysates can be stored for 12 months at –20°C or for 1 day at 4°C.

RT-PCR

In RT-PCR, the mRNA has to be transcribed into first strand cDNA before exponential amplification. There are two strategies for RT-PCR; a one-step/one-tube process or a two-step system (synthesis of cDNA in one tube followed by PCR in another tube). The one-step/one tube process used in this study reduces handling and saves time.

A Standard Curve

A standard curve is created by plotting the log of the intial template concentration vs the crossing point. The measured values should be close to the theoretically expected values. The error value should be lower than 10^{-3}.

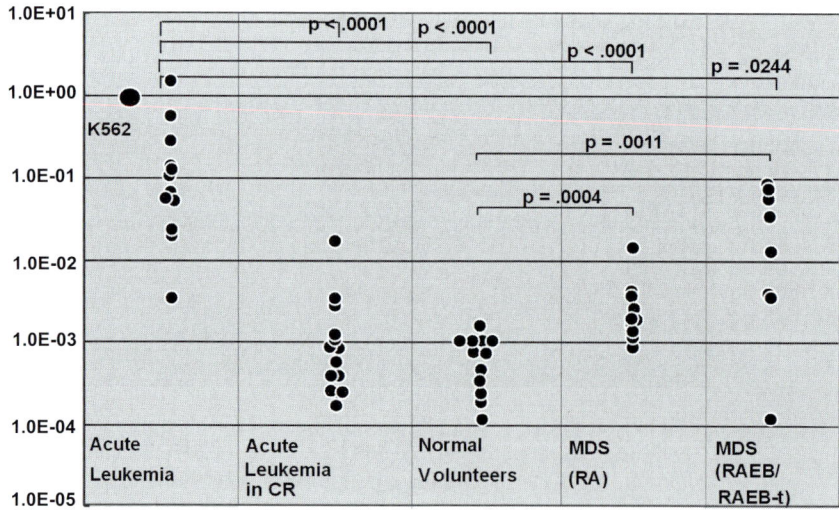

Fig. 5. WT1 expression in patients with acute leukemia and myelodysplatic syndrome (RA=refractory anemia, RAEB=refractory anemia with excess blasts)

Applications

WT1 expression of bone marrow cells in patients with acute leukemia and MDS (refractory anemia and refractory anemia with excess blasts) were quantified. Signifant differences in WT1 expression were found between acute leukemia at diagnosis and normal volunteers ($p<0.0001$), acute leukemia at diagnosis and during complete remission ($p<0.0001$), refractory anemia (RA) and normal volunteers ($p=0.0004$) and refractory anemia with excess blasts (RAEB) and normal volunteers ($P=0.0011$) as shown in Fig. 5.

References

1. Miwa H, Beran M, Saunders GF (1992) Expression of the Wilms' tumor gene (WT1) in human leukemias. Leukemia 6: 405–9
2. Inoue K, Sugiyama H, Ogawa H, Nakagawa M, Yamagami T, Miwa H, Kita K, Hiraoka A, Masaoka T, Nasu K, Kyo T, Dohy H, Nakauchi H, Ishidate T, Akiyama T, Kishimoto T (1994) WT1 as a new prognostic factor and a new marker for the detection of minimal residual disease in acute leukemia. Blood 84:3071–9
3. Caplin BE, Rasmussen RP, Bernard PS, Wittwer CT (1999) LightCycler Hybridization Probes. Biochemica 1: 5–8

Real-Time Detection of Minimal Residual Disease by Amplifying Immunoglobulin Genes in Acute Lymphoblastic Leukemia on the LightCycler

Makoto Nakao, Bart Janssen, Claus R. Bartram*

Introduction

Acute lymphoblastic leukemia (ALL) is the most frequent malignancy among children. ALL is characterized as the malignant transformation and clonal expansion derived from a single progenitor cell at early lymphoid differentiation. It has become possible to assess minimal residual disease (MRD) by PCR-based methods during the course of therapy in ALL [1]. The relative number of leukemic cells remaining after initial therapy reflects the responsiveness to treatment and therefore bears prognostic potential [2]. For the monitoring of MRD, immunoglobulin heavy chain (*IGH*) gene rearrangements have been widely used as PCR targets due to their high incidence (almost every B-lineage ALL) and the diversity of V_H-N-D_H-N-J_H junctional regions. The junction ('N-D_H-N') codes for an antibody complementarity determining region (CDR) and serves as a clone-specific marker for each ALL patient. Rearrangements of the kappa deleting element (Kde) within the immunoglobulin kappa (*IGK*) gene locus occur in 40%–50% of patients with B-lineage ALL. The rearrangement between the intron recombination signal sequence (RSS) and Kde is also characterized by clone-specific diversity. Synthesized oligonucleotides corresponding to these junctional regions can be utilized as hybridization probes or clone-specific primers for MRD monitoring.

In this section we describe the application of the LightCycler technology [3] for the detection of MRD in ALL. We analyzed *IGH*/*IGK* rearrangements from two patients by allele specific amplification on a block thermocycler and by SYBR Green I detection on the LightCycler and then compared the respective sensitivities.

Materials

Block Thermocycler (Primus 96, MWG, Ebersberg, Germany)   **Equipment**
LightCycler Instrument (Roche Diagnostics, Mannheim, Germany)

---

* Claus R. Bartram (✉) (e-mail: cr_bartram@med.uni-heidelberg.de)
  Institut für Humangenetik der Universität Heidelberg, Im Neuenheimer Feld 328, 69120 Heidelberg, Germany

Reagents	Taq DNA polymerase (Life Technologies, Karlsruhe) DNA polymerization mix (Amersham Pharmacia Biotech, Piscataway, NJ) Amplification Primers (BIG Biotech, Freiburg, Germany) LightCycler-DNA Master SYBR Green I

Procedure

Sample Preparation	High molecular weight DNA was extracted from 1–3 ml of bone marrow cells from two patients with ALL, using a standard procedure [5]. Buffy coat (BC) control DNA was prepared from peripheral blood of five healthy volunteers. 100 ng of each DNA sample was electrophoresed on a 0.6% agarose gel to determine its concentration and quality.
Characterization of IGH/IGK Gene Rearrangements	Rearrangements of *IGH/IGK* genes were screened by PCR amplification with rearrangement specific primers [4, 6]. Briefly, consensus primers were used to amplify V_H-J_H rearrangements (patient 1). Monoclonality was subsequently determined by a heteroduplex analysis on a polyacrylamide gel. V_H-J_H rearrangements are observed in over 90% of ALL patients. For the detection of an RSS-Kde rearrangement (patient 2), which is found in approximately 20% of ALL patients [6], we also applied rearrangement specific consensus primers (Table 1) for PCR amplification and confirmed the homoduplex formation by the heteroduplex analysis. Both PCR products were directly sequenced [4, 6]. Allele specific primers corresponding to each junctional region were designed (Table 1, Fig. 1).
First-Round PCR	A tenfold serial dilution of initial leukemic DNA in BC control DNA was prepared (10^{-1}-10^{-6}). In the first PCR the monoclonal (ALL-specific) and polyclonal (ALL-unrelated) *IGH* or *IGK* gene fragments were co-amplified in order to obtain a sufficient amount of template for the second (quantitative) round of PCR. One microgram of each dilution (10^{-1}-10^{-6}) was amplified in a 100 µl PCR mixture containing 1 U of Taq polymerase, 1×PCR buffer (Life Technologies), 2 mM $MgCl_2$, 40 pmol of each rearrangement-specific consensus primer (Fig. 1, Table 1) on a Primus 96 cycler (MWG) (first PCR). BC DNA and H_2O served as controls. After 40 cycles of amplification, the PCR products were resolved on a 1.5% agarose gel. In total, 10 µl of the first PCR product was diluted in 990 µl of 0.5×TE.
Conventional Allele Specific Amplification	For patient 1, a V_H4-outer primer and an allele specific primer (CDR in Table 1) were used to amplify a leukemia-specific fragment (Fig. 1A). Allele specific amplification of DNA from patient 2 was performed using an allele specific primer (N in Table 1) and a Kde-inner primer (Fig. 1B). PCR was performed on a Primus 96 cycler (MWG) in 50 µl consisting of 1 U of Taq polymerase, 1×PCR buffer (Life Technologies), 1.5 mM $MgCl_2$, 20 pmol of each primer, and 2 µl of the diluted first PCR product. Two H_2O reactions were included in order to check for cross-contamination; H_2O1 contained the diluted first-round PCR reaction of H_2O, and H_2O2 contained water prepared for second-round PCR (no template). The PCR protocol consisted of 4 min initial denaturation at 94°C, followed by

Table 1. Sequences of PCR primers used in the present study

Sequence (5'→3')	*Size	Name	Length	GC (%)	T_m (°C)
Patient 1					
V_H4-J_H rearrangement-specific consensus primers	376 bp				
GCC CAG GAC TGG TGA AGC		V_H4-outer	18	66.7	65.0
ACC TGA GGA GAC GGT GAC C		J_H	19	63.2	65.1
V_H4-J_H allele-specific amplification primers	177 bp				
ATC TAT TAT AGT GGG AGC ACC		V_H4-inner	21	42.9	57.9
AAC CCG TAC CAG CTG CCT CC		CDR	20	65.0	68.6
Patient 2					
RSS-Kde rearrangement-specific consensus primers	509 bp				
GTT ATT CCC AAA AGC TCA ATC TCA AAG		RSS	27	37.0	62.3
CCC TTC ATA GAC CCT TCA GGC AC		Kde-outer	23	56.5	66.1
RSS-Kde allele-specific amplification primers	161 bp				
TAG CCA GCT TTC CTT TTT ACC		N	21	42.9	60.0
GTT TAC AGA CAG GTC CTC AG		Kde-inner	20	50.0	59.2

*Size of amplification products

20–30 cycles of 30 s at 94°C, 30 s at 60°C, and 45 s at 72°C. PCR products were electrophoresed on a 3% agarose gel containing ethidium bromide.

Allele Specific Amplification by LightCycler (LC-PCR)

The following master mix was used:

	Volume [µl]	[Final]
LightCycler-DNA Master SYBR Green I	2	1×
$MgCl_2$ stock solution	1.6	3 mM
Inner Primer (20 µM)	0.5	0.5 µM
Allele-specific primer (20 µM)	0.5	0.5 µM
First PCR product	1	–
H_2O	14.4	–
Total volume	20	–

A total of 19 µl of master mix and 1 µl of diluted first PCR product was added to each capillary. Each reaction was prepared in duplicate. The sealed capillaries were centrifuged in a microcentrifuge and placed into the LightCycler rotor.

The following PCR protocol (software version 3.39) was used for the SYBR Green detection:

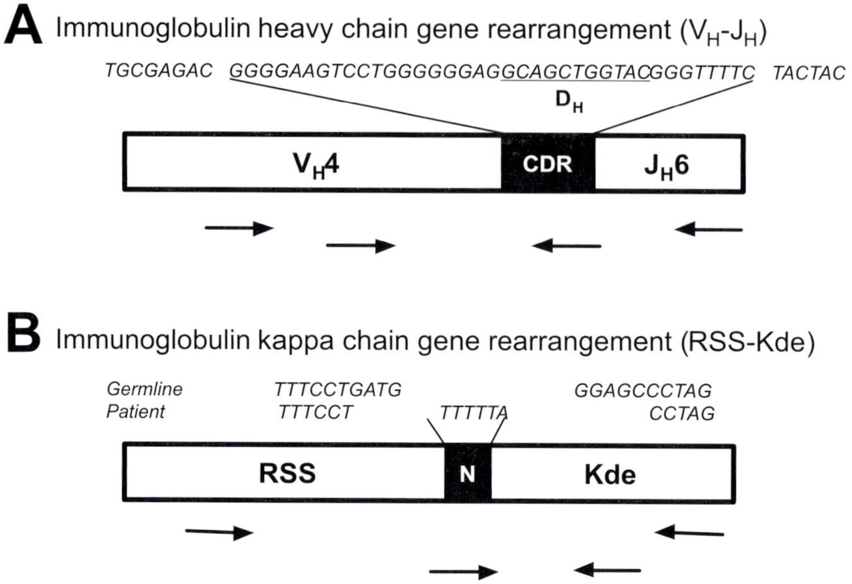

A Immunoglobulin heavy chain gene rearrangement (V_H-J_H)

TGCGAGAC GGGGAAGTCCTGGGGGGAGGCAGCTGGTACGGGTTTTC TACTAC

D_H

| V_H4 | CDR | J_H6 |

B Immunoglobulin kappa chain gene rearrangement (RSS-Kde)

Germline   TTTCCTGATG   GGAGCCCTAG
Patient   TTTCCT   TTTTTA   CCTAG

| RSS | N | Kde |

Fig. 1. Schematic representation of *IGH/IGK* rearrangements and primer positions. **A** *IGH* rearrangement of patient 1. The CDR consists of 38 bp including an 11 bp fragment of D_H. A 1 bp deletion was found at the 3' end of V_H4. No deletion occurred at the 5' end of J_H6. An allele specific primer was designed as an antisense primer ('CDR-primer') (Table 1). **B** *IGK* rearrangement of patient 2. The Kde was found to be rearranged to an intron recombination signal sequence (RSS). Deletions 4 and 5 bp were observed at the 3' end of the RSS and at the 5' end of the Kde, respectively. A sense allele specific primer was used ('N-primer')

- Denaturation for 2 min at 95°C
- Amplification

Parameter	Value		
Cycles	35		
Analysis mode	Quantification		
	Segment 1	Segment 2	Segment 3
Target temperature [°C]	95	60	72
Incubation time [s]	1	10	10
Temperature transition rate [°C/s]	20.0	20.0	20.0
Acquisition Mode	None	None	Single
Gains			F1=5

● Melting Curve Analysis

Parameter	Value		
Cycles	1		
Analysis mode	Melting curves		
	Segment 1	Segment 2	Segment 3
Target temperature [°C]	95	60	95
Incubation time [s]	1	10	1
Temperature transition rate [°C/s]	20.0	20.0	0.2
Acquisition Mode	None	None	Cont.
Gains			F1=5

Cooling was performed with a target temperature of 40°C and an incubation time of 15 s.

Results

The result of the conventional allele specific amplification is shown in Fig. 2A. Visual inspection and densitometer analysis showed that a bone marrow sample taken in complete remission 1 month after polychemotherapy (CR1) contained between 10^{-3} and 10^{-4} residual leukemic cells. The LC-PCR results are shown in Fig. 2B,C. As the melting curve of the BC control (not shown) revealed the same T_m as the 10^{-1}-10^{-6} dilution samples, the amplification signals obtained for samples BC and 10^{-6} in Fig. 2B must be due to the amplification of polyclonal (ALL-unrelated) rearrangements of *IGH*, late in the PCR reaction. Quantification indicated that the 10^{-6} and the control samples both started to increase around cycle 21, supporting the estimated detection limit of 10^{-5} in this patient. We determined an MRD level in the CR1 sample of 3.8×10^{-4} (Fig. 2B).

MRD Analysis Using IGH Rearrangement (Patient 1)

The melting curve analysis showed that melting temperatures of the BC control and dilution samples 10^{-5} and 10^{-6} differed from those of 10^{-1}-10^{-4} (Fig. 3A), indicating a sensitivity of 10^{-4} in this assay. Figure 3B demonstrates the reproducibility of the LC-PCR. Under these conditions, the MRD level of the bone marrow remission sample (CR1) was quantified (Fig. 3C) as 8.9×10^{-2}. Although the sensitivity using allele specific amplification on a block thermocycler was also 10^{-4} at 25 cycles, the conventional method failed to quantify the MRD level as all bands between 10^{-1} and 10^{-3} exhibited the same intensity (not shown).

MRD Analysis Using IGK Rearrangement (Patient 2)

Comments

Since a degraded sample may contain less of a target gene compared to a sample of adequate quality [7], the DNA should be checked before use. To prevent cross-contamination, PCR mixtures should be prepared in a room free of any PCR

General Recommendations for Quantitative PCR

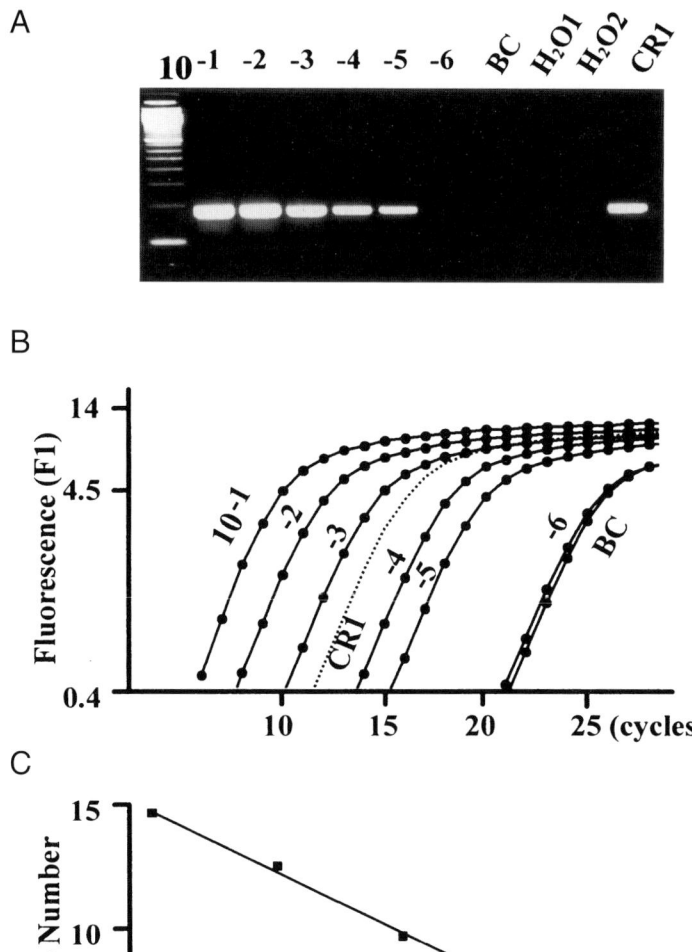

Fig. 2. MRD analysis using *IGH* rearrangement (Patient 1). **A** Conventional allele specific amplification. The detection limit was 10^{-5} at 25 cycles. Lane 1, 100 bp size marker; Lanes 2–7, dilutions of the patient's DNA obtained at the time of initial diagnosis; BC, buffy coat control DNA from healthy persons; H_2O1, negative control of first PCR; H_2O2, negative control of conventional allele specific amplification; CR1, a complete remission marrow sample of patient taken 1 month after initiation of polychemotherapy. **B** Allele-specific amplification on the LightCycler. The PCR products of 10^{-6} and BC showed a similar amplification pattern which started to increase at 21 cycles (crossing point), indicating polyclonal (ALL-unrelated) amplification. Melting peaks of all products showed the same T_m (not shown). **C** Standard curve at the range of 10^{-1}–10^{-5}

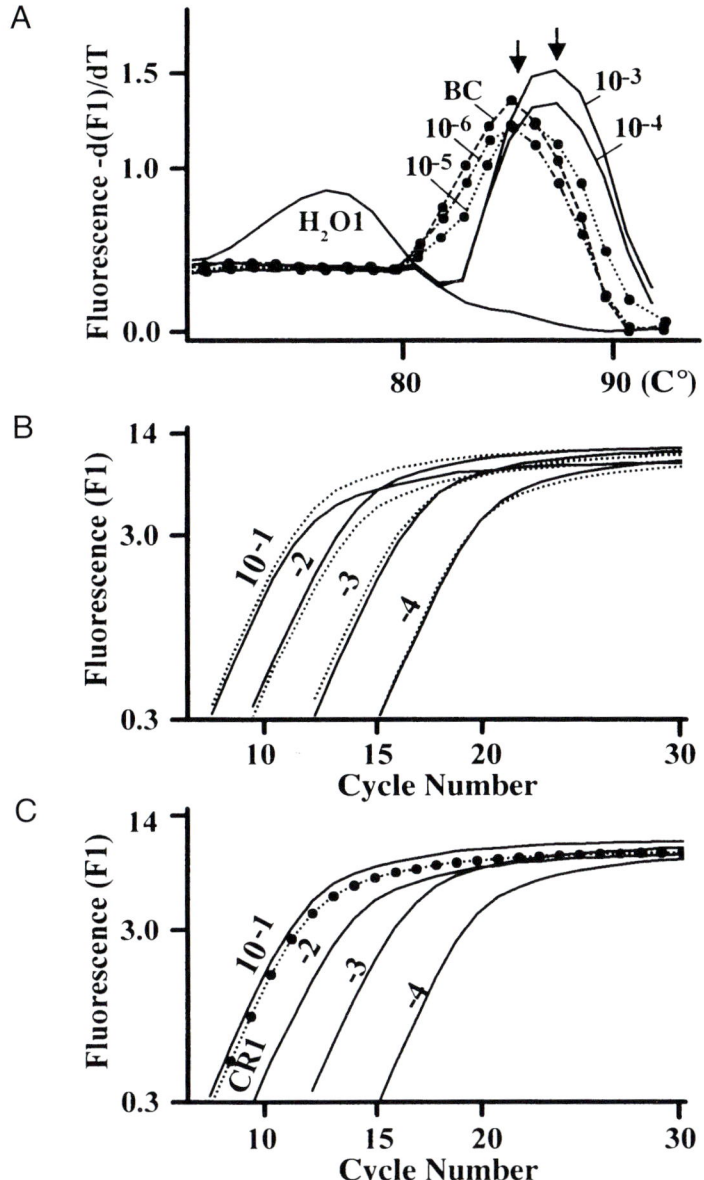

Fig. 3. MRD analysis using *IGK* rearrangement (Patient 2). **A** Melting curve analysis. The peaks of BC control as well as $10^{-5}/10^{-6}$ samples were different from those of 10^{-1}-10^{-4}, indicating a sensitivity of 10^{-4}. The peak of H_2O1 represents a primer dimer. **B** Reproducibility of the experiment. Each duplicate reaction showed an identical amplification pattern. **C** Bone marrow sample (CR1) obtained 1 month after the start of treatment was quantified on the LightCycler. The residual leukemic cell was calculated to be 8.9×10^{-2}, whereas it was difficult to assess an accurate MRD level with the conventional allele specific amplification method

products. Every experiment must include a control sample, and strict adherence to the handling procedure is necessary if quantitative results are to be obtained.

Allele Specific Primer Design

The CDR or N-regions at the rearranged *IGH/IGK* genes differ widely from patient to patient. A GC-rich sequence such as the CDR in patient 1, repeated sequences, and short allele specific sequences at junctional regions often make it difficult to design optimal allele specific primers. Primer dimer and internal hairpin formation should be avoided for the SYBR Green I detection. However, it is even more important to optimize the specificity of the primers. Ideally, the clone-specific sequence should anneal to the 3′ end of the allele specific primer. With short junctional regions, a limited number of extra bases at the 3′ end is acceptable.

Interpretation of the Results

Immunoglobulin genes are also rearranged in normal lymphocytes (BC control DNA) and are amplified when overcycled. Therefore, the melting curve of the control DNA should always be checked. In principle, the most dilute sample with the same melting peak as the more concentrated samples indicates the detection limit. Melting curve analysis can identify non-specific PCR products (patient 2) [4]. When the control DNA shows the same melting peaks (patient 1), a standard curve analysis (calibration graph) is also important for determining the detection limit. Curves of dilutions that show up near the control DNA and curves that do not fit on the calibration graph cannot represent a genuine detection limit. When it is difficult to interpret melting curve characteristics, the PCR product should be resolved on an agarose gel. Two-fold dilution series instead of tenfold dilutions may provide more accurate information on the detection level when an amplification curve of a remission sample is near the detection limit.

Availability of the LC-PCR as a Routine Test for MRD Detection

In the present study the total running time of LC-PCR was 31 min 40 s. This rapid method for MRD quantification on the LightCycler provides an attractive alternative to conventional methods. We routinely apply the method described here for our ongoing MRD studies on patients with ALL. Real-time quantification through LightCycler technology has been established for immunoglobulin (IGH, IGK) and T-cell receptor (TCRγ, TCRδ) rearrangements [4].

References

1. Bartram CR (1993) Detection of minimal residual leukemia by the polymerase chain reaction: potential implications for therapy. Clin Chim Acta 217:75–83
2. van Dongen JJ, Seriu T, Panzer-Grumayer ER, Biondi A, Pongers-Willemse MJ, Corral L, Stolz F, Schrappe M, Masera G, Kamps WA, Gadner H, van Wering ER, Ludwig WD, Basso G, de Bruijn MA, Cazzaniga G, Hettinger K, van der Does-van den Berg A, Hop WC, Riehm H, Bartram CR (1998) Prognostic value of minimal residual disease in acute lymphoblastic leukaemia in childhood. Lancet 352: 1731–1738
3. Wittwer CT, Ririe KM, Andrew RV, David DA, Gundry RA, Balis UJ (1997) The LightCycler: a microvolume multisample fluorimeter with rapid temperature control. Biotechniques 22: 176–181

4. Nakao M, Janssen JWG, Flohr T, Bartram CR (2000) Rapid and reliable quantification of minimal residual disease in acute lymphoblastic leukemia using rearranged immunoglobulin and T-cell receptor loci by LightCycler technology. Cancer Res 60: 3281–3289
5. Miller SA, Dykes DD, Polesky HF (1988) A simple salting out procedure for extracting DNA from human nucleated cells. Nucleic Acids Res 16: 1215
6. Nakao M, Janssen JWG, Bartram CR (2000) Duplex PCR facilitates the identification of immunoglobulin κ *(IGK)* gene rearrangements in acute lymphoblastic leukemia. Leukemia 14: 218–219
7. Cross NCP (1995) Quantitative PCR techniques and applications. Br J Haematol 89: 693–697

HER2/*neu* Gene Amplification Quantified by PCR and Melting Peak Analysis Using a Single Base Alteration Competitor as an Internal Standard

Elaine Lyon, Ph.D*, Alison Millson, MT(ASCP), Arminda Suli, B.Sc.

Introduction

The HER2/*neu* gene is amplified in 25–30% of primary breast cancers and correlates with relapse and shorter survival time (1,2,3,4,5). HER2/*neu* status may also impact therapeutic decisions since monoclonal antibodies against HER2/*neu* have shown inhibition of tumor growth when combined with traditional chemotherapy (6). Due to its prognostic significance and potential in therapeutic decisions, molecular methods for HER2/*neu* gene quantification are being developed. Recent advances in PCR technology using real-time fluorescent monitoring capabilities of the LightCycler allow mutation detection and quantification. Fluorescent hybridization probes have been designed to monitor product accumulation during PCR for real time detection (7). Hybridization probes can also distinguish a single base change by differences in melting temperature (Tm) as the probe melts from the wild-type or mutant allele (8,9,10,11).

Melting curve analyses and peak area determinations were adapted as a competitive PCR method to quantify HER2/*neu* gene copy. To reduce PCR efficiency differences, competitors were designed with a single base alteration from the wild-type sequences. The competitor and the target were amplified by the same primers and distinguished by melting curve analysis. The relative peak areas of target to competitor with known copy numbers allow quantification of the target. We describe this real-time PCR technique for quantifying HER2/*neu* using a single base change competitor and fluorescent hybridization probes.

Materials

LightCycler (Roche Diagnostics, Mannheim, Germany)
LightCycler capillary tubes (Roche Diagnostics, Mannheim, Germany)
Centrifugal concentrators (MicroCon 50, Millipore, Bedford, MA)
RapidCycler, (Idaho Technology, Idaho Falls, ID)

Equipment

---

* Elaine Lyon (✉) (e-mail: lyone@aruplab.com)
  ARUP Laboratories, 500 Chipeta Way, Salt Lake City, UT 84108, USA

Reagents

Taq polymerase (Roche Diagnostics, Mannheim, Germany)
TaqStart Antibody™ (Clontech Palo Alto, CA)
LightCycler $MgCl_2$: 25mM $MgCl_2$ (Roche Diagnostics, Mannheim, Germany)
$MgCl_2$ buffer: 30 mM $MgCl_2$, 50 mmol/L Tris, pH 8.3, 500 mg/L bovine serum albumin (Idaho Technology, Salt Lake City, UT)
dNTP's: 2 mM each in 10 mM Tris, pH 8.1, 0.1 mM EDTA (Sigma, St. Louis, MO)
LightCycler DNA Master Hybridization Probes (10X) (Roche Diagnostics, Mannheim, Germany)
Primers and Probes (Idaho Technology, Salt Lake City, UT)

Procedure

Sample Preparation

The assay was developed using SKBR3 (ATCC# HTB-30), a cell line with approximately 11 copies of HER2/*neu* (12,13) and MRC5, a fetal lung cell line with a single copy HER2/*neu* (ATCC# CCL-171). DNA was prepared from cell lines using phenol/chloroform extraction and ethanol precipitation (14). DNA concentration was determined by spectroscopy and adjusted to 50 ηg/μl.

Primers

The primer and probe sequences with fluorescent labels and relative positions are shown for HER2/*neu* (12) and β-globin (a single copy reference gene;15) in Table 1. Primer and probe synthesis and labeling was performed as previously described (16).

Competitor preparation

The competitors for HER2/*neu* and β-globin were constructed using site directed mutagenesis PCR and amplifying genomic DNA by rapid cycle PCR (17) thereby incorporating the alteration (Table 1). PCR was performed with the following master mix.

Master Mix	Volume [μl]	[Final]
$MgCl_2$ Buffer (30mM: see Reagents)	5	3.0mM
Primers (5μM each)	5	0.5μM each
dNTP (2mM each)	5	200μM each
Taq polymerase (0.4U/μl)	5	2U/50μl
DNA (50ng/μl)	5	250ng/50μl
H_2O	25	
Total Volume	50	

Table 1. Oligonucleotides

HER2/*neu* Gene (Genebank Accession # M11730)				
	Position	Length	GC (%)	T_m (°C)
Primers				
CCTCTGACGTCCATCGTCTC	2098	20	60.0	61.7
CGGATCTTCTGCTGCCGTCG	2198R	20	65.0	65.6
CCTCTGACGTCCATCGTCTCTGCGGTGG-TTAGCATTCTGCTGGT (competitor)*	2098	44	56.8	78.5
Product	2098–2198	100		
Hybridization Probes				
CTTGATGAGGATCCCAAAGACCACCCCC-AAGACCAC-F	2142	36	55.6	73.9
LCRed640-ACCAGCAGAATGCCAACCA-P	2162	19	52.6	62.2
β-Globin Gene (Genebank Accession # AF 117710)				
Primers				
ACACAACTGTGTTCACTAGC	(–)7	20	45.0	58
CAACTTCATCCACGTTCACC	103R	20	50.0	59.1
CAACTTCATCCACGTTCACCTTGCCCC-ACAGGGTAGTAACGG (competitor)*	103	42	54.8	76.4
Product	(–)7–103	110		
Hybridization Probes				
CCACAGGGCAGTAACGG-F	61	17	64.7	59.8
LCRed705-AGACTTCTCCTCAGGAGTCAGG-TGCACCATG-P	59	31	54.8	71.3

* Altered base is shown in bold

PCR for competitor templates was performed on a rapid air thermocycler using thin-walled PCR microtubes with the following conditions.

Parameter	Value		
Cycles	30		
	Segment 1	Segment 2	Segment 3
Target temperature (°C)	94	50	75
Incubation time (s)	30	30	30

PCR products were visualized as single bands by agarose gel electrophoresis and purified by phenol/chloroform extraction and ethanol precipitation. Primers were removed from PCR products by micro-concentrators and 5 minutes of centrifugation at 10,000 x g. To confirm the base alterations, competitors were sequenced by dye-terminator chemistry (ABI 377 sequencer, Applied

Competitive PCR Serial two-fold dilutions of competitors from 250,000 copies to 4000 copies were mixed with 100 ηg genomic DNA. PCR mixtures were added to the samples, transferred into capillary tubes and briefly centrifuged to spin the reaction mixture to the bottom of the capillary and loaded into the LightCycler. Competitive PCR was performed with the following master mix.

Master Mix	Volume [µl]	[Final]
LightCycler-DNA Master Hybridization Probes (10X)*	2µl	1X
Taqstart Antibody (1.1 µg/µl)	0.32µl	0.352µg/20µl
MgCl$_2$ (25mM)	2.4µl	4mM**
Primers (5 µM both forward primer, 2µM both reverse primer)	2µl	0.5 +0.2 µM
HER2/*neu* fluorescein probe (2 µM)	2µl	0.2µM
β-globin fluorescein probe (2 µM)	2µl	0.2µM
HER2/*neu* LCRed 640 probe (2 µM)	2µl	0.2 µM
β-globin LCred 705 probe (2 µM)	2µl	0.2 µM
Competitor templates (2x10^3-1.25x10^5 copies/µl each)	2µl	4x10^3-2.5x10^5 copies/20µl
DNA (50 ηg/µl)	2µl	100ηg/20µl
H$_2$O	1.28µl	
Total Volume	20	

* LightCycler-DNA Master Hybridization Probes mix contains *Taq* DNA polymerase, reaction buffer, dNTP mix (with dUTP instead of dTTP), and 10mM MgCl$_2$
** Includes MgCl$_2$ in master mix

PCR for competitive PCR was performed on the LightCycler using the following conditions.

Parameter	Value		
Cycles	35		
Type	None		
	Segment 1	Segment 2	Segment 3
Target temperature (°C)	94	53	72
Incubation time (s)	3	10	0
Temperature Transition rate (°C/s)	20	20	1
Acquisition mode	None	Single	None
Gains	F1 = 1	F2 = 13	F3 = 23

Immediately after cycling, samples were heated slowly with continual monitoring using the following parameters.

Parameter	Value			
Cycles	1			
Type	Melting Curve			
	Segment 1	Segment 2	Segment 3	Segment 4
Target temperature (°C)	94	65	40	80
Incubation time (s)	3	0	120	0
Temperature Transition rate (°C/s)	20	20	1	0.1
Acquisition mode	None	Single	Single	Continuous
Gains	F1 = 1	F2 = 13	F3 = 23	

Five additional cycles and a second melt with identical parameters were performed for β-globin melting cuves.

Color Compensation

Due to emmision spectra overlaps of the fluorophores, each channel was corrected for fluorescence crosstalk of the other fluorophores as previously described (16) or by using a commercially available kit (Roche Diagnostics).

Analysis

Melting curves were obtained for HER2/*neu* after 35 cycles of amplification. β-globin melting curves were analyzed after 40 cycles. The Tm at which the probe melts from the template is seen as a loss of fluorescence. With the color compensation enabled, the data analysis program converts the melting curves into melting peak plots, (derivative curves with respect to temperature). The competitor and target PCR products are distinguished by differences in melting peak Tm. Peaks are analyzed by polynomial estimation of the derivative with background subtraction and a sliding window of 8°C. The software calculates areas for each peak. Two peaks are fit, one to the competitor and one to the target for the each sample. The peak areas are calculated by non-linear least square analysis of multiple Gaussians by the LightCycler analysis software. An example of the Gaussian fit is shown in Figure 1.

Results

Melting curve analysis easily distinguishes between the competitor and target amplicons that differ by a single base. The HER2/*neu* and β-globin wild-type and competitor Tms are shown in Table 2. Derivative melting curves of DNA co-amplified with varying amounts of competitors are shown in Figure 2. The areas of the competitor and target peaks were determined at each competitor concentration. A ratio of the competitor area to target area (C/T) was calculated for each competitor concentration.

To determine the copy number, the log C/T versus the log of the competitor copy number [C] was plotted for each cell line and a second order polynomial

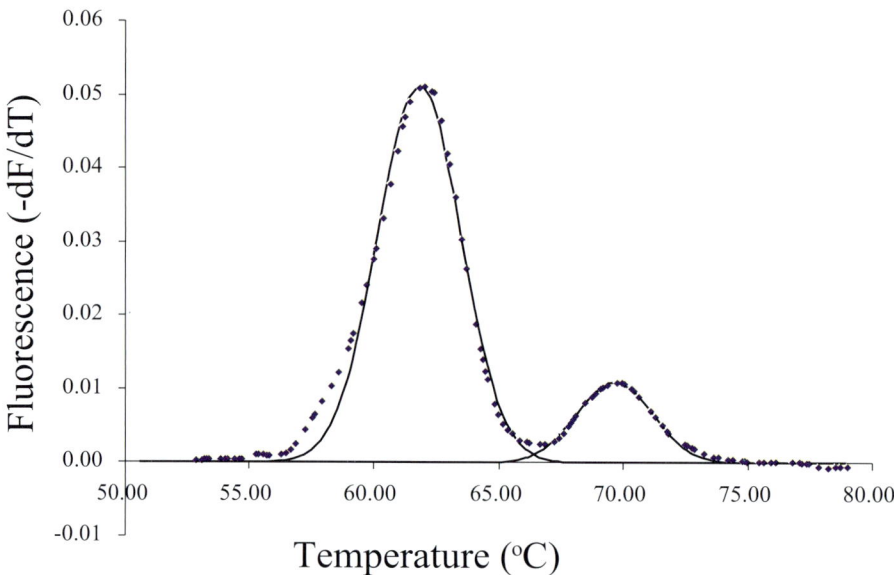

Fig. 1. Gaussian fit for HER2/*neu* and competitor. Genomic DNA was co-amplified with 30,000 copies of competitor and distinguished by melting peak analysis (♦). The gaussian fit is shown by the solid line

Table 2. Melting Temperatures of HER2/*neu* and β-globin Target and Competitor

Gene		Pairing	T_m (observed)
HER2/*neu*	Target	G-C match	65.9°C
	Competitor	A-C mismatch	55.2°C
β-globin	Target	G-C match	61.9°C
	Competitor	A-C mismatch	50.6°C

curve was fit (Figure 3). Assuming that the amount of competitor required for equal areas of competitor and target peaks (i.e. the x-intercept of the log(C/T):log[C] plot) is equal to the amount of target, the copy number of each sample was determined by calculating the x-intercept for each curve. The MRC5 cell line had 25,000 copies and SKBR3 had 250,000 copies of HER2/*neu*.

Copy numbers for β-globin were also calculated. The MRC5 cell line had 25,000 copies of β-globin and SKBR3 had 32,000 copies. To compare our results to those reported in the literature, we calculated a gene dose, or the amount of amplification compared to the single copy control gene β-globin (a gene dose of one). This comparison adjusts for concentration differences of amplifiable DNA between samples. When adjusted for β-globin (copies HER2/*neu*/copies β-globin), MRC5 had a gene dose of one and SKBR3 had a gene dose of 7.8. (Table 3). These results are comparable to the reported gene dose of one for MRC5 and eleven for SKBR3 (12,13).

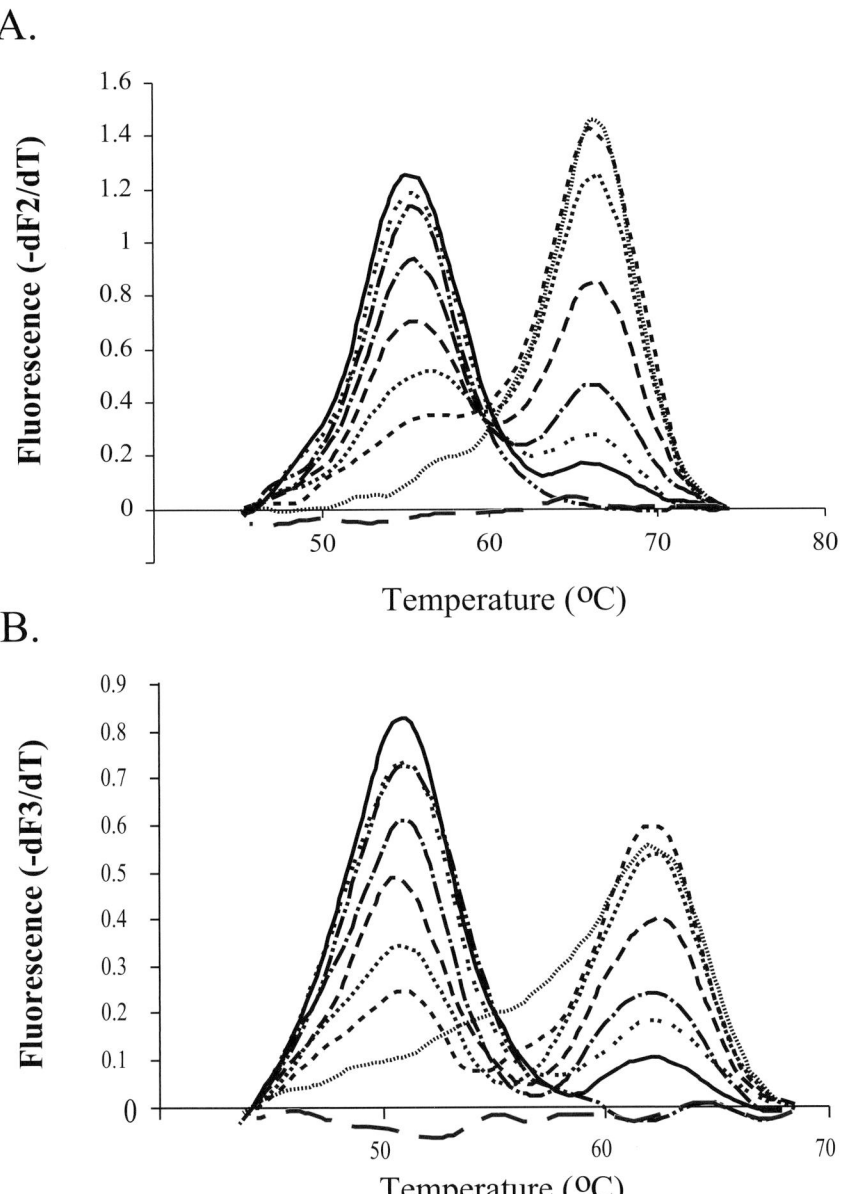

Fig. 2. Melting peaks for targets and competitors. A. HER2/*neu*. B. β-globin with color compensation. MRC5 DNA (100ηg) was co-amplified with varying competitor concentrations; 250,000 copies (———), 125,000 copies (-- -- --), 62,000 copies (—·—), 31,000 copies (— — —), 15,000 copies (·········), 7,500 copies (— — —). Controls: No template (— —), MRC5 (100ηg) (············), HER2/*neu*/β-globin competitor 31,000 copies (——··)

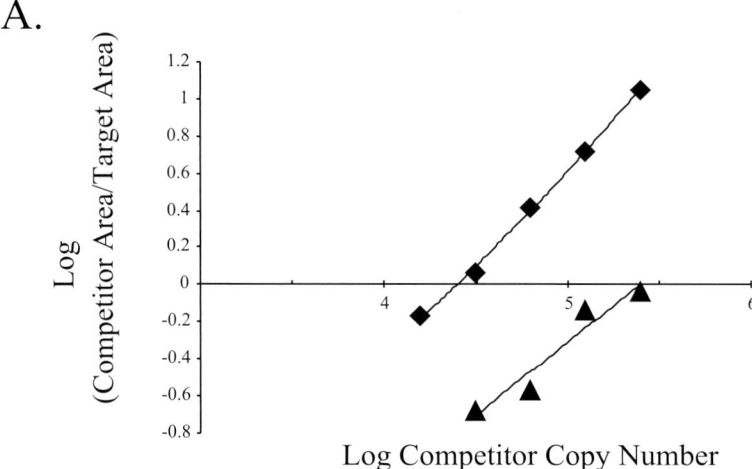

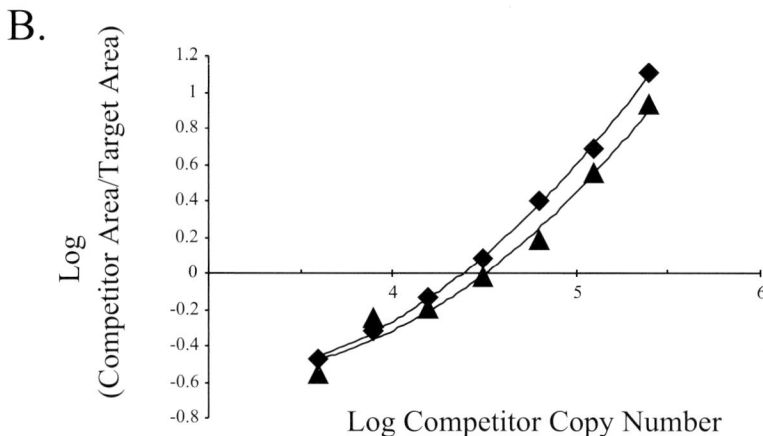

Fig. 3. Co-amplification of cell lines with varying competitor concentrations ranging from 250,000 to 4000 copies. Curves were fit to a second degree polynomial to determine copy numbers. A. HER2/*neu* ; MRC5 (◆) y = 0.106x² + 0.017x − 2.117, R² = 0.998: SKBR3 (▲) y = -0.0209x² + 0.9906x − 4.7398, R² = 0.9333. B. β-globin; MRC5 (◆) y = 0.270x² − 1.566x + 1.672, R² = 0.999: SKBR3 (▲) y = 0.2678x² − 1.6436x + 1.9723, R² = 0.9822

To reduce the number of competitor concentrations needed per sample, a gene dose can be caluclated directly from the ratios of HER2/*neu* and β-globin using the following formula:

$$\frac{(\text{Target Area/Competitor Area})_{HER2/neu}}{(\text{Target Area/Competitor Area})_{\beta\text{-globin}}}$$

Table 3. Copy numbers and gene doses for MRC5 and SKBR3 cell lines

	MRC5	SKBR3
β-globin copy number	25,000	32,000
HER2/*neu* copy number	25,000	250,000
HER2/*neu*/β-globin (gene dose)	1.0	7.8
Reported HER2/*neu* gene dose	1.0	11

When calculated as single point ratios at each competitor concentration, gene doses for MRC5 ranged from 0.76–1.6. Since a gene dose greater than two is considered amplified, the MRC5 cell line would not be classified as "amplified." The SKBR3 cell line had calculated gene doses ranging from 7.4–11.7 at different competitor concentrations.

Comments

Quantitative PCR methods employ either external or internal standards. Advantages of external standards are wide dynamic ranges and results from a single tube per sample. Disadvantages include the necessity of using a complete standard curve for each run. Additionally, if a reference gene is amplified along with the target gene, a separate standard curve must be included, or the two reactions must have equal efficiencies. Using external standards, analysis is best performed during the log phase of PCR.

Advantages of internal standards include control of PCR efficiency between samples and end-point analysis. However, internal standard systems typically have a dynamic range of 2 logs (18). Although some external standard systems may have greater dynamic ranges, HER2/*neu* breast tumors are amplified between 2 and 20 copies (12,18), so a range of two logs is more than adequate for this application.

We have developed a fluorescent quantitative PCR assay for HER2/*neu* using single base alteration competitors for internal standards. This technique employs several controls within a single reaction. An internal competitor that differs by a single base alteration controls for sample to sample efficiency differences, since any efficiency difference will affect both the competitor and the target equally. Also included in the reaction is a single copy reference gene. The reference gene controls for sample to sample variation in the amount of amplifiable DNA. This is particularly important for sub-optimal samples such as paraffin-embedded, formalin-fixed samples that may be partially degraded. Since spectroscopy cannot distinguish between high molecular weight DNA and degraded DNA, DNA concentration based on spectroscopy may not accurately reflect amplifiable DNA. Since HER2/*neu* and β-globin are both controlled for sample to sample efficiency differences by their own competitors, the amplification efficiencies of the two reactions do not need to be equalized. To reduce the number of competitor concentrations per sample, a single point determination can be used to directly calculate gene dose.

Fluorescent PCR monitoring offers several advantages over existing PCR methods for quantification. PCR amplification and analysis are performed in a single capillary tube, eliminating manual manipulation for post-amplification analysis. This simplifies sample tracking and reduces the chance of PCR product contamination. Rapid cycle PCR also reduces amplification time and increases specificity over traditional PCR (17). Thirty-two samples can be amplified and analyzed in approximately 45 minutes. Furthermore, computerized programs are available to rapidly calculate peak areas and further automate this technique.

References

1. Slamon DJ, Godolphin W, Jones LA, Holt, JA, Wong SG, Keith DU, Levin WJ, Stuart SG, Udonve J, Ullrich A, Press MF. (1989) Studies of the HER-2/*neu* proto-oncogene in human breast and ovarian cancer. Science 244:707–712.
2. Borg A, Tandon AK, Sigurdsson H, Clark GM, Ferno M, Fuqua SA, Killander D, McGuire WL. (1990) HER-2/*neu* amplification predicts poor survival in node-positive breast cancer. Cancer Res 50:4332–4337.
3. Ciocca DR, Fujimura FK, Tandon AK, Clark GM, Mark C, Lee-Chen GJ, Pounds GW, Vendely P, Owens MA, Pandian MR et al. (1992) Correlation of HER-2/*neu* amplification with expression and with other prognostic factors in 1103 breast cancers. J Natl Cancer Inst 84:1279–1281.
4. Parkes HC, Lillycrop K, Howell A, Craig RK. (1990) C-erbB-2 mRNA expression in human breast tumours: comparison with c-erbB2 DNA amplification and correlation with prognosis. Br J Cancer 61: 39–45.
5. Slamon DJ, Clark GM, Wong SG, Levin WJ, Ullrich A, McGuire WL. (1987) Human breast cancer: Correlation of relapse and survival with amplification of the HER-2/*neu* oncogene. Science 253:177–182.
6. Baselga J, Norton L, Albanell J, Kim YM, Mendelsohn J. (1998). Recombinant humanized anti-HER2 antibody (Herceptin) enhances the antitumor activity of paclitaxel and doxorubicin against HER2/*neu* overexpressing human breast cancer xenografts. Cancer Res 58:2825–2831.
7. Wittwer CT, Herrmann MG, Moss AA, Rasmussen RP: (1997) Continuous fluorescent monitoring of rapid cycle DNA amplification. Biotechniques 22:130–138.
8. Lay MJ, Wittwer CT. (1997) Real-time fluorescence genotyping of factor V Leiden during rapid-cycle PCR. Clin Chem 43:2262–2267.
9. Bernard PS, Lay MJ, Wittwer CT. (1998) Integrated amplification and detection of the C677T point mutation in the methylenetetrahydrofolate reductase gene by fluorescence resonance energy transfer and probe melting curves. Anal Biochem 255:101–107.
10. Bernard PS, Ajoika RS, Kushner JP, Wittwer CT. (1998) Homogenous multiplex genotyping of hemochromatosis mutations with fluorescent hybridization probes. Am J Pathol. 153:1055–1061.
11. Lyon E, Millson A, Phan T, Wittwer CT. (1998) Detection and identification of base alterations within the region of factor V Leiden by fluorescent melting curves. Molec Diag 3:203–210.
12. Sestini R, Orlando C, Zentilin L, Donatella L, Gelmini S, Pinzani P, Giacca M, Pazzagli M. (1995) Gene amplification for c-erbB-2, c-myc, epidermal growth factor receptor, int-2, and N-myc measured by quantitative PCR with a multiple competitor template. Clin Chem 41:826–832.
13. Hynes, NE, Geber HA, Saurer S, Groner B. (1989) Overexpression of the c-erbB-2 protein in human breast tumor cell lines. J Cell Biochem 39:145–51.

14. Thomas SM, Moreno RF, Tilzer LL. (1989) DNA extraction with organic solvents in gel barrier tubes. Nucleic Acids Res 17:5411.
15. Saiki RK, Scharf S, Faloona F, Mullis KB, Horn GT, Erlich HA, Arnheim N. (1985). Enzymatic amplification of β-globin genomic sequences and restriction site analysis for diagnosis of sickle cell anemia. Science 230:1350–1354.
16. Bernard PS, Pritham GH, Wittwer CT. (1999) Color multiplexing hybridization probes using the apolipoprotein E locus as a model system for genotyping. Anal Biochem 273:221–228.
17. Wittwer, CT, Reed GB, Ririe KM: (1994) Rapid cycle DNA amplification. In Mullis KB, Ferré F, Gibbs RA: The polymerase chain reaction. Birkhauser, Boston, pp174–181.
18. Diviacco S, Norio P, Zentilin L, Menzo S, Clementi M, Biamonti G, Riva S, Falaschi A, Giacca M. (1992) A novel procedure for quantitative polymerase chain reaction by coamplification of competitive templates. Gene 122:313–320.

Quantification of Residual Tumor Cells in Monoclonal B-cell Lymphoma

THOMAS PFITZNER*, ANDREAS ENGERT, STEFAN BARTH

Introduction

With the advent of more effective therapeutic modalities, better methods to evaluate and quantify minimal residual disease (MRD) in patients with malignant lymphoma are needed. These methods should be highly sensitive in detecting very low amounts of malignant cells and should be specific for the malignant clone. In addition, these methods should allow the quantification of residual tumor cells. At present, the highest sensitivity is reached with PCR, allowing the detection of one malignant cell in 10^6 normal cells [1]. The clone-specific hypervariable complementarity determining regions (CDRs) of the immunoglobulin heavy chain locus (IgV$_H$) provide a useful marker for monitoring MRD in B-cell lymphoma during and after treatment [2, 3].

Reports on various malignancies support the hypothesis that residual tumor cells may only lead to relapse once a threshold cell number is reached. Easy methods for accurate quantification of residual tumor cells must be available.

This protocol describes a strategy to develop PCR systems for quantification of residual tumor cells in monoclonal B-cell lymphoma. The assay is based on an immunoglobulin heavy chain (IgV$_H$) specific PCR with tumor specific primers complementary to hypervariable CDRII and CDRIII regions. A set of framework region III (FRIII) specific hybridization probes (Hyb-probes) can be used for the detection of the specific amplification product. A schematic illustration of the test set-up is shown in Fig. 1.

As an example, sample DNA from peripheral mononuclear cells and from the bone marrow of one patient with B-CLL was analyzed before and after four cycles of therapy with Fludarabin and the monoclonal antibody IDEC-C2B8 (Rituximab).

---

* Thomas Pfitzner (✉) (e-mail: Thomas.Pfitzner@medizin.uni-koeln.de)
  Medizinische Klinik I der Universitaet zu Köln, Labor für Immuntherapie, LFI, E4, R 703, Joseph-Stelzmann-Str. 9, 50931 Köln, Germany

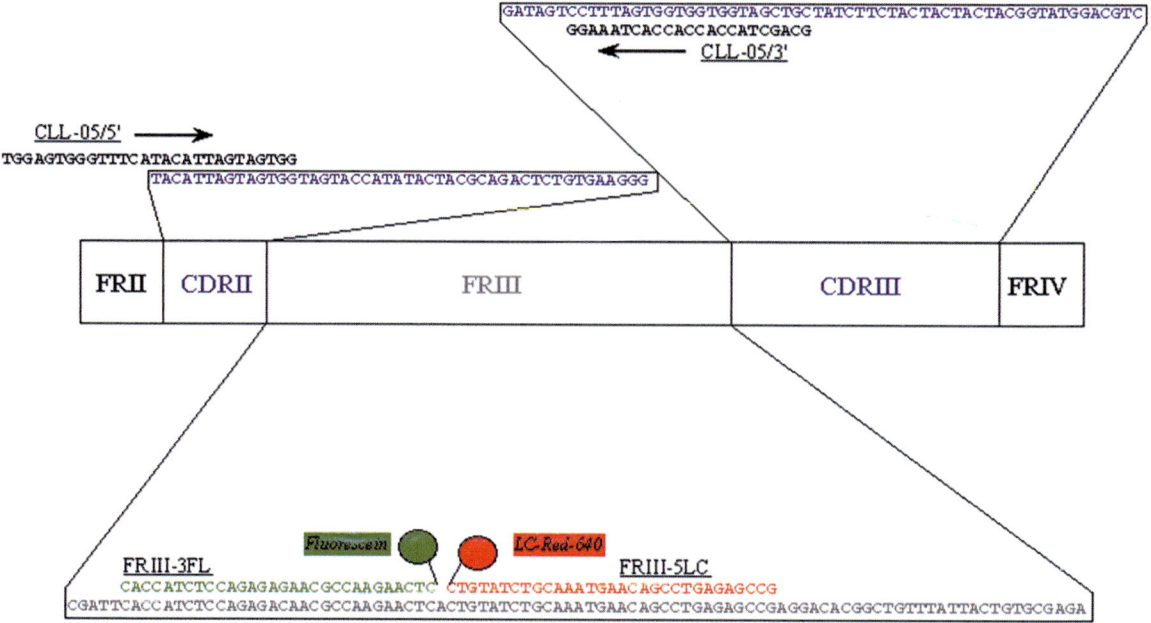

Fig. 1. A schematic illustration of the IgV$_H$ segment of one patient together with primers and hybridization probes used for the minimum residual disease quantification with the LightCycler. The sequence of CDRII and CDRIII is shown, with the primers, specific for these segments. The sequence of FRIII is also shown, together with the hybridization probes [11]

Materials

Equipment LightCycler Instrument (Roche Diagnostics, Mannheim, Germany)
Biometra Trio-Thermoblock
ABI PRISM Automated Sequencer (Perkin Elmer Corp., Norwalk, USA)
OLIGO 4.0 Primer analysis software. (National Biosciences, Plymouth, MN, USA)

Reagents Ficoll Hypaque (Pharmacia Biotech, Uppsala, Sweden)
Amplification Primers (Eurogentec, Brussels, Belgium)
Hybridization Probes (TIB MOLBIOL, Berlin, Germany)
Taq-Polymerase (Pharmacia)
Amplitaq Gold DNA Polymerase, dNTPs (containing dUTP instead of dTTP; Perkin Elmer Corp.)
T/A cloning kit (Invitrogen Corporation, San Diego, CA)
TaqStart antibody (Clontech, Palo Alto, USA).
  The following reagents were purchased from Roche Diagnostics:
DNA Isolation Kit for Blood/Bone Marrow/Tissue
LightCycler-DNA Master Hybridization probes
Uracil-N-Glycosylase (heat labile)
Human Genomic DNA, *Hin*dIII

Procedure

Mononuclear cells were isolated from EDTA blood and bone marrow aspirates by Ficoll density gradient centrifugation. Cells were washed three times in sterile 0.9% NaCl solution. DNA was extracted following a standard protocol which has previously been used for detection of MRD [4]. At the same time, 1×10^6 cells were resuspended in 160 µl PCR buffer, containing 50 mM KCl, 10 mM Tris-HCl (pH 8.3), 2.5 mM $MgCl_2$, 0.1 mg/ml gelatin, 0.45% NP40, 0,45% Tween 20 and 60 µg/ml proteinase K. After incubation at 55°C for 1 h and heat inactivation of proteinase K at 95°C, samples were stored at –80°C.

In an alternative procedure, mononuclear cells were resuspended in lysis buffer (DNA Isolation Kit for Blood/Bone Marrow/Tissue) and stored at –80°C until further preparation of DNA according to the kit protocol.

The OD_{260} of samples was determined to estimate the DNA concentration [5].

Sample Preparation

In order to design tumor-clone CDR-specific primers, the IgV_H segment of the tumor clone has to be sequenced. In a first step, a set of VH consensus primers was used for amplification of IgV_H (Table 1). Primers were described previously [6]. PCR was performed with 0.2 µg genomic DNA, 0.2 µM primers, 200 µM each dNTP (with dUTP instead of dTTP), 1.0 mM $MgCl_2$, 2.5 U Taq-Polymerase and appropriate 10× buffer in a 50-µl volume. After 5 min at 95 °C and a hot start, the

Development of a Tumor-Specific PCR

Table 1. VH consensus primers [6], tumor-specific primers, and fluorescent hybridization probes used for minimum residual disease detection [11]

	Region	Length	GC (%)	T_m (°C)
Consensus primers:				
CCTCAGTGAAGGTYTCCTGCAAGGC	FRI (VH1)	25	60/56	64.7/63.6
GTCCTGCGCTGGTGAAACCCACACA	FRI (VH2)	25	60	67.2
GGGGTCCCTGAGACTCTCCTGTGCAG	FRI (VH3)	26	65.4	67.0
GACCCTGTCCCTCACCTGCRCTGTC	FRI (VH4)	25	64/68	66.1/68.3
AAAAAGCCCGGGGAGTCTCTGARGA	FRI (VH5)	25	52/56	63.4/64.9
ACCTGTGCCATCTCCGGGGACAGTG	FRI (VH6)	25	64	67.7
ACCTGAGGAGACGGTGACCAGGGT	J_H 1,2,4,5	24	62.5	66.3
TACCTGAAGAGACGGTGACCATTGT	J_H 3	25	48	61.2
ACCTGAGGAGACGGTGACCGTGGT	J_H 6	24	62.5	66.8
Tumor-Specific Primers:				
TGGAGTGGGTTTCATACATTAGTAGTGG	FRII-CDRII	28	42.9	67.6
TAGCAGCTACCACCACCACTAAAGG	CDRIII	25	52	69.6
Product		189		
Hybridization probes:				
CACCATCTCCAGAGACAACGCCAAG-AACTC-F	FRIII	30	53.3	73.2
LCRed640-CTGTATCTGCAAATGAAC-AGCCTGAGAGCCG-P	FRIII	31	51.6	73.5

following parameters were used: 40 cycles with 1 min at 95°C, 1 min at 65°C, and 2 min at 72°C. For the patient described in this protocol, we obtained a specific product with primers VH3 and J_H6, from both blood and bone marrow samples. The PCR product was directly ligated into pCRII, recombinant clones were analyzed, and five IgV_H containing plasmids were sequenced.

Corradini et al. [7] introduced a tumor-specific PCR, which utilized primers specific for the CDRII and CDRIII regions of the IgV_H segment to specifically amplify tumor cell DNA in a background of normal cells. Our test is based on this approach. CDRII and CDRIII specific primers were designed with OLIGO 4.0 primer analysis software.

We determined optimal PCR conditions for tumor-specific PCR as described previously [8] with 10 copies of plasmid DNA diluted into 1 μg of genomic DNA of a healthy individual. Optimal PCR conditions were 250 μM each dNTP (with 500 μM dUTP instead of dTTP), 3 mM $MgCl_2$, 1 μM primers, 2.5 U AmpliTaq gold DNA polymerase, and 1 U uracil-N-glycosylase (UNG) in 50 μl volume. After 10 min incubation at room temperature for elimination of "carry-over" contamination by UNG, PCR was performed on a Trio-Thermoblock (Biometra). The following parameters were used: 10 min at 95°C and 40 cycles with 1 min at 95°C, 1 min at 65°C, and 1 min at 72°C. No specific amplification was observed from 1 μg of genomic DNA from a polyclonal B-cell population of a healthy individual.

Generation of Plasmid Standards for Real-Time Quantitative PCR

A standard plasmid preparation of pCRII containing the IgV_H sequence of the patients tumor clone was performed. DNA was resuspended in PCR grade water, and DNA concentration and copy number were estimated from the OD_{260}.

Plasmid DNA was digested with 1 U of *Hin*dIII per 50 μl for 1 h at a concentration of 1×10^8 copies/μl. The enzyme was inactivated at 95°C for 10 min. DNA was diluted into 1.5 ng/μl human genomic DNA in log steps from 1×10^8 down to 1 copy per 2 μl volume. For evaluation of standard concentration, a limiting dilution assay based on Poisson analysis was performed [9]. The copy number estimate based on absorbance (1 copy in 2 μl). was in good correlation with the copy number estimate based on limiting dilution (1.6 copies in 2 μl). Standards were aliquoted and stored at −20°C.

LightCycler PCR

Two hybridization probes binding in the FRIII segment of IgV_H, were chosen the T_ms about 5 degrees higher than the primer T_ms (Fig. 1, Table 1).

Quantification of IgV_H copy number was based on a set of seven external standards, containing 1 to 1×10^6 plasmid copies per 2 μl volume. PCR with these standards was performed in the presence of 1 μg genomic DNA of a healthy individual. The DNA samples from patient with B-CLL were analyzed in parallel with standards.

Hybridization Probe Master Mix:

	Volume [µl]	[Final]
LightCycler-DNA Master Hybridization Probes	2	1x
TaqStart antibody (5 min incubation at room temperature)	0.35	
MgCl$_2$ (25 mM)	3.2	5 mM
Primers (2.5 µM each)	1	0.125 µM
Hybridization probes (1 µM each)	4	0.20 µM
Heat-labile uracil-DNA glycosylase (1 U/µl)	1	1 U/tube
Genomic DNA of a healthy donor (250 ng/µl)	4	1 µg/tube
Standards in appropriate dilution (1 to 10^6 copies)	2	
H$_2$O (PCR grade)	2.45	
Total volume:	20	
For unknown samples, replace the healthy donor and standard DNA with patient genomic DNA (1 µg/tube) and make up to 20 µl with PCR grade H$_2$O.		

The sealed capillaries were incubated for 10 min at room temperature for elimination of "carry over" contamination by uracil-DNA glycosylase.

The following PCR protocol was used:

- Denaturation for 2 min at 95°C
- Amplification

Parameter	Value		
Cycles	60		
Type	Quantification		
Temperature targets	Segment 1	Segment 2	Segment 3
Target temperature (°C)	95	67	72
Incubation time [s]	0	20	3
Temperature transition rate [°C/s]	20	20	20
Acquisition mode	None	Single	None
Gain	F2=10		

- Melting curve analysis

Parameter	Value		
Cycles	1		
Type	Melting curve		
Temperature targets	Segment 1	Segment 2	Segment 3
Target temperature (°C)	95	40	95
Incubation time [s]	0	10	0
Temperature transition rate [°C/s]	20	20	0.2
Acquisition mode	None	None	Step
Gain	F2=10		

Data analysis

After background subtraction to normalize fluorescence between various samples, a noise band was defined, and 3 points in the exponential phase of PCR were used to define log-linear lines. The intersections (crossing points) of these lines with the noise band were plotted against the "log" of the concentrations of the standards to generate a standard curve. Sample concentration was interpolated from this standard curve. All samples were analyzed at least 3 times. Mean concentration, standard deviation and coefficient of variation were calculated.

Results

Generation of a Standard Curve for Quantification of IgV_H Copy Number

Standards from 1 to 1×10^6 copies of a plasmid containing the patient's IgV_H segment were used. This range was chosen to bracket the amount of IgV_H DNA which could theoretically be contained within 1 µg of genomic DNA [10]. The fluorescence curves are shown in Fig. 2. Four tubes containing an average of one copy were run in parallel. Since the probability of a tube containing a certain number of molecules is described by the Poisson distribution [9], not all tubes containing an average of one copy gave positive results. The log-linear lines, corresponding to the exponential portion and a standard curve are shown in Fig. 3.

Patient samples were analyzed with 1 µg of genomic DNA per reaction; 1 µg of genomic DNA from a healthy individual was added to the standard curve tubes to achieve identical reaction conditions between standards and samples. This was

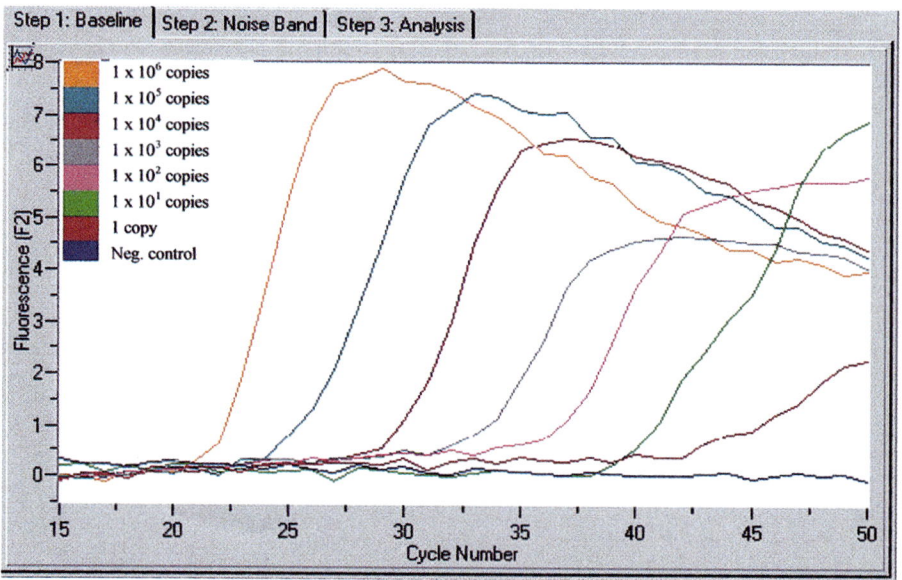

Fig. 2. Amplification curves of IgV_H plasmid standards [11]

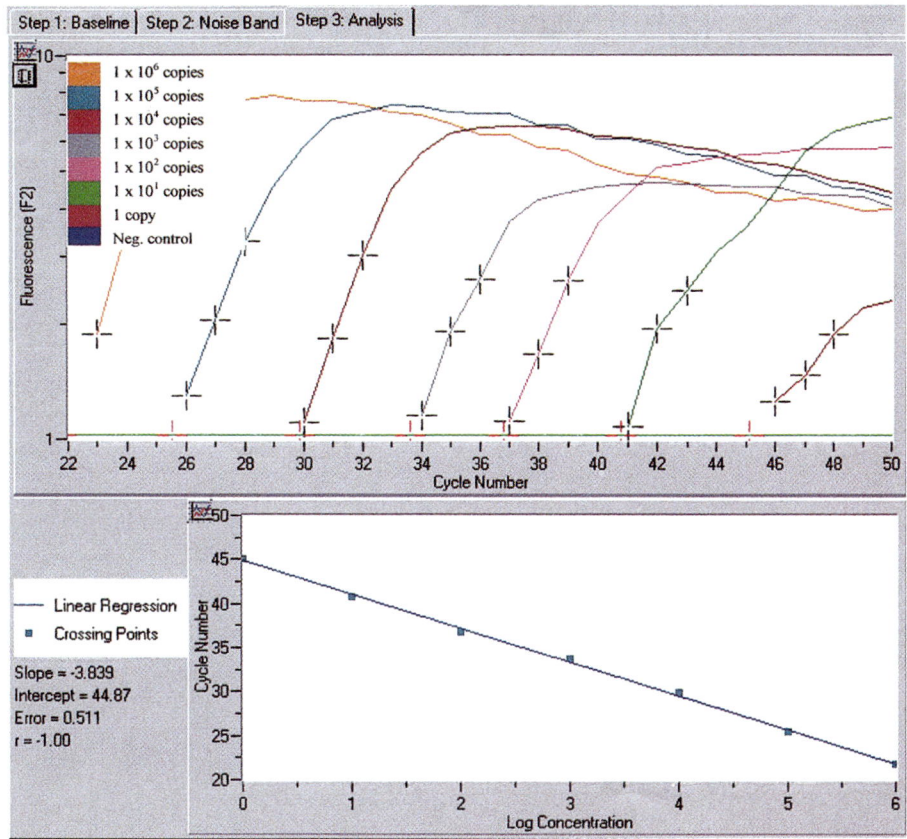

Fig. 3. Log-lines of amplification curves from Fig. 2 were used to define a standard curve for quantification of samples with unknown concentration [11]

necessary because different copy numbers were quantified when standards without 1 µg human genomic background DNA were analyzed.

The standard amplifications were analyzed on a 1% agarose gel to demonstrate that unspecific amplification products were not present. (Fig. 4).

Determination of IgV$_H$ Copy Number

Samples from the patient were analyzed in parallel with standards under identical conditions. Figure 5 shows standard curves and three replicate runs of DNA extracted from peripheral blood mononuclear cells before therapy. The IgV$_H$ copy number was calculated by interpolation from the standard curve as described above.

Copy numbers of tumor IgV$_H$ from peripheral blood and bone marrow before and after therapy were estimated in multiple parallel reactions. Mean values, standard deviations, and coefficients of variation were calculated. Representative results are documented in Table 2 [11] and Figure 6.

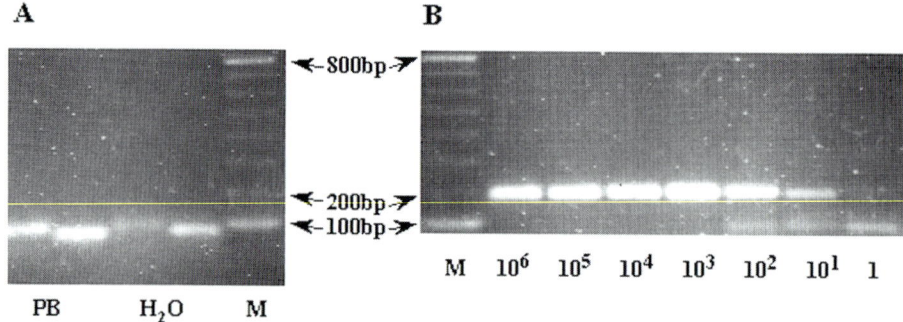

Fig. 4. A, B. Analysis of standard amplification (documented in Fig. 2) with 1% agarose gel electrophoresis. **A** Negative controls with genomic DNA from a negative donor and water alone (PB: DNA from a negative donor). **B** A specific PCR product is visible at approximately 200 bp for standards from 1×10^6 to 1 copy. M: 100bp marker [11]

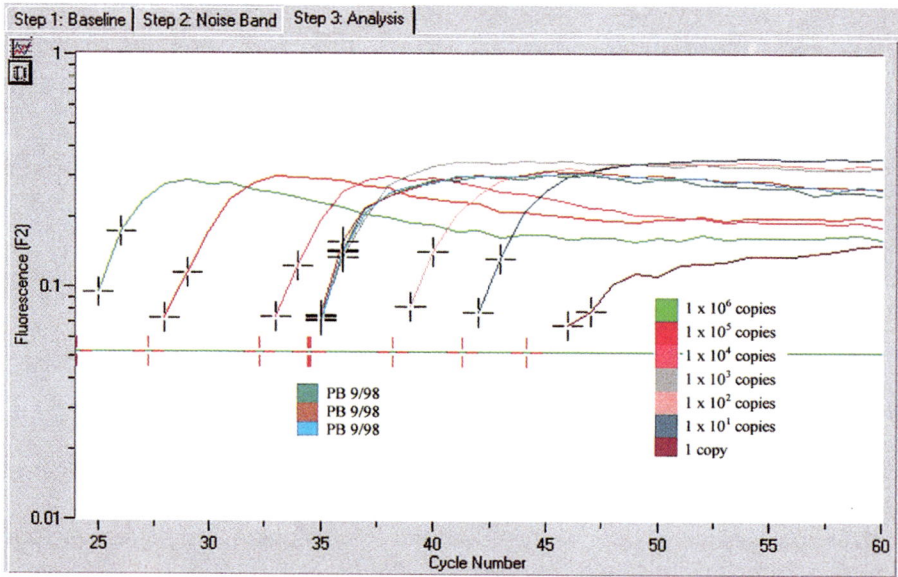

Fig. 5. Log-lines of standards and 3 replicates of DNA extracted from the peripheral blood before treatment. [11]

A statistical analysis was performed to check whether the estimated copy-numbers before and after therapy [11] were significantly different. The Kolmogorov-Smirnov test, a test for two independent samples, showed that the difference in IgV_H copy number in peripheral blood ($P<0.0001$), and bone marrow ($P=0.003$) were highly significant.

Table 2. Quantification of tumor-specific immunoglobulin heavy chain in 1 µg of DNA from peripheral blood before (PB 9/98) and after (PB 3/99) therapy, and from bone marrow before (BM 9/98) and after (BM 3/99) therapy [11]

Sample ID:	Copy-no.:	Mean value	Standard deviation	Coefficient of variation
PB 9/98	1500			
	2000			
	2600			
	1900	2010	310	15%
PB 3/99	110			
	180			
	150			
	130			
	100			
	250			
	170	155	30	18%
BM 9/98	5300			
	5600			
	6000	5650	230	4.12%
BM 3/99	2800			
	2350			
	3600			
	3800			
	5400	3600	670	18.8%

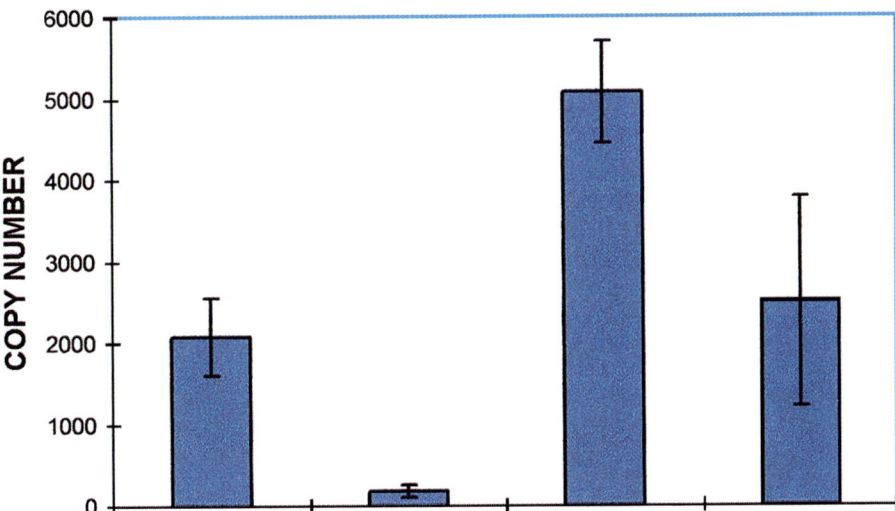

Fig. 6. Mean copy number and standard deviation of samples before (PB 9/98; BM 9/98) and after therapy (PB 3/99; BM 3/99) [11]

Comments

Design of the tumor-specific PCR

The CDRII specific primer used binds in both FRII and CDRII. This allowed a reasonable Tm, G/C-content and lack of secondary structures. The primer set was tested for sensitivity and specificity. A specific PCR product of 189 bp was detected only with genomic DNA from blood and bone marrow of the patient, but not from genomic DNA from polyclonal B-cells of a healthy individual.

Contamination Prevention

Quantification of DNA molecules by PCR is very contamination sensitive. To prevent contamination, sample preparation, PCR set-up, analysis of PCR product, cloning of the IgV_H segment, and preparation of plasmid standards was performed in physically separated areas. In addition to these precautions, we used uracil-N-glycosylase to eliminate "carry-over" contaminations.

Minimal Residual Disease Detection in B-cell Lymphoma

Minimal residucal disease (MRD) status in B-cell malignancies has been evaluated by various approaches. Several immunological methods have been used, but PCR has proven to be the most sensitive assay. A typical PCR application in MRD diagnosis is the detection of tumor specific chromosomal translocations. The tumor clone specific immunoglobulin heavy chain gene rearrangement has been described as an alternative marker for MRD monitoring. Corradini et al. [7] established an approach with tumor-specific CDRII/CDRIII primers, which was transferred to the LightCycler PCR system, allowing the quantification of residual tumor cells [11].

The described method is applicable to a variety of monoclonal B-cell malignancies. We do not know whether a CDRII/CDRIII specific LightCycler PCR might be applicable to tumors with a high rate of ongoing mutations. To answer this question, further studies will be performed.

References

1. Roberts W, Estrov Z, Zipf T (1995) Detection of minimal residual disease in acute lymphoblastic leukemia [letter; comment]. Blood 86: 237–9
2. Maloum K, Pritsch O, Dighiero G (1997) Minimal residual disease detection in B-cell malignancies by assessing IgH rearrangement. Hematol Cell Ther 39: 19–24
3. Schultze JL, Donovan JW, Gribben JG (1999) Minimal residual disease detection after myeloablative chemotherapy in chronic lymphytic leukemia. J Mol Med 77: 259–265
4. Zwicky C, Maddocks A, Andersen N, Gribben J (1996) Eradication of polymerase chain reaction detectable immunoglobulin gene rearrangement in non-Hodgkin's lymphoma is associated with decreased relapse after autologous bone marrow transplantation. Blood 88: 3314–22
5. Sambrook J, Fritsch EF, Maniatis T (1989) Molecular Cloning: A laboratory manual. Cold Spring Harbor Laboratory, Cold Spring Harbor NY
6. Kueppers R, Hansmann ML, Rajewsky K (1996) Micromanipulation and PCR analysis of single cells from tissue sections. In: Herzenberg LA, Weir D, Blackwell D (eds) Handbook of Experimental Immunology, 5th edn., Blackwell Science Ltd., Oxford 206.1-206.4.
7. Corradini P, Voena C, Astolfi M, Ladetto M, Tarella C, Boccadoro M, Pileri A (1995) High-dose sequential chemoradiotherapy in multiple myeloma: residual tumor cells are detectable in bone marrow and peripheral blood cell harvests and after autografting. Blood 85: 596–602

8. Boleda M, Briones P, Farres J, Tyfield L, Pi R (1996) Experimental design: a useful tool for PCR optimization. Biotechniques 211: 34–40
9. Wang Z, Spadoro J (1998) Determination of target copy number of quantitative standards used in PCR-based diagnosticassays. In Ferre F (ed) Gene Quantification. Birkhäuser, Boston 33–43
10. Jeffreys AJ, Wilson V, Neumann R, Keyte J (1988) Amplification of human minisatellites by the polymerase chain reaction: towards DNA fingerprinting of single cells. Nucleic Acids Res 16: 10953–71
11. Pfitzner T, Engert A, Wittor H, Schinköthe T, Oberhäuser F, Schulz H, Diehl V, Barth S (2000): A real-time PCR assay for the quantification of residual malignant cells in B-cell chronic lymphatic leukemia. Leukemia (in press)

Development of PCR-Based Assays for the Detection of Chromosomal Translocations Using SYBR Green I

SANDRA D. BOHLING, KOJO S. J. ELENITOBA-JOHNSON*

Introduction

The majority of human hematopoietic malignancies carry non-random chromosomal translocations. The deregulation of specific genes that occurs as a consequence of translocations has been shown to be central to the pathogenesis of such diseases. In addition, each translocation is often characteristic of a particular subtype of neoplasm, thus making chromosomal translocations valuable as diagnostic markers of disease.

Molecular detection of chromosomal translocations has traditionally been achieved using one of three techniques: Southern blot hybridization, the polymerase chain reaction (PCR) or Fluorescence In-Situ Hybridization (FISH). The LightCycler's integration of rapid PCR and fluorescence technology can be applied to the detection of chromosomal translocations using the double stranded DNA (dsDNA) binding dye SYBR Green I. When a dsDNA binding dye is included in a PCR, a cycle-dependent increase in fluorescence which correlates with product accumulation is observed. If product identity can be reliably established by a specific property of the PCR product such as its melting temperature (T_m), fluorescence monitoring by dsDNA binding dyes can effectively supplant conventional methods requiring postamplification analysis by gel electrophoresis.

The t(14;18) (q32;q21) chromosomal translocation is the most common specific recurrent karyotypic abnormality in non-Hodgkin's lymphomas, and is detectable in up to 95% of a histologic subtype known as follicular lymphoma [1]. This translocation juxtaposes the bcl-2 gene on band 18q21 to the immunoglobulin heavy chain gene joining region (JH) on 14q32, and results in overexpression of bcl-2 protein with consequent increased longevity of the follicle center cells. The majority of the chromosome 18 breakpoints (50–60%) occur within a 150 bp cluster on exon 3 of the bcl-2 gene known as the major breakpoint region (MBR) [1]. A significant proportion of the t(14;18) breakpoints (20%) do not fall within the MBR region of chromosome 18 [2]. These breakpoints occur at a second site distant from the bcl-2 gene known as the minor cluster region (MCR). Most breakpoints within the MCR tend to tightly

---

* Kojo S. J. Elenitoba-Johnson (✉) (e-mail: kojo.elenitobaj@path.utah.edu)
  University of Utah Medical Center, Department of Pathology, 50 N. Medical Drive, Salt Lake City, UT 84132, USA

cluster within 500 base pairs of each other, thus making them amenable to PCR. The LightCycler technology can be used for the detection of various translocations including the t(14;18) MBR and MCR breakpoints. Using the MBR/JH fusion as a model, we describe a strategy to develop PCR-based assays for the detection of specific recurrent chromosomal translocations. This method addresses the detection of specific translocations only. In several instances, the quantification of these fusions is desirable and can be accomplished by implementing the use of known standards and monitoring cycle thresholds [3, 4].

Materials

Equipment LightCycler instrument (Roche Diagnostics, Mannheim, Germany)
LightCycler capillary tubes (Roche Diagnostics, Mannheim, Germany)

Reagents Oligonucleotide primers (Operon Technologies, Alameda, CA)
SYBR Green I (Molecular Probes, Eugene, OR)
Taq DNA Polymerase (Promega, Madison, WI)
TaqStart Antibody (ClonTech, Palo Alto, CA)
Deoxynucleotide triphosphates (SIGMA, St. Louis, MO)
PCR buffer (50 mM Tris pH=8.3 (25°C), 30 mM $MgCl_2$, 2.5 mg/ml bovine serum albumin [BSA])
Enzyme Diluent (10 mM Tris, pH 8.3, 250 µg/ml BSA)

Procedure

Primer Design The LightCycler can be used to detect various chromosomal translocations using PCR and RT-PCR-based formats (Table 1). To detect chromosomal translocations via PCR, oligonucleotide primers are designed complementary to the DNA sequences flanking the breakpoints of the chromosomes participating in the translocation. In DNA samples containing the translocation, an amplification product will be detected, whereas no product is obtained from the DNA of normal tissue in which the sequences complementary to the primers remain on separate chromosomes. The chromosomal breakpoints are not identical from sample to sample of the same tumor type, but the majority of the breakpoints are "clustered" within a fairly defined region or regions.

Conventional PCR is most consistently performed if the DNA target ranges from 100–1000 base pairs (bp). For rapid-cycle PCR applications, we recommend that the amplification products be 500 bp or less if possible. In certain instances, a simple amplification across the translocation breakpoint, as occurs with the t(14;18), will not be feasible due to the widely dispersed locations of the translocation breakpoints in the genomic DNA (i.e., spanning a distance of 200 kb or more). However, RNA splicing frequently results in the precise juxtaposition of exons such that the chimeric transcript is smaller in size and amenable to reverse

Table 1. Chromosomal Translocations Characteristic of some Human Neoplasms

Clinical Disease	Translocation(s)	% Disease with Translocation Detectable by PCR/RT-PCR	Method of Detection
Acute myelogenous leukemia M2	t(8;21)(q22;q22)	40%	RT-PCR
Acute myelogenous leukemia M3	t(15;17)(q22;q12–21)	95%	RT-PCR
Acute myelogenous leukemia M4εo	Inv (16)(p13;q22)	~100%	RT-PCR
B-precursor acute lymphoblastic leukemia	t(9;22)(q32;q11), t(1;19)(q23;p13)	25%, 5%	RT-PCR
Chronic myelogenous leukemia	t(9;22)(q34;q11)	95%	RT-PCR
Follicular lymphoma	t(14;18)(q32;q21)	85–95%	PCR
Mantle cell lymphoma	t(11;14)(q13;q32)	40%	PCR
Diffuse large B-cell lymphoma	t(14;18)(q32;q21)	20–30%	PCR
Marginal zone lymphoma	t(11;18)(q21;q21)	10–20%	RT-PCR
Anaplastic large cell lymphoma	t(2;5)(p23;q35)	30–65%	RT-PCR
Ewing's sarcoma/PNET	t(11;22)(q24;q12), t(21;22)(q22;q12)	85%, 7%	RT-PCR
Alveolar rhabdomyosarcoma	t(2;13)(q35;q14), t(1;13)(p36;q14)	35%, 12%	RT-PCR
Synovial sarcoma	t(X;18)(p11.2,q11.2)	95%	RT-PCR
Desmoplastic small round cell tumor	t(11;22)(p13;q12)	90%	RT-PCR
Myxoid liposarcoma	t(12;16)(q13;p11)	80–90%	RT-PCR

PNET, primitive neuroectodermal tumor; RT-PCR, reverse transcription-polymerase chain reaction; t, translocation; Inv, inversion

transcription PCR (RT-PCR). RT-PCR entails initial RNA isolation, reverse transcription of the RNA to cDNA and then subsequent PCR using the cDNA as a template. Since the breakpoints on MBR are clustered within a confined region, the MBR/JH fusion is detectable via a DNA-based PCR assay.

Sample Preparation

DNA was isolated from a total of 30 samples consisting of 7 peripheral blood, 6 bone marrow, 8 fresh/frozen tissues and 9 fixed paraffin-embedded tissue samples using standard procedures [5, 6]. DNA quantity and quality was assessed by spectrophotometric analysis. All samples were diluted in TE buffer (10 mM Tris, 0.1 mM EDTA, pH=8.0) to a concentratin of 50 ng/μl prior to PCR amplification.

LightCycler PCR

For detection of the MBR/JH fusion product, we utilized an oligonucleotide primer flanking the MBR and a consensus oligonucleotide primer complementary to the IgH joining region as previously described [7]. PCR amplification was also performed under identical conditions with primers specific to the human β-globin gene to confirm the amplifiability of the DNA sample (Table 2).

Table 2. Oligonucleotides

Sequence	Genbank Accession #	Position	Length	GC (%)	T_m (°C)
MBR/JH Oligonucleotides					
GAG TTG CTT TAC GTG GCC TG	M14745	2997	20	55	62.4
ACC TGA GGA GAC GGT GAC C	X60830	63R	19	63	64.1
Product Size=120–500 bp					
β-globin Oligonucleotides					
GGC TTC CTA GAG ACC AAT CA	U01317	47693	20	50	59.4
AAC CAA GAC AGC CAG TTC AC	U01317	47799R	20	50	61.3
Product Size=107 bp					

The mastermix was made in sufficient quantity to amplify n+2 samples and contained the following:

Component	Volume [µl]	[Final]
10 ×PCR buffer (50 mM Tris, pH=8.3, 30 mM MgCl$_2$, 2.5 mg/ml BSA)	1	1×
SYBR Green I (diluted 1:3,000 in dH$_2$O)	1	1:30,000 dilution
Primers (5 µM each)	1 + 1	0.5 µM each
dNTP's (2 mM)	1	0.2 mM
Diluted Enzyme + TaqStart antibody*	1	0.4 U Taq and 8.8 ng per reaction TaqStart antibody
H$_2$O (Molecular Grade)	3	
Total volume	9	

* Diluted enzyme was prepared by adding 1 µl of Promega Taq (5 U/µl) to 1 µl of ClonTech TaqStart Antibody (110 ng/µl). This mixture was incubated at room temperature for 5 min prior to the addition of 10.5 µl enzyme diluent. This solution was then added to the mastermix.

In total, 9 µl of master mix and 1 µl of sample were added to each capillary. Capillaries were centrifuged in a microcentrifuge, sealed and placed into the LightCycler rotor. The following PCR protocol was used for amplification and detection using SYBR Green I:

- Denaturation at 95°C for 90 s
- Amplification

Parameter	Value		
Cycles	45		
Type	Quantification		
	Segment 1	Segment 2	Segment 3
Target temperature (°C)	94	68	74
Incubation time [s]	10	0	20
Temperature transition rate [°C/s]	20	20	3
Acquisiton mode	None	None	Single
Gains	F1=4		

- Melting Curve

Parameter	Value			
Cycles	1			
Type	Melting Curve			
	Segment 1	Segment 2	Segment 3	Segment 4
Target temperature (°C)	94	45	72	100
Incubation time [s]	10	20	0	0
Temperature transition rate [°C/s]	20	20	0.2	0.2
Acquisition mode	None	None	None	Step
Gains	F1=4			

Analysis of PCR Products

During initial optimization of cycling conditions, PCR products were analyzed using agarose gels to ensure correct product size. Samples amplified on the LightCycler were recovered from the capillaries by reverse centrifugation into microcentrifuge tubes, mixed with sucrose/bromphenol blue loading dye and loaded on 1.5% agarose gels in 1×TBE buffer containing ethidium bromide. Once the product size was verified (120–500 bp), the T_m of the products using SYBR Green I was used for product identification. In our study, the MBR/JH positive samples demonstrated melting temperatures of 88.85±1.15°C (n=19) on a graph plotting the negative derivative of fluorescence over temperature ($-d(F1)/dT$) versus temperature (T) (Fig. 1). No amplification product was detected in the negative samples using melting temperature or gel electrophoresis. Replicates (n=9) of the RL (ATCC # CRL-2261) MBR/JH positive cell line demonstrated little variation in melting temperature (standard deviation=0.057)

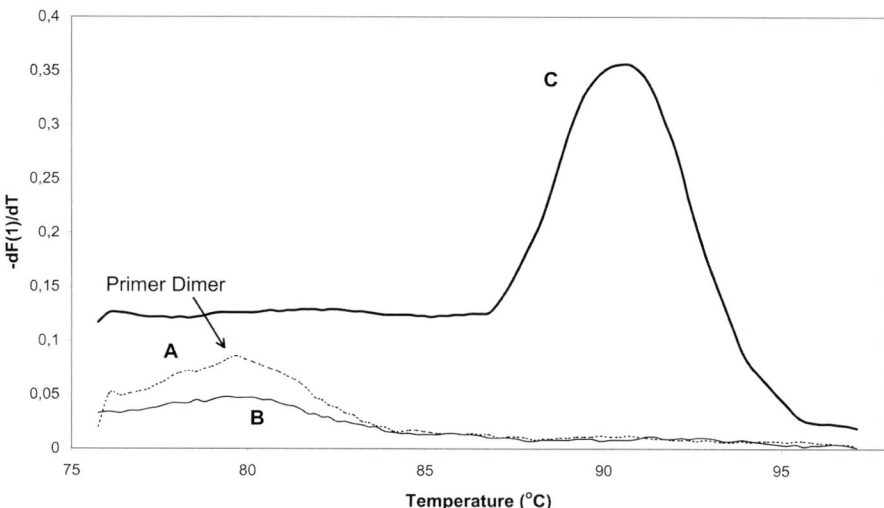

Fig. 1. Fluorescence melting curve for MBR/JH translocation product. *Curve A*, fluorescence profile of the template-free control (H$_2$O) (note the broad nonspecific melting peak from the primer dimer); *curve B*, melting curve obtained from amplification of human placental DNA; *curve C*, characteristic melting peak (90.5°C) obtained for a 256-bp JH/bcl-2 translocation product

Results

Product Detection

Because the chromosomal breakpoints involved in translocations do not occur at the same site in every instance, the sizes of the PCR products obtained will vary. In traditional PCR, this is apparent by the range of band sizes seen between positive samples when analyzed on an agarose gel. Using SYBR Green I and the LightCycler, the varying products result in a range of melting temperatures for positive samples (Fig. 2). In our application, this observation is accounted for by the fact that distinct breakpoints in the MBR of the bcl-2 locus are juxtaposed to JH region sequences, in each case yielding MBR/JH fusion products of different lengths and sequence compositions, which consequently exhibited slightly different melting temperatures.

Primer Dimer Formation

Because SYBR Green I binds indiscriminately to all double-stranded DNA species, it is necessary to determine the presence of a specific product in the amplification reaction. During melting, primer dimers are easily distinguished from authentic product by their low melting temperature and broad peaks (Fig. 1).

Sensitivity

Derivative melting curve analysis for the MBR/JH product allowed product identification at initial template concentrations as low as 0.05 ng in a 10-µl reaction (Fig. 3). This represents an increase in sensitivity over traditional PCR by up to two orders of magnitude [7, 9]. This increased sensitivity is not unique to the detection of the MBR/JH product. We have seen similar improvements in sensi-

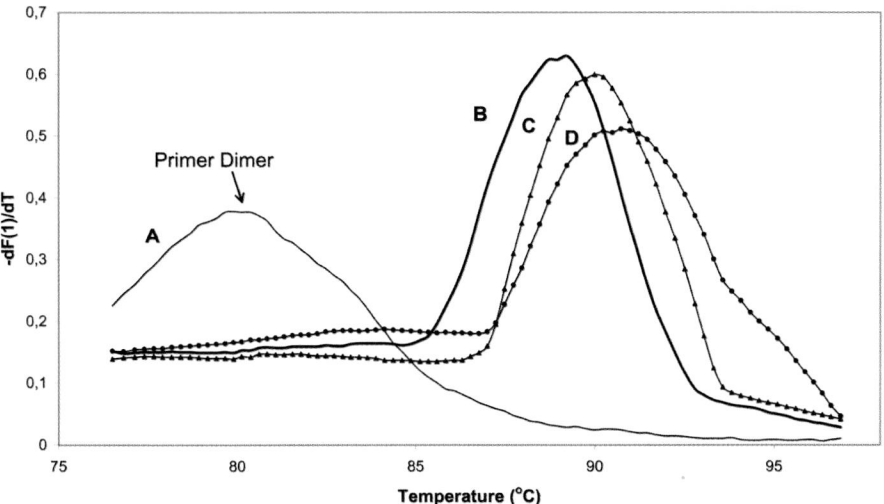

Fig. 2. Fluorescence melting peaks for different MBR/JH products showing slightly different melting temperatures (T_m). *Curve B*, T_m=89.3°C (184 bp); *curve C*, T_m=90.0°C (192 bp); *curve D*, T_m=90.5°C (207 bp). Product sizes were determined by DNA sequencing analysis. The fluorescence profile of the negative control (human placental DNA) is represented by *curve A* and shows only a primer dimer peak

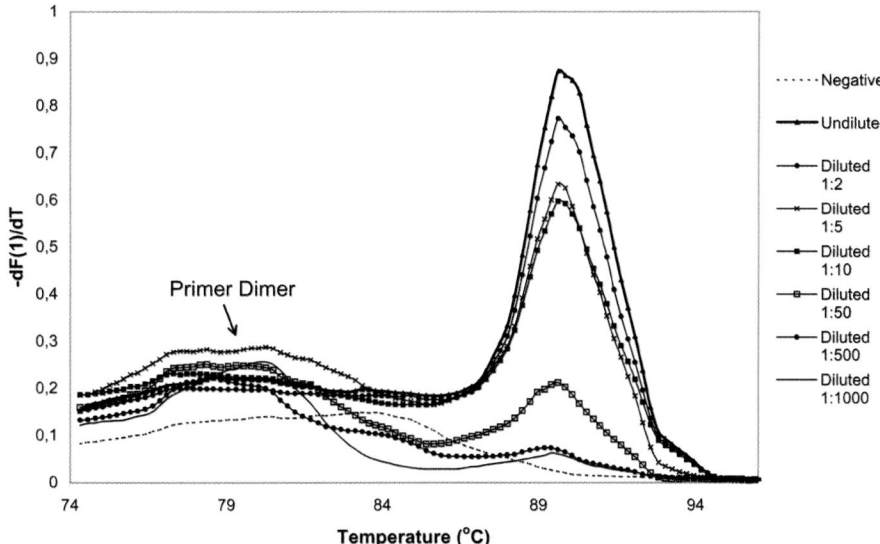

Fig. 3. Dilutional analysis by fluorescence melting peak analysis: fluorescence melting peaks are shown for serial dilutions of DNA extracted from the SUDHL-6 (Stanford University Diffuse Histiocytic Lymphoma-6) cell line using an initial template concentration of 50 ng per 10 μl PCR reaction. MBR/JH product melting at 90.5°C is evident at dilutions as low as 1:1000

tivity using other translocation targets [10]. We also performed dilutions of a MBR/JH positive cell line (RL, ATCC# CRL-2261) into placental DNA. Our assay detected 10% cell line DNA in the background of placental DNA.

Comments

Sample Preparation

Sample quantity and quality are of the utmost importance when utilizing the LightCycler. For optimal results, DNA should be used at a starting concentration of 50 ng/µl whenever possible. If samples are too concentrated or too dilute, a clear melting profile cannot be obtained. Inhibitors present in the sample can also influence the PCR, but can often be overcome by diluting the sample.

LightCycler PCR

Our particular target, the MBR/JH translocation, contains significant sequence homology with the Epstein Barr Virus (EBV) genome. Consequently, EBV positive samples could be misinterpreted as MBR/JH positive with some amplification assays [8]. To avoid amplification of an unwanted target, primer sequences should be designed to minimize mis-priming. BLAST homology searches can be used to identify potential mispriming sites. If some sequence homology cannot be avoided, other parameters can be altered to ensure specific amplification:
- Limit cross-hybridization to one primer such that mispriming can only generate linear amplification.
- Increase stringency of PCR by increasing annealing temperature and decreasing salt concentration.

In our MBR/JH assay, an annealing temperature of 68°C and an extension temperature of 74°C were employed to increase the stringency of the reaction [7].

When detecting translocations via the LightCycler, amplifiability of DNA must be verified to avoid scoring samples that failed to amplify as negative for the translocation. To reduce false negatives, we multiplexed our MBR/JH assay with a β-globin amplification. Both targets were amplified in the same reaction vessel using two separate primer sets. Products were differentiated using their melting temperatures (MBR/JH ~ 90°C, β-globin=81.2°C) (data not shown). If multiplexing is technically unsuccessful or undesirable, amplifiability of DNA must be ensured using a separate PCR directed at amplifying a house-keeping gene such as β-globin or GAPDH. In this application, β-globin PCR is used as an amplification control only. However, multiplexing of the target of interest and a second gene, such as a housekeeping gene, can be used for quantification of the target. Target amplification is normalized against the amplification of the housekeeping gene and is normally expressed as a ratio of cycle thresholds.

Sensitivity

At very low initial template concentrations, the sensitivity of fluorescence monitoring of PCR product accumulation by dsDNA binding dyes is limited by non-specific amplification, and this background fluorescence increases as excessive amplification cycles are performed. The specificity of dsDNA binding dyes is dependent on the inherent specificity of the primers and the adequacy of the optimization of the amplification reaction. Quantification of very low copy numbers of template by fluorescence analysis may thus be better achieved by using sequence-specific fluorescent probes that confer a greater specificity to the reaction. It is important to recognize that the t(14;18) has been detected in normal individuals in low copies [11]. It is thus imperative that clinical and pathologic data be integrated with molecular studies in the evaluation of the individual patient.

References

1. Tsujimoto Y, Cessman J, Jaffe E, Croce CM (1985) Involvement of the *bcl-2* gene in human follicular lymphoma. Science 228 (4706): 1440–1443
2. Ngan BY, Jamison N, Cleary ML (1989) Detection of chromosomal translocation t(14;18) within the minor cluster region of bcl-2 by polymerase chain reaction and direct genomic sequencing of the enzymatically amplified DNA in follicular lymphomas. Blood 73(7): 1759–1762
3. Luthra R, McBride JA, Cabanillas F, Sarris A (1998) Novel 5' exonuclease-based real-time PCR assay for the detection of t(14;18)(q32;q21) in patients with follicular lymphoma. Am J Pathol 153(1): 63–68
4. Olsson K, Gerard CJ, Zehnder J, Jones C, Ramanathan R, Reading C, Hanania EG (1999) Real-time t(11;14) and t(14;18) PCR assays provide sensitive and quantitative assessments of minimal residual disease (MRD). Leukemia 13(11) 1833–1842
5. Miller SA, Dykes DD, Polesky HF (1988) A simple salting out procedure for extracting DNA from human nucleated cells. Nucleic Acids Res 16(3): 1215
6. Forsthoefel KF, Papp AC, Snyder PJ, Prior TW (1992) Optimization of DNA extraction from formalin-fixed tissue and its clinical application in Duchenne muscular dystrophy. Am J Clin Pathol 98(1): 98–104
7. Bohling SD, King TC, Wittwer CT, Elenitoba-Johnson KSJ (1999) Rapid simultaneous amplification and detection of the MBR/JH chromosomal translocation by fluorescence melting curve analysis. Am J Pathol 154(1): 97–103
8. Bohling SD, Wittwer CT, King TC, Elenitoba-Johnson KSJ (1999) Fluorescence melting curve analysis for the detection of the bcl-1/JH translocation in mantle cell lymphoma. Lab Invest 79(3):337–345
9. Segal GH, Scott M, Jorgensen T, Braylan RC (1993) Primers frequently used for detecting the t(14;18) major breakpoint also amplify Epstein-Barr viral DNA. Diagn Mol Pathol 3(1): 15–21
10. Segal GH, Jorgensen T, Scott M, Braylan RC (1994) Optimal primer selection for clonality assessment by polymerase chain reaction analysis: II. Follicular Lymphomas. Hum Pathol 25(12): 1276–1282
11. Limpens J, de Jong D, van Krieken JH, Price CG, Young BD, van Ommen GJ, Kluin PM (1991) Bcl-2/JH rearrangements in benign lymphoid tissues with follicular hyperplasia. Oncogene 6(12): 2271–2276

Relative Quantification of the HER2/*neu* Oncogene Using SYBR Green I

Rachel Woods*

Introduction

The HER2/*neu* oncogene is amplified in 10%–34% of breast tumors [1]. HER2/*neu* gene amplification has been correlated to poor patient prognosis and resistance to conventional therapies [2]. Current methods for detection of HER2/*neu* amplification include immunohistochemistry (IHC), fluorescence in-situ hybridization (FISH), southern blotting, and quantitative polymerase chain reaction (PCR). Quantitative PCR can be performed easily and rapidly on the LightCycler, making it an excellent alternative to the other methods of quantification.

SYBR Green I is a dsDNA dye that can be used to detect amplification on the LightCycler [3]. Unlike sequence-specific probes, it is generic and can detect any PCR product. In addition, it is generally easier to optimize a primer set with SYBR Green I than with sequence-specific probes. It is a good choice for laboratories that routinely use a variety of primer sets, as it eliminates the need to design, synthesize, and optimize a separate probe set for each primer set.

There are a variety of methods for quantification using SYBR Green I on the LightCycler. A standard curve can be generated from DNA of known concentrations and used to quantify unknown samples by interpolation [4]. To compare the target gene to a housekeeping or reference gene, two separate standard curves can be generated, one for the target and another for the reference gene. Any number of targets can be quantified in this manner, and the amplification efficiencies of the different reactions do not need to be matched. However, a standard curve is needed for each target and reference gene.

For this study, a simpler, relative quantification method was performed, using SYBR Green I. Relative quantification compares a target of interest to a reference gene (presumed to be constant between all samples) instead of to a standard curve [5]. The position of an amplification curve along a fluorescence vs. cycle number plot contains quantitative information. The greater the concentration of starting template for a reaction, the earlier the curve's position will be along the cycle number axis. The position of the HER2/*neu* amplification curve was compared to the position of the reference gene (albumin) amplification curve. The position was described by examining the shape of the curve, specifically, the place where the acceleration of the curve was the greatest. This point is where the

---

* Rachel Woods (✉) (e-mail: rachel@idahotech.com)
 Idaho Technology, 390 Wakara Way, Salt Lake City, Utah 84108, USA

second derivative of the curve is at a maximum (Maximum Second Derivative). The difference between the Maximum Second Derivative for HER2/*neu* and albumin was calculated and used to determine the amount of HER2/*neu* relative to the amount of albumin. If there was no difference, the amounts of HER2/*neu* and albumin were the same, indicating that the HER2/*neu* gene was not amplified in that sample. If there was a difference, the fold increase of HER2/*neu* over albumin was calculated by raising the efficiency (E) of the reaction to the difference between the Maximum Second Derivatives for each reaction. Because reactions are compared directly, temperature cycling conditions of the amplifications were adjusted so that the efficiencies were matched. This relative quantification method was demonstrated on three cell lines: two with known HER2/*neu* gene amplifications, and one control cell line with no gene amplification.

Materials

Equipment LightCycler Instrument (Roche Diagnostics, Mannheim, Germany)
LightCycler Reaction Cuvettes (Roche Diagnostics, Mannheim, Germany)
Centrifuge (Roche Diagnostics, Mannheim, Germany)

Reagents 50 mM 10× PCR Buffer (500 mM Tris, pH 8.3; 5 mg/ml BSA; 50 mM $MgCl_2$)
20 mM 10× PCR Buffer (500 mM Tris, pH 8.3; 5 mg/ml BSA; 20 mM $MgCl_2$)
dNTPs (Roche Diagnostics, Mannheim, Germany)
Amplification primers (IT Biochem, Salt Lake City, UT)
Taq DNA Polymerase (Roche Diagnostics, Mannheim, Germany)
TaqStart Antibody (Clontech Laboratories, Palo Alto, CA)
SYBR Green I (Molecular Probes, Eugene, OR)
PureGene DNA Isolation Kit (Gentra Systems, Minneapolis, MN)
LSI HER2/*neu* SpectrumOrange/CEP 17 SpectrumGreen DNA Probe kit (Vysis Inc. Downers Grove, IL)

Procedure

Sample Preparation DNA was extracted from three cell lines: SKBR3 (ATCC number HTB-30), T47D (ATCC number HTB-133), and MRC5 (ATCC number CCL-171), using the Puregene DNA Isolation kit according to the manufacturers instructions. After extraction, the DNA was boiled for 10 min, and absorbance readings at 260 nm were taken. The DNA was diluted to 50 ng/μl for standardization between samples.

Primer Design Oligonucleotides were designed from the published sequences for the reference gene albumin (Table 1). The HER2/*neu* primer sequences (Table 1) were derived from Sestini et al. [7], but the A in position 16 was listed incorrectly as an A by Sestini and was changed to a G when synthesized for this project.

Table 1. Oligonucleotides

HER2/neu (Genebank Accession #M11730)				
	Position	Length	GC (%)	T_m (°C)
CCTCTGACGTCCATCGTCTC	2098	20	60	67.5
CGGATCTTCTGCTGCCGTCG	2198 R	20	65	71.4
Product	2098–2198	100		
Albumin (Genebank Accession #M12523)				
CCGTGGTCCTGAACCAGTTA	15538	20	55	67.4
GTCGCCTGTTCACCAAGGAT	15633 R	20	55	68.4
Product	15538–15633	96		

Master Mix Preparation

Master Mix for each 10 µl reaction:

	Volume [µl]	[Final]
PCR Buffer (10×)	1	1×
Taq Polymerase (0.4 Units/µl)	1	0.04 Units/µl
SYBR Green I (1:3,000)	1	1:30,000
Primers (5 µM each)	1	0.5 µM
H_2O (PCR Grade)	5	
Total Volume	9	

If a hot start is desired, add 2 µl of concentrated TaqStart Antibody to 2 µl of concentrated Taq polymerase and incubate at room temperature for 5 min. Then dilute mixture to a final Taq polymerase concentration of 0.4 units/µl.

Master mixes were prepared for the HER2/neu (using 50 mM 10X PCR Buffer) and albumin (using 20 mM 10X PCR Buffer) primer sets. 9 µl of master mix and 1 µl of DNA were added to each capillary. Sealed capillaries were centrifuged and placed into the LightCycler rotor.

Amplification on the LightCycler

The following amplification protocol was used for both primer sets:

Parameter	Value			
Cycles	40			
Type	Quantification			
	Segment 1	Segment 2	Segment 3	Segment 4
Target temperature (°C)	94	64	74	79
Incubation time [s]	0	0	4	5
Temperature transition rate [°C/s]	20	20	2	5
Acquisition mode	None	None	None	Single
Gains	F1=5			

Product Confirmation on the LightCycler

The temperature where a PCR product denatures (melts) is characteristic of both its length and GC content. This melting temperature can be used to determine if a specific product has been made. After amplification, the melting temperature of the product is determined by slowly raising the temperature of the sample while monitoring fluorescence continuously. The following protocol was used for both primer sets:

Parameter	Value		
Cycles	1		
Type	Quantification		
	Segment 1	Segment 2	Segment 3
Target temperature [°C]	94	50	97
Incubation time [s]	0	30	0
Temperature transition rate [°C/s]	20	20	0.2
Acquisition mode	None	None	Step
Gains	F1=5		

Data Analysis

Background subtraction and determination of the position of the amplification curves was performed using the LightCycler 3.0 analysis software (Roche Diagnostics, Mannheim). The Maximum Second Derivative for HER2/*neu* was subtracted from the Maximum Second Derivative for albumin to determine the cycle shift (Δn) between the curves. The relative copy number was calculated by raising the efficiency (E) to the power Δn. The efficiency was determined by non-linear fitting of the log linear portion of the curve, defined by the equation $F=C*E^n$, where F is fluorescence, C is a proportionality constant, E is the efficiency of the reaction, and n is the cycle number.

FISH

Chromosome preparations were performed on the cell lines as outlined in the ACT Cytogenetics Laboratory Manual [9]. The FISH was performed according to manufacturers instructions for the LSI HER2/*neu* SpectrumOrange/CEP 17 SpectrumGreen DNA Probe kit.

Results

Two cell lines, SKBR3 and T47D were chosen as a model for relative quantification of HER2/*neu*. SKBR3 is a cell line that is highly amplified for HER2/*neu*, while T47D is a more moderately amplified cell line. DNA from the cell lines was split between two tubes, one containing master mix with HER2/*neu* primers, and the other containing master mix with the reference gene (albumin) primers. Reactions containing the two primer sets were amplified side-by-side using the same protocol on the LightCycler. Figure 1 shows representative amplification curves for SKBR3. MRC5, a cell line with no HER2/*neu* amplification was used as a control (Fig. 2).

Table 2 compares the HER2/*neu* copy numbers from relative quantification on the LightCycler to copy numbers we observed by FISH. The copy numbers that we

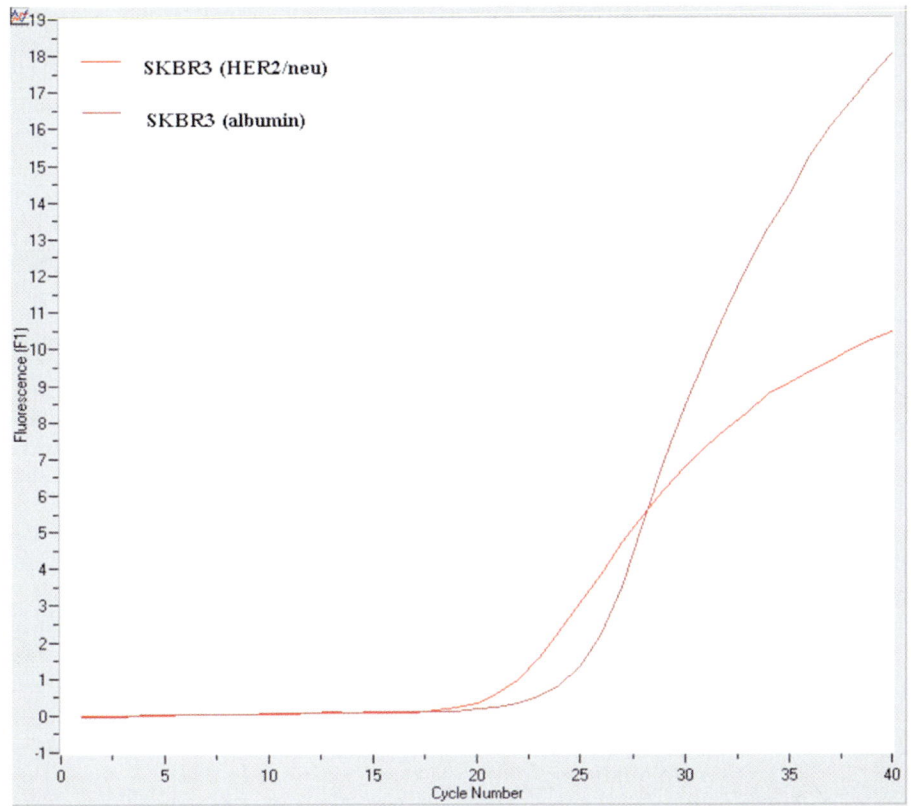

Fig. 1. Quantification of SKBR3, a cell line with known HER2/*neu* gene amplification. The Maximum Second Derivative of the amplification curve for the reaction with HER2/*neu* primers (*light brown line*) is earlier than the Maximum Second Derivative of the amplification curve for albumin primers (*dark brown line*). This is true even though the final fluorescence for the albumin primers is higher than that of the HER2/*neu* primers. The Maximum Second Derivative for the HER2/*neu* reaction for the SKBR3 cell line is 3.8 cycles earlier than the Maximum Second Derivative for the albumin reaction, meaning that the HER2/*neu* gene is amplified in this cell line

determined for the cell lines by FISH are similar to those reported by other investigators [6–8]. The relative quantification copy numbers can be seen to closely match those observed by FISH.

Troubleshooting

The accuracy of relative quantification depends on how well the efficiencies of the target and reference gene are matched. For these experiments, efficiencies were matched by comparing the HER2/*neu* and albumin Maximum Second

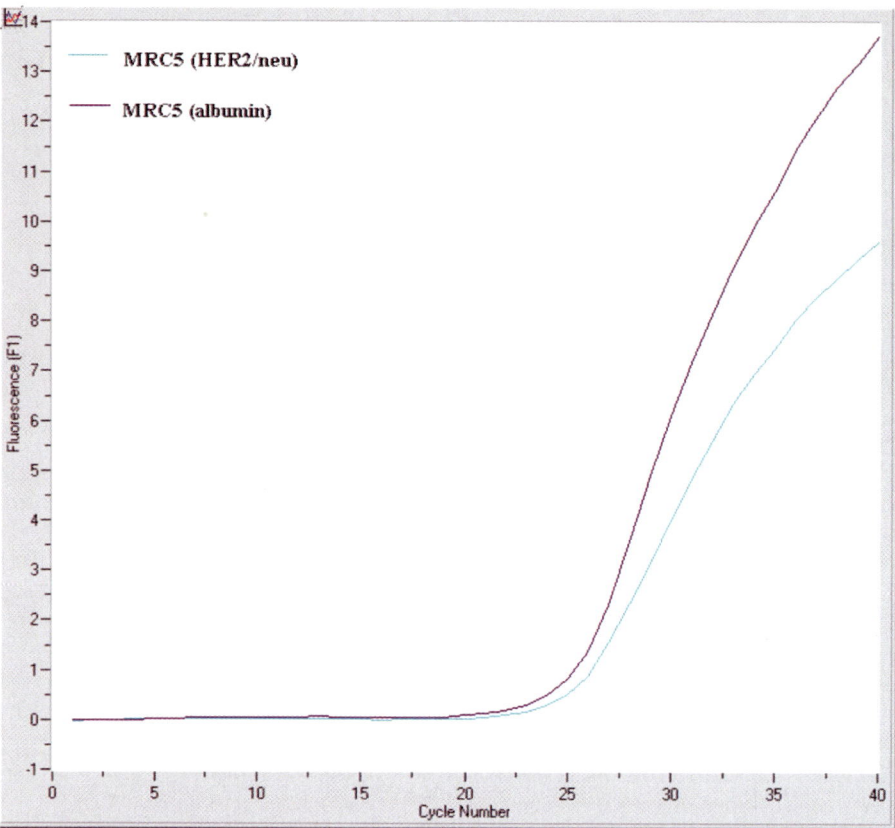

Fig. 2. Quantification of MRC5, a control cell line with no HER2/*neu* gene amplification. The Maximum Second Derivative for MRC5 with HER2/*neu* primers (*green line*) is nearly equivalent (0.2 cycles earlier) to the Maximum Second Derivative for MRC5 with albumin primers (*purple line*), meaning that there is no HER2/*neu* gene amplification in this cell line. The Maximum Second Derivative assesses maximum acceleration, so, as in this case, two amplification curves may have equivalent Maximum Second Derivatives even though the two curves are not directly on top of one another

Derivatives for MRC5, the control cell line. The annealing temperature was varied from 55°C to 64°C, and the $MgCl_2$ concentration was varied between 2 and 5 mM.

Comments

Sample Preparation

Following extraction, the genomic DNA in this study was boiled in a water bath for 10 min. This is a one-time denaturation step that is very important for reducing background fluorescence in SYBR Green I reactions on the LightCycler.

Table 2. Comparison of copy numbers obtained by relative quantification and FISH. The cycle shift (Δn) and the efficiency (E) are listed for each cell line. The relative quantification copy number was calculated by raising E to the power Δn. The copy numbers determined for the cell lines by FISH are given in the final column for comparison.

Cell line	Cycle Shift (Δn)	Efficiency (E)[1]	Relative quantification copy number[2]	Copy number by FISH
SKBR3	3.8	1.7	7.5	8.0
T47D	2.0	1.7	2.9	2.6
MRC5	0.2	1.7	1.1	1.0

[1] Average of four amplification curves
[2] Calculated by raising efficiency (E) to the power Δn

Efficiency Calculations

For these experiments, the efficiencies of the reactions were calculated by curve fitting. This method generates an efficiency value from a single curve by non-linear fitting of the log linear change in fluorescence. Although this reduces the number of samples that need to be run, the method may be limited in that it calculates efficiency from only a small window of amplification (3–4 cycles) and assumes that efficiency is not changing during the course of the reaction. An alternative would be to use a dilutional series that covers a range of starting copy number concentrations [10]. Nevertheless, the method used gave results in good agreement with FISH data. The log linear portions of the curves were estimated by eye and efficiencies were calculated in DeltaGraph.

Relative Quantification vs Standard Curve Quantification

Relative quantification requires more optimization work initially than standard curve quantifications do, because reactions should have similar efficiencies for relative quantification. However, relative quantifications do not require the preparation and purification of PCR product to be used as standards. Furthermore, relative quantification allows two PCR reactions to be compared directly, eliminating potential problems from comparing amplification of purified PCR product with amplification of genomic DNA. Amplification from genomic DNA introduces a variety of alternative primer binding sites which may alter the efficiency of the reaction.

Applications

The method described here can be applied to a variety of systems. It is amenable to monitoring with SYBR Green I, Hybridization probes, or TaqMan probes. The method is applicable to any system where quantification can be expressed in relative amounts. PCR reactions for the target of interest and the reference or housekeeping gene need only to be similar in efficiency. Relative quantification is ideal for researchers who wish to quantify large numbers of samples on the LightCycler. Since no standard curve is needed, every rotor position can be filled with clinical or experimental samples for maximum throughput.

References

1. Ross JS, Fletcher JA (1998) The HER-2/neu Oncogene in Breast Cancer: Prognostic Factor, Predictive Factor, and Target for Therapy. Stem Cells 16: 413–428
2. Slamon DJ, Clark GM, Wong SG, Levin WJ, Ullrich A, McGuire WL (1987) Human breast cancer: correlation of relapse and survival with amplification of the HER2/*neu* oncogene. Science 235: 177–182
3. Wittwer CT, Herrmann MG, Moss AA, Rasmussen RP (1997) Continuous fluorescence monitoring of rapid cycle DNA amplification. Biotechniques 22: 130–138
4. Wittwer C, Ririe K, Rasmussen R (1997) Fluorescence monitoring of rapid cycle PCR for quantification. In: Ferre F (ed) Gene Quantification. Birkhauser, New York
5. Boulay J, Reuter J, Ritschard R, Terracciano L, Herrmann R, Rochlitz C (1999) Gene dosage by quantitative real-time PCR. Biotechniques 27: 228–232
6. Hynes NE, Gerber HA, Saurer S, Groner B (1989) Overexpression of the c-erbB-2 protein in human breast tumor cell lines. J Cell Biochem 39: 167–173
7. Dati C, Antoniotti S, Taverna D, Perroteau I, De Bortoli M (1990) Inhibition of c-erbB-2 oncogene expression by estrogens in human breast cancer cells. Oncogene 5: 1001–1006
8. Sestini R, Orlando C, Zentilin L, Lami D, Gelmini S, Pinzani P, Giacca M, Pazzagli M (1995) Gene amplification for c-*erb*B-2, c-*myc*, epidermal growth factor receptor, *int*-2, and N-*myc* measured by quantitative PCR with a multiple competitor template. Clin Chem 41: 826–832
9. Barch MJ (ed) (1991) The ACT Cytogenetics Laboratory Manual, p 126 Raven Press, New York
10. Fink L, Seeger W, Ermert L, Hanze J, Stahl U, Grimminger F, Kummer W, Bohle R (1998) Real-time quantitative RT-PCR after laser assisted cell picking. Nat Med 4: 1329–1333

Receptors and Mediators

Development of Quantitative RT-PCR Tests for the Expression of Cytokine Genes on the LightCycler 251
Thomas Giese

Quantitative RT-PCR for the Detection of T Cell Receptor Transcripts in T Lymphocytes Populations Using LightCycler Technology 263
E. Jouvin-Marche, I. Vigan, V. Leroy, P. N. Marche

Rapid, Homogeneous Genotyping of Human Platelet Antigen 1 by Fluorescence Resonance Energy Transfer and Probe Melting Curves 273
Markus S. Nauck, Hedi Gierens, Matthias A. Nauck, Winfried März, Heinrich Wieland

Development and Validation of an Externally Standardised Quantitative Insulin-like Growth Factor-1 RT-PCR Using LightCycler SYBR Green I Technology .. 281
Michael Pfaffl

An Application of Melting Curve Analysis to Large-Scale Genetic Analysis in Atherosclerotic Disease: Two Linked Polymorphisms of Glycoprotein Ia Gene and Myocardial Infarction in Japanese 293
H. Morita, H. Kurihara, Y. Yazaki and R. Nagai

Development of Quantitative RT-PCR Tests for the Expression of Cytokine Genes on the LightCycler

Thomas Giese*

Introduction

Cytokines and chemokines are a major focus in biomedical research because of their numerous functions and disease associations. Direct evaluation of these mediators as proteins by enzyme immunoassay, immunohistochemistry or flow cytometry requires specific antibodies and assay sensitivity is variable. Since cytokine gene expression (mRNA) is usually quantitatively correlated to the level of produced protein, RT-PCR is widely used as an alternative method to describe cytokine profiles. Quantification of mRNA does not require the accumulation of measurable amounts of protein and hence is a more sensitive and dynamic monitor of gene expression.

Although highly sensitive, conventional RT-PCR is labor intensive and difficult to use for quantification. Many semi-quantitative RT-PCR methods that use external standards and competitors are available [1]. However, all require considerable time and effort. Amplification is usually followed by manual gel analysis and band intensities are compared visually or by densitometry.

Many of the difficulties of conventional semi-quantitative RT-PCR are overcome with the newly available LightCycler system. Continuous detection of accumulating PCR product each cycle allows rapid and reliable quantification of gene expression. We were interested in developing RT-PCR systems to estimate the expression level of cytokine genes. Methods for the T-lymphocyte growth factors interleukin-2 (IL-2) and interleukin-4 (IL-4) are detailed here.

Materials

Equipment

LightCycler Instrument (Roche Diagnostics, Mannheim, Germany)
Oligo Primer analysis software (6.3 for Macintosh, MBI, Cascade,CO)
Fluorescence imager FLA-2000 (Fuji Photo Film CO., LTD. Tokyo, Japan)

Kits

RNA isolation kit (High-Pure total RNA, Roche Diagnostics)
First Strand cDNA synthesis kit for RT-PCR (AMV, Roche Diagnostics)
LightCycler-DNA Master SYBR Green I (Roche Diagnostics)
LightCycler-DNA Master Hybridization Probes (Roche Diagnostics)

---

* T. Giese (✉) (e-mail: Thomas.Giese@urz.uni-heidelberg.de)
  Institute of Immunology, School of Medicine, University of Heidelberg
  Im Neuenheimer Feld 305, 69120 Heidelberg, Germany

Reagents

Histopaque1077 (Sigma, Deisenhofen, Germany),
Phorbol 12-myristate 13-acetate (PMA, Sigma),
Ionomycin (Sigma)
Fetal calf serum (Sigma)
RPMI 1640 (Life Technologies, Karlsruhe, Germany)
Amplification Primers (Life Technologies)
Hybridization Probes (TIB MOLBIOL, Berlin, Germany)

Procedure

Sample Preparation

Mononuclear cells (MNC) were isolated on a Histopaque1077 density gradient using Leuco Sep tubes (Greiner Labortechnik, Frickenhausen, Germany) and 2×10^6 cells were resuspended in RPMI 1640 with 10% FCS. Cultures were stimulated with 10 ng/ml PMA and 0.5 µg/ml ionomycin for various time points at 37°C in 7% CO_2. Cells were harvested, resuspended in 200 µl PBS and 400 µl of High-Pure lysis solution was added. Resulting lysates were stored at −70°C. After thawing at 37°C for 10 min, RNA was extracted (RNA isolation kit) and eluted from the spin column in a volume of 50 µl. An aliquot of 8.2 µl RNA was reverse transcribed using AMV-RT and oligo-(dT) as primer (cDNA synthesis kit). As a control, a reaction was performed without reverse transcriptase (no-RT control). After termination of cDNA synthesis the reaction mix was diluted to a final volume of 500 µl and stored at −20°C until PCR analysis.

Primer Design

For both IL-2 and IL-4, three sets of primers and a single probe pair were used (Table 1). The first primer set (Set1) in each case consisted of previously published primers [2]. Sets 2 and 3 were selected with the aid of Oligo primer analysis software. The search stringency was set to, "very high", with length adjustment set to, "match priming efficiency points (P.E.# s)".

For IL-4, Set 1 binds to exons 1 and 4, and will amplify the IL-4 -Δ2 splicing variant [3]. In contrast, Sets 2 and 3 use a primer that binds to exon 2 and therefor will not amplify this deletion variant. Hybridization probes were designed as published [4]. The primer and probe sequences and characteristics are shown in Table 1.

LC-PCR

SYBR Green I Master Mix for each 20 µl reaction:

	Volume [µl]	[Final]
LightCycler-DNA Master SYBR Green I	2	1 x
$MgCl_2$ (25 mM)	1.2	2.5 mM
Primers (15µM each)	0.8	0.6 µM
H_2O (PCR grade)	6	
Total volume	10	

Table 1. Oligonucleotides

Interleukin-2 (Genebank Accession # E 00210)					
	Position	Exon	Length	GC (%)	T_m (°C)
Set 1					
ACAGCTACAACTGGAGCATT	136	1	20	45	59.8
TGCTGTCTCATCAGCATATT	442 R	4	20	40	57.2
Product	136-442		307		
Set 2					
CACAGCTACAACTGGAGCATTA	135	1	23	43.5	61.0
AGAAATTCTACAATGGTTGCTGTC	459 R	4	24	37.5	59.6
Product	135-459		325		
Set 3					
TAATTACAAGAATCCCAAACTCAC	193	1	24	33.3	56.9
TGATGCTTTGACAAAAGGTAATCC	491 R	4	24	37.5	59.8
Product	193-491		299		
Hybridization Probes					
ACCAGGATGCTCACATTTAAGTTTT-ACAT-F	215	2	29	34.5	61.5
LCRed640-CCCAAGAAGGCCACAGAA-CTGAAACATCT-P	245	2/3	29	48.3	66.6
Interleukin-4 (Genebank Accession # M 13982)					
	Position	Exon	Length	GC (%)	T_m (°C)
Set 1					
GCGATATCACCTTACAGGAG	143	1	20	50.0	57.8
TTGGCTTCCTTCACAGGACA	430 R	4	20	50.0	61.9
Product	143-430		307		
Set 2					
AGTTGACCGTAACAGACATCTTTGC	212	2	25	44.0	63.4
ACTCATAAATTAAAATATTCAGCTCG	518 R	4	26	26.9	56.1
Product	212-518		332		
Set 3					
AGAAGACTCTGTGCACCGAGTTGA	194	1/2	24	50.0	65.7
CTCTCATGATCGTCTTTAGCCTTT	476 R	4	24	41.7	60.5
Product	194-476		306		
Hybridization Probes					
ACAGGCACAAGCAGCTGATCCGA-TTCCT-F	356	3	28	53.6	69.7
LCRed640-AAACGGCTCGACAGGAA-CCTCTGGGGCC-P	385	3	28	64.3	73.1

Hybridization Probe Master Mix for each 20 μl reaction:

	Volume [μl]	[Final]
LightCycler-DNA Master Hybridization Probes	2	1 x
MgCl$_2$ (25 mM)	1.2	2.5 mM
Primers (15μM each)	0.8	0.6 μM
Probes (2 μM each)	2+2	0.2 μM
H$_2$O (PCR grade)	2	
Total volume	10	

If a hot start is desired, add 0.16 μl/reaction concentrated TaqStart antibody (Clontech, Palo Alto, CA) directly to the 10 x DNA Master solutions and incubate at room temperature for 5 min. Then add MgCl$_2$, primers, (probes), and water (less 0.16 μl/reaction).

10 μl of master mix and 10 μl of cDNA template were added to each capillary. Sealed capillaries were centrifuged in a microcentrifuge and placed into the LightCycler rotor.

The following PCR protocol was used for the SYBR Green I detection:

- Denaturation at 95°C for 2 min
- Amplification

Parameter	Value		
Cycles	40		
Type	Quantification		
	Segment 1	Segment 2	Segment 3
Target temperature [°C]	95	68	72
Incubation time [s]	1	10	16
Temperature transition rate [°C/s]	20	20	20
Secondary target temperature [°C]	0	58	0
Step size	0	0.5	0
Step delay	0	1	0
Acquisition mode	None	None	Single
Gains		F1 = 5	

- Melting Curve Analysis

Parameter	Value		
Cycles	1		
Type	Melting Curve		
	Segment 1	Segment 2	Segment 3
Target temperature [°C]	95	58	95
Incubation time [s]	0	10	0
Temperature transition rate [°C/s]	20	20	0.1
Acquisition mode	None	None	Cont.

For hybridization probes, the amplification protocol was modified as follows:

- Amplification

Parameter	Value		
Cycles	50		
Type	Quantification		
	Segment 1	Segment 2	Segment 3
Target temperature [°C]	95	68	72
Incubation time [s]	1	15	16
Temperature transition rate [°C/s]	20	20	20
Secondary target temperature [°C]	0	58	0
Step size	0	0.5	0
Step delay	0	1	0
Acquisition mode	None	Single	None
Gains		F1 = 1 ; F2 = 10	

Gel Analysis of PCR Product

PCR products amplified in the SYBR Green I format were recovered from the capillaries by reverse centrifugation into microtiter V-bottomed plates. Samples were diluted in gel loading buffer, heated to 65°C for 5 min and loaded on 1.5% Agarose gels in 0.5 x TBE buffer containing SYBR Green II at a dilution of 1:20,000. Gels were analyzed on a fluorescence imager.

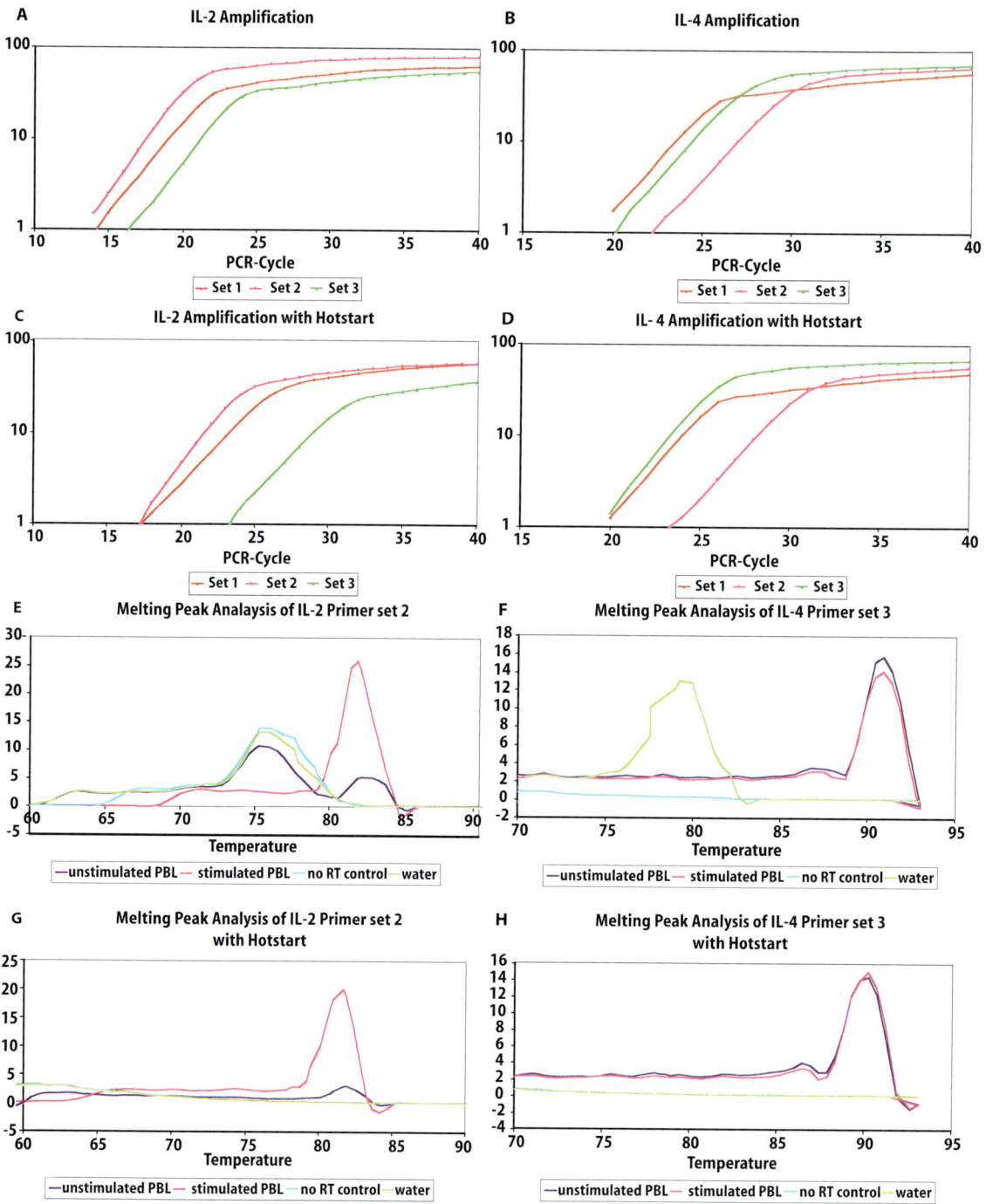

Results

Although computer programs can eliminate some poor primer pairs, finding primers with optimal specificity remains mostly trial and error. Besides single product specificity, amplification sensitivity is important. If amplification sensitivity for different products is equal, a single external standard curve can be used for quantification of multiple parameters. In real-time PCR, sensitivity depends on the reaction efficiency (slope) and the plateau fluorescence. The greater the efficiency, the greater the slope and the more the curve shifts to the left on log fluorescence vs cycle number plots (Fig 1).

With IL-2 amplification, primer Set 2 had the highest efficiency and plateau. Primer Set 3 demonstrated poor efficiency compared to the other sets. With IL-4, primer Set 3 was selected. Primer Set 1 co-amplifies a truncated form of IL-4 missing exon 2 [3] and is shown in Figure 2. Primer Set 2 showed relatively low amplification efficiency.

The effect of using a hot start antibody is evident in Figures 1 and 2. Without the antibody, low melting, non-specific products are often formed that appear to be primer dimers on gel analysis. The low melting products and dimers are most often present when the template copy number is low. These undesired products shift the amplification curve to the left and can give false positive results when SYBR Green I is used, especially if melting curve analysis is not performed. In almost all cases, the formation of these undesired products can be inhibited by using the hot start antibody.

One way to assess the linearity and accuracy of a quantification method is by titration as shown in Table 2 and Figure 3. Using primer Set 2 for IL-2 and primer Set 3 for IL-4, cDNA from PMA/ionomycin stimulated cells was diluted with the no RT control and amplified. All amplification curves showed approximately the same slope. As expected, samples that differed in concentration by a factor of 5 were separated by a little more than 2 cycles. Table 2 quantifies the deviation of the measured value from the expected value. Deviations are generally less than 20 %, with hot start runs more accurate than in the absence of antibody, particularly when SYBR Green I is used as an indicator.

Fig. 1. Selection of optimal primer pairs. cDNA from cells cultured alone, or stimulated with PMA/ionomycin for 3h, no-RT control and water were amplified in the LightCycler using the SYBR Green I format. Three primer pairs were compared for their amplification without (A,B) and with (C,D) TaqStart antibody. Melting curve analysis for primer set 2 (IL-2) and 3 (IL-4) is shown without (E,F) and with (G,H) the antibody

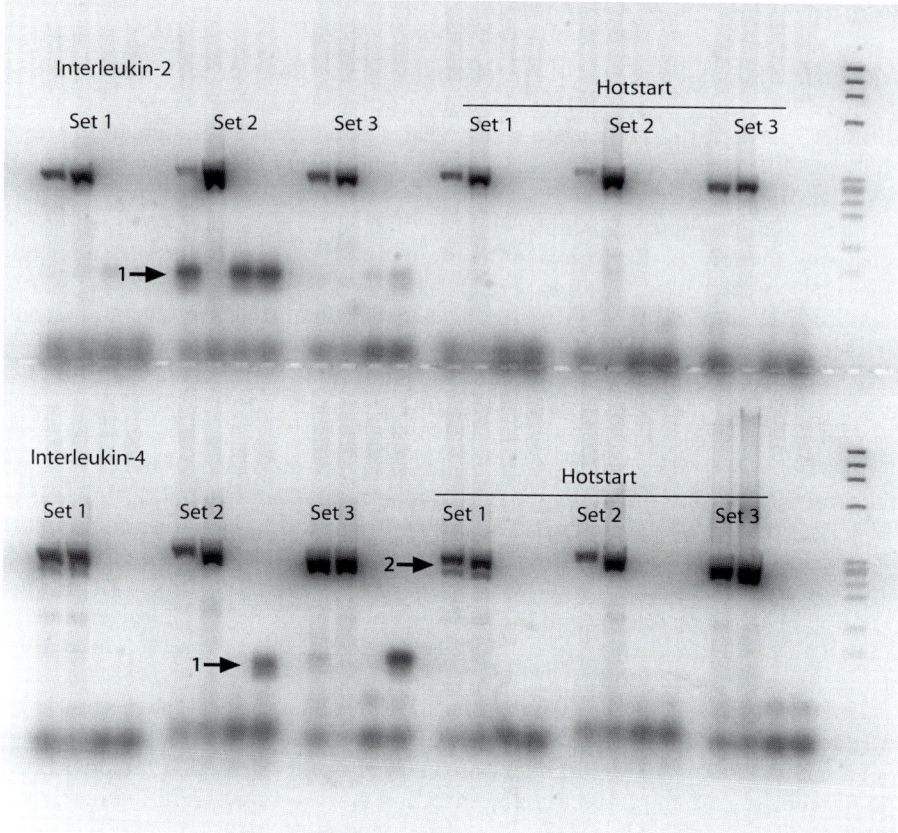

Fig. 2. Gel analysis of PCR products amplified in the LightCycler. cDNA from cells cultured alone, or stimulated with PMA/ionomycin for 3 h, no-RT control and water were amplified in the LightCycler using the SYBR Green I format. Three primer pairs were compared for their amplification with and without TaqStart antibody. The products were recovered and visualized on a SYBR Green II Agarose gel using a fluorescence imager.
[1]) Typical dimer
[2]) IL-4 Δ2 amplification using primer set 1

Comments

Sample Preparation

Sample preparation is one of the most critical aspects of quantitative PCR. Isolation of high quality RNA is an important first step for the quantification of gene expression. This step has to be easy, quick and reliable. For high sample throughput a microformat is preferred. Total RNA is sufficient for analysis and contamination of DNA should be minimal.

RNA has to be transcribed into first strand cDNA. There are two strategies that can be followed. Using a one-step or one-tube RT-PCR reduces handling and time, but requires separate RT-reactions for every parameter studied and house-

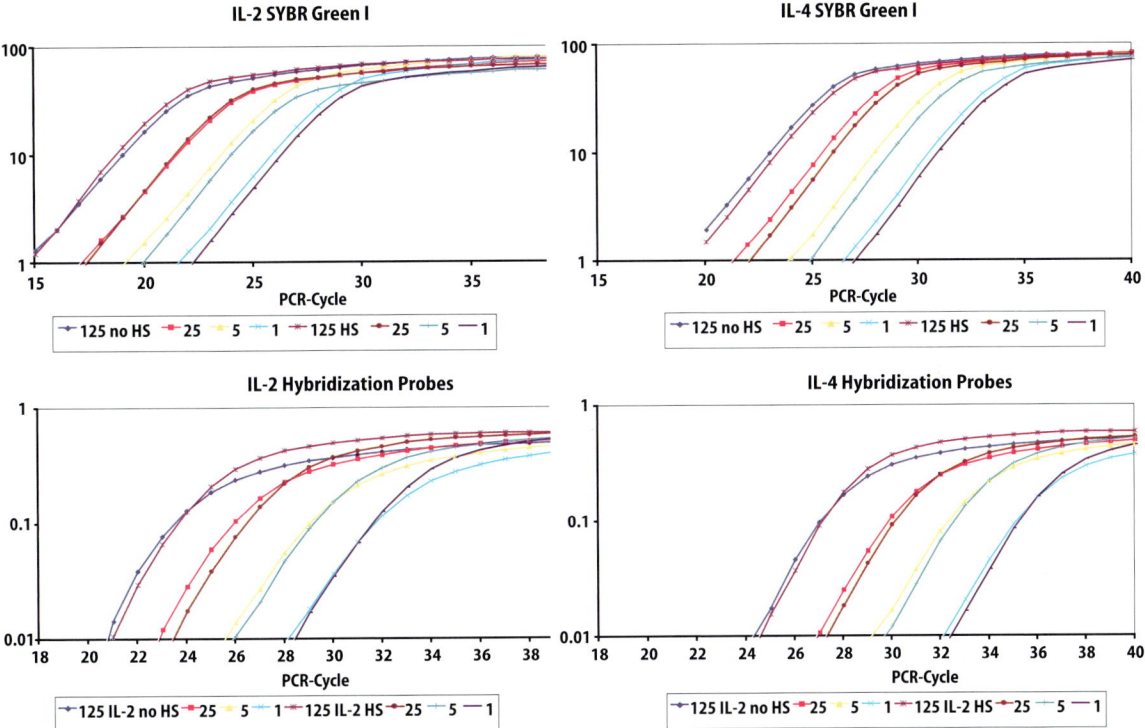

Fig. 3. Quantitative PCR of titrated cDNA samples. cDNA form PMA/ionomycin stimulated cells was serially diluted in 1:5 steps with the no RT control. PCR was performed for 40 cycles using the SYBR Green and Hybridization Probe format described in procedure. Each curve represents the mean of two amplifications

Table 2. Quantitative PCR of titrated cDNA samples. cDNA from PMA/ionomycin stimulated cells was serially diluted in 1:5 steps with no RT control. PCR was done for 40 cycles using the SYBR Green I and Hybridization Probe format described in Procedure. The expected value was calculated from the dilution of the measured amount of message in the undiluted sample.

	IL-2				IL-4			
	Hyb Probes		SYBR Green		Hyb Probes		SYBR Green	
Dilution	–	Hotstart	–	Hotstart	–	Hotstart	–	Hotstart
1 : 5	121[1]	97	103	91	97	87	94	103
1 : 25	84	87	140	91	104	94	93	108
1 : 125	93	104	175	117	96	92	96	117

[1] Percent of expected value, mean of two amplifications

keeping gene used to normalize sample concentration. In a two step system, first strand synthesis is performed in a separate reaction, followed by parameter-specific PCR analysis. Usually cDNA obtained from 1 µg total RNA is sufficient to analyze 50 parameters. We obtained good and reliable results with a two step system using total RNA isolation and first strand cDNA synthesis with AMV.

Primer Design

Good primer design is absolutely critical for the successful development of reliable quantitative RT-PCR tests in the LightCycler. The amplicon should be smaller than 600 bp (optimal between 250 and 350 bp) and no genomic sequences should be amplified. This is verified by PCR "with no RT" controls (cDNA reaction in the absence of reverse transcriptase) and / or with genomic DNA. Genomic products above 1500 bp are usually not amplified in the LightCycler using standard rapid cycling protocols.

It is important, that only one specific product is amplified especially in the SYBR Green I format. Amplification primers should have a high threshold for mispriming. Searching different databases with the sequence of the selected primer can be very helpful. Increasing the stringency of the amplification (e.g. touchdown PCR) can also enhance the specific amplification of a target. A computer aided search for primer sequences with highly stringent binding properties is recommended.

Applications

Quantitative RT-PCR tests have been developed in our laboratory for the LightCyler for the following cytokine and chemokine related parameters.

CHEMOKINES	CYTOKINES
CCR-4	G-CSF
CCR-5	GM-CSF
ELC	IL-12R β1
ENA-78	IL-12R β2
Eotaxin	IL-1α IL-1β
I-309	IL-2, 3, 4, 5, 6, 7, 8, 9
IP-10	IL-10, 12, 13, 14, 15
MCP-1	IL-16, 17, 18
MCP-3	CD25
MDC	Interferon-α
MIP-1α	Interferon-γ
MIP-1β	M-CSF
NAP-2	MIF
RANTES	sIL-1RA
TARC	TGF-β1
MIG	TNF-α
I-TAC	TNF-β

References

1. Benveniste O, Martin M, Villinger F and Dormont D (1998) Techniques for quantification of cytokine mRNAs. Cytokines.Cell Mol Ther 4:207-214
2. Zhou L J, Tedder TF (1995) A distinct pattern of cytokine gene expression by human CD83+ blood dendritic cells. Blood 86:3295-3301
3. Atamas S P (1996) An alternative splice variant of human IL-4, IL-4 delta 2, inhibits IL-4-stimulated T cell proliferation. J Immunol 156:435-41
4. Caplin B E, Rasmussen R P, Bernard P S and Wittwer C T (1999) LightCycler Hybridization Probes. Biochemica 1:5-8

Quantitative RT-PCR for the Detection of T Cell Receptor Transcripts in T Lymphocytes Populations Using LightCycler Technology

E. Jouvin-Marche, I. Vigan, V. Leroy, P. N. Marche*

Introduction

Various approaches have been developed to identify the frequency of distinct subsets of lymphoid cells such as flow cytometry analysis or competitive polymerase chain reaction (PCR). However, these procedures are difficult to apply and somewhat imprecise when the sample to analyse contains a very low number of cells and when multiple parameters are analyzed. We took advantage of the sensitivity and reproducibility of the LightCycler system to develop an assay for the determination of the frequencies of lymphocyte sub-populations and the expression of cytokine genes. These studies were carried out with peripheral blood lymphocytes and with tissues from human liver biopsies.

In the human, $\alpha\beta$ and $\gamma\delta$ T cell receptor (TCR) heterodimers are expressed in classical T lymphocytes and in natural killer T cells (NKT), the latter coexpress the TCR and natural killer cell markers [1]. T and NKT cells, are both present in liver tissue. $\alpha\beta$ is a clonotypic TCR that recognizes antigenic peptides complexed with products of the major histocompatibility complex (MHC) on presenting cells, whereas the ligands recognized by $\gamma\delta$ TCR have not yet been identified [2]. Like the immunoglobulins, the TCR proteins are encoded by several gene segments, namely variable (V), diversity (D – β and δ chains only), joining (J) and constant (C) regions.

In this report, to follow expression of $\alpha\beta$ and $\gamma\delta$ heterodimers we have designated a set of oligonucleotide primers specific to the TCRαC and TCRδC genes, encoding TCR α and β constant region respectively. The primers were chosen in two independent exons in order to avoid amplification of putative DNA contamination in RNA preparations. The amounts of TCRα and TCRδ transcripts in a given sample were determined by performing an amplification of appropriate series of TCRα and TCRδ standard DNA dilutions. Standards were generated by classical PCR from TCRαC and TCRδC genes respectively. Furthermore, since RNA extraction and reverse transcription often have variable yields, we have developed an assay for housekeeping gene transcripts in order to normalize the results obtained from different RNA and DNA samples. We have chosen the G3PDH gene, which like other housekeeping genes is

---

* P. N. Marche (✉) (e-mail: patrice.marche@cea.fr)
 Laboratoire d'Immunochimie, CEA-Grenoble/DBMS, INSERM U238, 17, rue des Martyrs, 38054 Grenoble Cedex 9, France

expressed approximately at constant levels in most cell types. Thus, the results of each amplified product are given as a number of copies per 1000 copies of G3PDH cDNA.

Materials

Equipment
RoboCycler gradient 96 temperature cycler (Stratagene)
DNA Sequencer 725 (Amersham, Molecular Dynamics)
LightCycler Instrument (RocheDiagnostics, Mannheim)
LightCycler Capillaries (RocheDiagnostics, Mannheim)
Primers Select software (DNA STAR)

Reagents
RNA-NOW kit (Biogentex-Ozyme)
SuperScript RNase H- Reverse transcriptase kit (Gibco BRL, Life Technologies)
LightCycler-DNA Master SYBRGreen I (RocheDiagnostics, Mannheim)
TaqStart Antibody (Clontech)
Primers (Genset, France)
QIAEX II gel extraction kit (QIAGEN)

Procedure

Sample Preparation
Liver biopsies were provided by the department of hepato-gastro-enterology (CHRU-Grenoble) and were obtained with Hepafix biopsy 16 G needles. Liver biopsies were 1 cm long and contained an average of 1×10^5 mononuclear cells. Liver biopsies were from healthy donors and from chronic hepatitis C patients, and were stored at –80°C until RNA extraction. Peripheral blood mononuclear cells (PBMC) were isolated by Ficoll density gradient separation. After washing with PBS, PBMC were pelleted and used for RNA extraction. This study was approved by the local ethical committee and written informed consent was obtained for each patient.

RNA Extraction and cDNA Synthesis
Total RNA was extracted from liver biopsy specimens or PBMC using the RNA-NOW kit (Biogentex-Ozyme), following the manufacturer's instructions. Single-strand cDNA synthesis was performed with 1 μg of total RNA using the SuperScript RNase H- Reverse transcriptase kit (Gibco BRL, Life Technologies). cDNA synthesis were performed in a final volume of 20 μl and stored at –20°C until PCR analysis [3].

Primer Design
Primers were designed with the Primers Select software (Lasergene, DNA STAR) and all primers were synthesized by Genset. Primers sequences and characteristics are shown in Table 1.

DNA External Standard Synthesis
The quantifications of TCRα, TCRδ and G3DPH transcripts were made with a DNA external standard generated by PCR. Briefly, the target sequences were amplified using PBMC cDNA preparations and each primer pair. PCR reactions were carried

Table 1. Primers for the amplification of TCRA, TCRD and G3PDH transcripts

TCR Alpha (Genebank Accession # L 02424)					
	Position	Exon	Length	GC (%)	T_m (°C)
Primers					
ATGAGGTCTATGGACTTCAAGAGCAACAG	154	1	29	44.8	67.1
ATTCGGAACCCAATCACTGACAGG	365 R	3	24	50.0	66.3
Product	154–365		212		
TCR Delta (Genebank Accession # Y00736)					
Primers					
AAGTCAGCCTCATACCAAACCATCC	121	1	25	48.0	66.1
AGCATTCGTAGCCCAAGCACTGT	525 R	3	23	52.2	68.4
Product	121–525		405		
G3PDH (Genebank Accession # NM_002046)					
Primers					
AGCAATGCCTCCTGCACCACCAAC	517	7	24	58.3	63.8
CCGGAGGGGCCATCCACAGTCT	653 R	8	22	68.2	65.6
Product	517–653		136		

out in a RoboCycler gradient 96 temperature (Stratagene). PCR reaction mixture contained: 0.2 mM of each dNTP, 0.4 µM of each primer, 1 U of Taq DNA Polymerase (Roche), 1× PCR buffer (Roche), and 1 µl cDNA preparation in a final volume of 50 µl. The PCR cycle started with an initial denaturation step of 1 min at 94°C, followed by 35 cycles consisting of 1 min at 94°C, 1 min at 60°C, and 1 min at 72°C and ended with an elongation step of 10 min at 72°C. PCR products were run on an electrophoresis gel and purified using the QIAEX II gel Extraction Kit (QIAGEN). The concentration of the purified PCR product was determined by absorbance at 260 nm and was used to make a set of standards dilutions. The amount of TCRα or δ transcripts in each sample is given in number of copies. The results of each amplified product are given as a number of copies per 1000 copies of G3PDH.

The analysis on the LightCycler were performed in a reaction volume of 20 µl and a Master Mix of the following composition was used:

Master Mix Preparation

	Volume [µl]	[Final]
LightCycler DNA Master SYBRGreen I	2	1×
TaqStart antibody (5 min incubation at room temperature)	2	
MgCl$_2$ (25 mM)	1.6	3 mM
Primers (10 µM each)	1 + 1	0.5 µM
H$_2$O (PCR grade)	10.4	
Total volume	18	

A total of 2 µl of cDNA template (samples or DNA external standards in appropriate dilution : 10–10^6 copies) were added to 18 µl of master mix in pre-cooled capillary. Sealed capillaries were centrifuged (5 s at 1000 rpm) in a microcentrifuge and placed into the LightCycler rotor.

PCR Protocol for TCRα Amplification

- Denaturation at 95°C for 2 min
- Amplification

Parameter	Value		
Cycles	40		
Type	Quantification		
	Segment 1	Segment 2	Segment 3
Temperature target [°C]	95	60	72
Incubation time [s]	5	10	9
Temperature transition rate [°C/s]	20	20	20
Acquisition mode	None	None	Single
Gain	FL1=5		

- Melting Curve Analysis

Parameter	Value		
Cycles	1		
Type	Melting curve		
	Segment 1	Segment 2	Segment 3
Temperature target [°C]	95	65	95
Incubation time [s]	0	120	0
Temperature transition rate [°C/s]	0	0.1	0
Acquisition mode	None	None	Cont.
Gain	FL1=5		

Cooling at 40°C for 1 min

PCR protocol for TCRδ amplification

- Denaturation at 95°C for 2 min
- Amplification

Parameter	Value		
Cycles	36		
Type	Quantification		
	Segment 1	Segment 2	Segment 3
Temperature target [°C]	95	65	72
Incubation time [s]	5	10	17
Temperature transition rate [°C/s]	20	20	20
Acquisition mode	None	None	Single
Gain	FL1=5		

- Melting Curve Analysis

Parameter	Value			
Cycles	1			
Type	Melting curve			
		Segment 1	Segment 2	Segment 3
Temperature target [°C]		95	70	95
Incubation time [s]		0	120	0
Temperature transition rate [°C/s]		0	0.1	0
Acquisition mode		None	None	Cont.
Gain			FL1=5	
Cooling at 40°C for 1 min				

- Denaturation at 95°C for 2 min
- Amplification

PCR protocol for G3PDH amplification

Parameter	Value			
Cycles	35			
Type	Quantification			
		Segment 1	Segment 2	Segment 3
Temperature target [°C]		95	59	72
Incubation time [s]		5	10	7
Temperature transition rate [°C/s]		20	20	20
Acquisition mode		None	None	Single
Gain			FL1=5	

- Melting Curve Analysis

Parameter	Value			
Cycles	1			
Type	Melting curve			
		Segment 1	Segment 2	Segment 3
Temperature target [°C]		95	64	95
Incubation time [s]		0	120	0
Temperature transition rate [°C/s]		0	0.1	0
Acquisition mode		None	None	Cont.
Gain	FL1=5			
Cooling at 40°C for 1 min				

PCR product gel analysis and sequencing

PCR products were recovered from the capillaries by reverse centrifugation into eppendorf tubes. Samples were diluted in classical bromophenol blue loading buffer, loaded on 1.5% agarose gels containing 0.5 µg/ml ethidium bromide and run in 0.5× TBE buffer. Gels were analyzed on a fluorescence ImageMaster VD (Pharmacia Biotech, FujiFilm FTI-500). Product amplification specificity was further determined by direct sequencing of PCR products using the Thermo Sequenase core sequencing kit (Vistra DNA system), following the manufacturer's instructions. Sequences were read on a DNA Sequencer 725 (Vistra DNA system) using the DNA Sequencher software (Vistra DNA system).

Results and discussion

The selected primers for amplification of G3PDH, TCRα and TCRδ transcripts produced a single product of the expected size without observable primer dimers (Fig. 1). A single and readable sequence was also obtained with each PCR product. For the DNA external standard amplification as depicted in Fig. 2, linearity was maintained over at least four orders of magnitude, allowing a reliable reference scale. To check the sensitivity and the reproducibility of our assays, we have quantified TCRα and TCRδ transcripts in PBMC and in healthy and hepatitis C livers samples either two or three times during the same run or during different runs of amplification. The experimental results were highly reproducible and are summarized in Table 2 for TCRδ amplification. Similar reproducible results were obtained for TCRα amplification, indicating that the LightCycler system allows quantification of TCRα and TCRδ transcripts from samples containing a limited number of cells (i.e., $0.1-1 \times 10^5$ T cells) and to analyze these two transcripts in a large variety of pathological situations. Furthermore, the amount of cDNA used in each amplification run is very low (0.01–0.1 μg of RNA) allowing the analysis of several parameters from a single cDNA sample.

The number of TCRA cDNA copies, obtained from the reverse transcription of 1 μg of total RNA, ranged from 50 to 250×10^5 copies in PBMC. Taking into account that the TCR αβ+ lymphocytes represent 40% of the total PBMC and that our yield of RNA extraction was 1 μg per 1×10^6 PBMC, thus the number of TCRα cDNA copies varied from 10 to 50 copies per T lymphocyte. For TCRδ, assuming that TCR γδ+ lymphocytes represent about 1% of PBMC [2], we have detected 100–200 TCRD cDNA copies per T lymphocyte. The number of TCRα copies detected is consistent with the fact that the transcription of TCR α chain is low in mature T cells [4].

Table 2. Quantification of TCRδ transcripts in liver biopsies and PBMC

	TCRδ copy number/1000 copies G3PDH
Healthy liver	10,300±1200
Hepatitis C liver	523±56
PBMC	94,000±11,000
Results are shown as mean±SD for two separate experiments performed in triplicate	

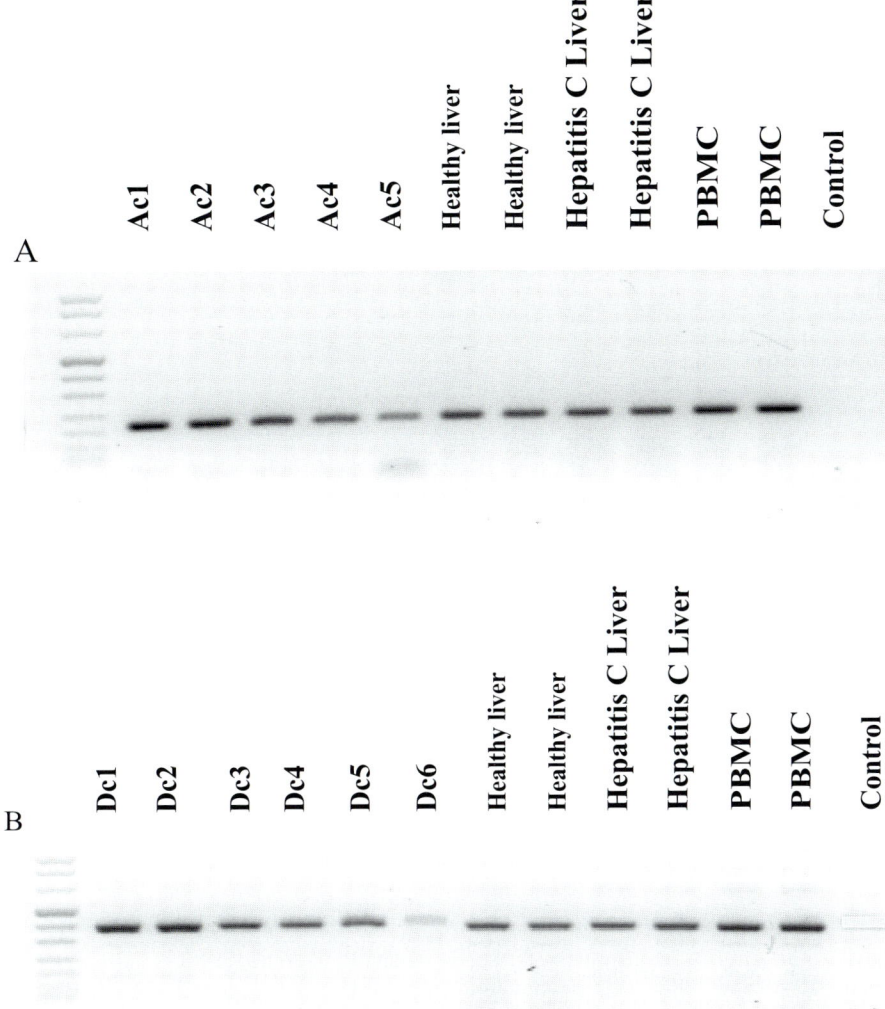

Fig. 1. PCR products analysis. DNA external standards (Ac-1 to Ac-5 for TCR alpha and Dc1 to Dc6 for TCR delta) and cDNA from an healthy liver, an hpatitis C liver and peripheral blood lymphocytes (PBMC) were amplified in the LightCycler, using SYBR Green I dye as detection format and TaqStart Antibody. As negative control, the template DNA was repaced with PCR grade water. The panels **A** and **B** represent respectively TCR alpha and TCR delta PCR products

Altogether, the LightCycler technology allows easy quantification of several transcripts present in biological sample and seems useful for analyzing a panel of markers in different pathologies. In our experience the primer design remains the crucial step.

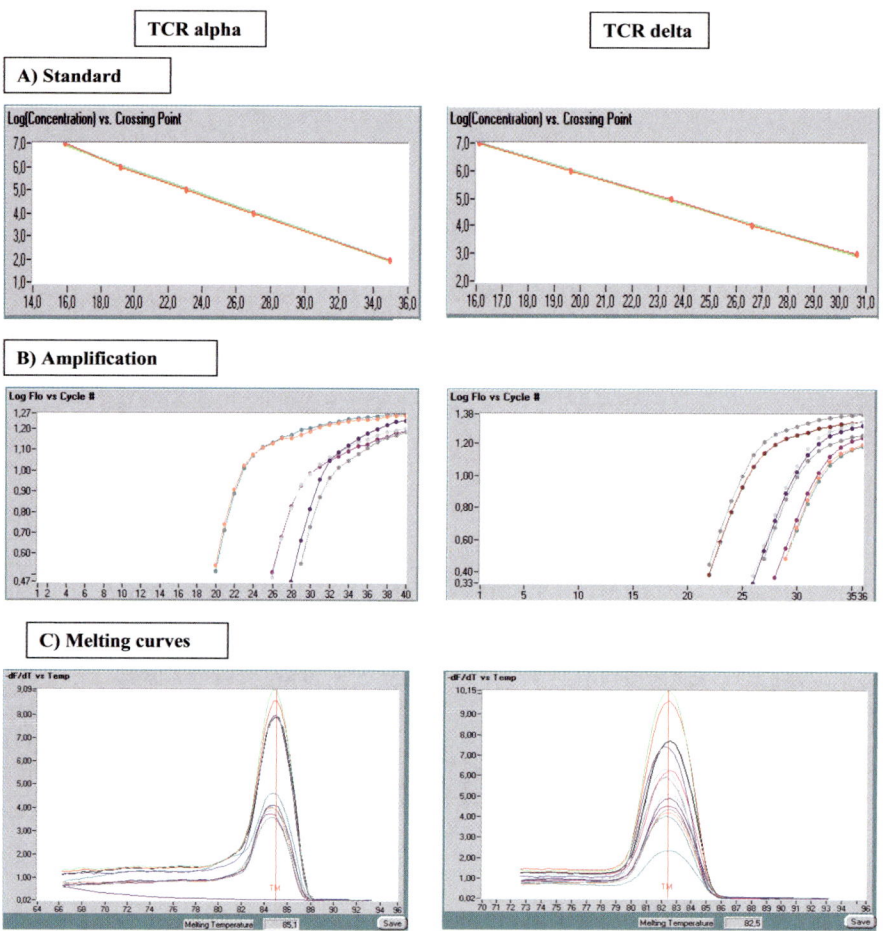

Fig. 2. TCRαC and TCR δC quantitative LC-PCR of standard and titrated cDNA samples. **A)** Detection of TCRαC and TCR δC amplification from different starting amount of TCRαC and TCRδC standard cDNA. **B)** Detection of TCRαC and TCR δC amplification from samples of liver and PBL cDNA with SYBRGreen I dye. **C)** Melting curve analysis of TCRαC and TCR δC amplification products

Applications

The following quantitative RT PCR assays have been developed in our laboratory for the analysis of lymphocyte populations in biological samples using the LightCycler system:

Lymphocytes markers	House keeping genes	Chemokines
TCRα chain	G3PDH	RANTES
TCRβ chain	HPRT	IP-10
TCRδ chain	Actine	MCP-1
CD3ε chain	**Cytokines**	SDF-1
CD4	IL-2	DC-CK1
CD8α chain	IL-4	**Cytotoxicity marker**
CD8β chain	TNF-α	Perforin
CD56	**Chemokine receptors**	**Virus sequence detection**
CD69	CCR3	Hepatitis C virus
	CXCR3	
	CXCR4	

References

1. Spits H, Lanier LI, Phillips JH (1997) Development if human T and Natural killer cells. Blood 85: 2654–2670
2. Norris S, Collins C, Doherty DG, Smith F, Mcentee G, Traynor O, Nolan N Hegarty J, O'Farrelly C (1998) Resident human hepatic lymphocytes are phenotypically different from circulating lymphocytes. J Hepatol 28: 84–90
3. Jouvin-Marche E, Aude-Garcia C, Candeias S, Borel E, Hachemi-Rachedi S, Gahery-Segard H, Cazenave PA, Marche PN (1998) Differential chronology of TCRADV2 gene use by alpha and delta chains of the mouse TCR. Eur J Immunol 28: 818–827
4. Kearse KP, Roberts JP, Wiest DL, Singer A (1997) Developmental regulation of αβ T cell antigen receptor assembly in immature CD4+ CD8+ thymocytes. Bioessays 17: 1049–1054

Rapid, Homogeneous Genotyping of Human Platelet Antigen 1 by Fluorescence Resonance Energy Transfer and Probe Melting Curves

Markus S. Nauck*, Hedi Gierens, Matthias A. Nauck, Winfried März, Heinrich Wieland

Introduction

Alloantigenic determinants on platelet membrane glycoproteins can be targets for alloimmune antibody responses that cause bleeding disorders such as neonatal alloimmune thrombocytopenic purpura, post-transfusion purpura (PTP), and refractoriness to platelet transfusion therapy [1]. In PTP, a transfused patient forms alloantigen-specific antibodies that, for still unknown reasons, may cause destruction of autologous platelets often resulting in life-threatening thrombocytopenia. In feto-maternal alloimmune thrombocytopenia, designated as neonatal alloimmune thrombocytopenic purpura (NATP), fetal thrombocytopenia is caused by maternal alloimmunization against one or more paternal platelet alloantigens with similar life-threatening complications. Severe NATP may lead to either prenatal or neonatal intracranial haemorrhage that may cause fetal death or psychomotor impairment.

The alloantigenic determinant implicated in the pathogenesis of most reported cases of PTP and NATP is human platelet antigen-1a (HPA-1a, PI^{A1}). The HPA-1 epitope is found on glycoprotein IIIa (GPIIIa). A single nucleotide transition from T to C at nucleotide 12548 [4] of the genomic DNA of GPIIIa causes a leucine to proline substitution at amino acid residue 33 of GPIIIa, converting HPA-1a (PI^{A1}) to HPA-1b (PI^{A11}) [2]. Although the function of GPIIIa is not profoundly affected, this change alters the antigenic properties of the molecule. Both immuno- and genotyping of human platelet alloantigens (HPA) have become important procedures in the diagnosis and prevention of these bleeding disorders.

Immunophenotyping of the HPA-1 system has several drawbacks, including the need for a sufficient number of platelets, shortages of typing antisera, and a time-consuming procedure. Therefore, genotyping of platelet antigens with the possibility of using any type of cellular material containing DNA as a source has become a preferred procedure, in combination with the various detection methods currently used. Polymerase chain reaction with allele-specific oligonucleotide hybridization, allele-specific restriction enzyme analysis (ASRA), amplification with sequence-specific primers, single strand conformation polymorphism, pref-

---

* M.S. Nauck (✉) (e-mail: msnauck@med1.ukl.uni-freiburg.de)
  University Hospital Freiburg, Department of Clinical Chemistry, Hugstetter Strasse 55, 79106 Freiburg i. Br., Germany

erential homoduplex formation, and oligonucleotide ligation assay have been applied to genotyping of human platelet alloantigens. However, all these methods have disadvantages which hinder a fast molecular diagnosis in a clinical laboratory setting, due to their requirement of extensive post-amplification processing, including electrophoresis or ELISA-based detection of allele-specific ligation-products.

We describe here a single-step protocol for HPA-1 genotyping on the LightCycler that is fast and robust and is thus ideally applicable to routine HPA typing in a clinical setting.

Materials

Equipment LightCycler Instrument

Reagents Amplification primers (MWG-Biotech, Eberswalde, Germany)
Hybridization probes (TIB MOLBIOL, Berlin, Germany)

The following reagents were purchased from Roche Diagnostics, Mannheim:
High-Pure PCR Template Preparation Kit,
LightCycler-DNA Master Hybridization Probes
LightCycler glass capillary cuvettes

Procedure

Sample Preparation Genomic DNA was isolated from whole blood or buffy coats using the High-Pure PCR Template Preparation Kit. The DNA was resuspended in 10 mM/l Tris (pH 7.4), containing 0.1 mM/l EDTA at a concentration of 100 ng/µl.

Design of Primers and Fluorogenic Probes The amplification primers were designed using the Primer3 input program from the Whitehead Institute for Biomedical Research (website: http://www-genome.wi.mit.edu/cgi-bin/primer/primer3www.cgi). The amplified fragment harboring the polymorphic site at nucleotide 12548 of the genomic DNA of GPIIIa is 181 bp in size (Table 1) [4]. The primers were synthesized by standard phosphoramidite chemistry (MWG-Biotech, Eberswalde, Germany).

The LCRed640-labeled detection probe is complementary to the antisense strand of the HPA-1b allele with the polymorphic nucleotide 8 bases from the 3´-end (underlined nt.) (Table 1, Fig. 1). The resulting C–A mismatch between detection probe and the HPA-1a allele is the most unstable mismatch possible with this polymorphism, ensuring a maximum difference in the melting temperatures (T_m) between both alleles. The 3'-fluorescein labeled anchor probe binds with a distance of one base 5´ to the detection probe. Both fluorophore-labeled probes were synthesized and purified by reverse-phase HPLC by TIB MOLBIOL, Berlin, Germany.

Rapid, Homogeneous Genotyping of Human Platelet Antigen 1 by Fluorescence Resonance Energy Transfer

Table 1. Oligonucleotides

	GP IIIa, exon 3 (Genebank Accession # M32672)			
	Position	Length	GC (%)	T_m (°C)
Primers				
TGCTCCAATGTACGGGGTAAAC	1374	22	50	66.9
CTGGGGCACAGTTATCCTTCAG	1554 R	22	54.5	67.0
Product	1374-1554	181		
Probes				
LCRed640-CCTGCCTCCGGGCTCAC-P	1490	17	76.5	68.4
GACTTCTCTTTGGGCTCCTGTCTTACAGG-F	1460	29	51.7	71.2

Experimental Protocol

Amplification on the LightCycler was performed in a reaction volume of 20 μl with 50 ng of genomic DNA and a master mix of the following composition:

	Volume [μl]	[Final]
LightCycler-DNA Master Hyb Probes	2.0	1×
MgCl$_2$ (25 mM)	3.2	5 mM
Primers (3 μM each)	2.0	0.3 μM each
Probes (2 μM each)	2.0	0.2 μM each
H$_2$O (PCR grade)	8.8	
Total volume	18.0	

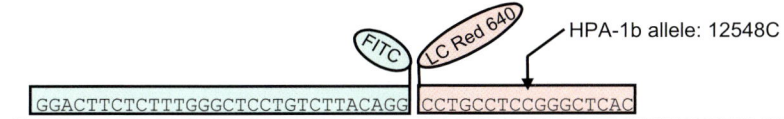
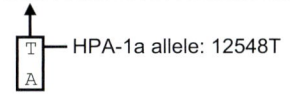

Fig. 1. Relative orientation of the fluorophore-labeled anchor and detection probe. The HPA-1 polymorphism is a result of a T–C substitution at nucleotide 12548 [4]. This polymorphism creates an A–C mismatch between the antisense strand of the HPA-1a allele and the detection probe. This mismatch destabilizes the hybrid which results in a decrease of the melting temperature of the probe. In contrast, complete matching of detection probe and antisense strand of the HPA-1b allele results in a higher melting temperature of the hybrid

After loading of the samples into the glass capillary cuvettes, the capillaries were sealed, briefly centrifuged and then placed into the LightCycler rotor. For amplification, the following thermocycling protocol was used:

- Denaturation at 94°C for 45 s
- Amplification

Parameters	Value		
Cycles	45		
Type	Quantification		
	Segment 1	Segment 2	Segment 3
Target temperature (°C)	94	55	72
Incubation time (s)	0	6	1
Temperature transition rate (°C/s)	20	20	3
Acquisition mode	None	Single	None
Gain	F2=15		

- Melting Curve Analysis

Parameters	Value		
Cycles	1		
Type	Melting curve		
	Segment 1	Segment 2	Segment 3
Target temperature (°C)	94	45	80
Incubation time [m:s]	1:0	3:0	0
Temperature transition rate (°C/s)	20	10	0.2
Acquisition mode	None	None	Cont.
Gain	F2=15		

Results

A total of 45 cycles of amplification were performed with genomic DNA of each genotype and a template-free control using the fluorescence resonance transfer detection system outlined in Fig. 1. The fluorescence signal was measured at the end of each annealing phase and increased as the product accumulated. Under our conditions, the fluorescence signal appeared above background levels after 25 cycles (Fig. 2). No increase in fluorescence signal was observed in the absence of template. When analyzed on an agarose gel we found a pure amplification product of the expected size.

The process of hybridization and melting of the detection probe to the target was monitored by melting curve analysis. Melting of the sample homozygous for the HPA-1a allele produced a rapid decrease in fluorescence at 55–56°C (Fig. 3a).

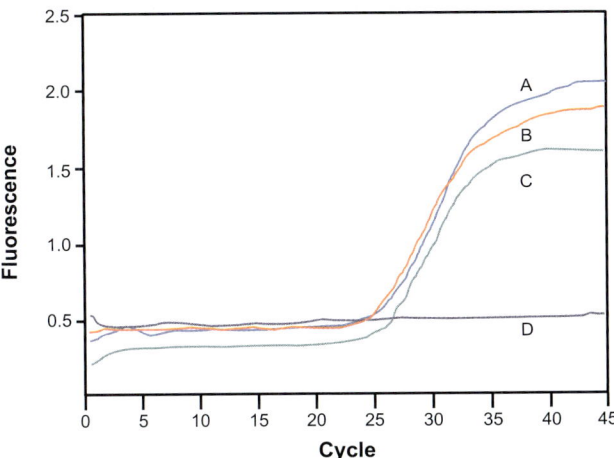

Fig. 2. Intensity of the fluorescence signal (*F2*) vs cycle number plot. A 181 bp fragment of the GPIIIa gene was amplified from genomic DNA of three different genotypes: homozygous for the HPA-1a allele *A*, heterozygous *B*, and homozygous for the HPA-1b allele *C*. Amplification of a no template control *D* was also attempted. The fluorescence signal was acquired once each cycle at the end of the annealing period

In contrast, in the sample homozygous for the HPA-1b allele, the transition occured at 65–66°C. The heterozygous sample exhibited two distinct decreases in fluorescence, corresponding to the presence of amplicons derived from both alleles. By plotting the negative derivative of the fluorescence signal with temperature versus temperature ($-(dF/dT)$ vs T), peaks are obtained at the respective melting temperatures (Fig. 3b, Table 2). With samples showing different amplification efficiencies the derivative melting curves were highly reproducible, with melting peaks differing by less than 1.0°C for the same allele, allowing easy and unambiguous assignment of genotypes to the respective melting curves.

To evaluate the reliability of the fluorescence genotyping, 100 human DNA samples were genotyped for the HPA-1 polymorphism by both the conventional ASRA method [3] and the homogenous fluorescence assay. The genotypes determined with both methods were in 100% concordance. The genotyping of the 100 samples on the LightCycler was completed within 2 hours, while the ASRA protocol took 8 hours and required several manual sample processing steps. Comparison of the costs for reagents and disposable material showed that both methods are about equally expensive with costs of about $ 2 US per typing reaction.

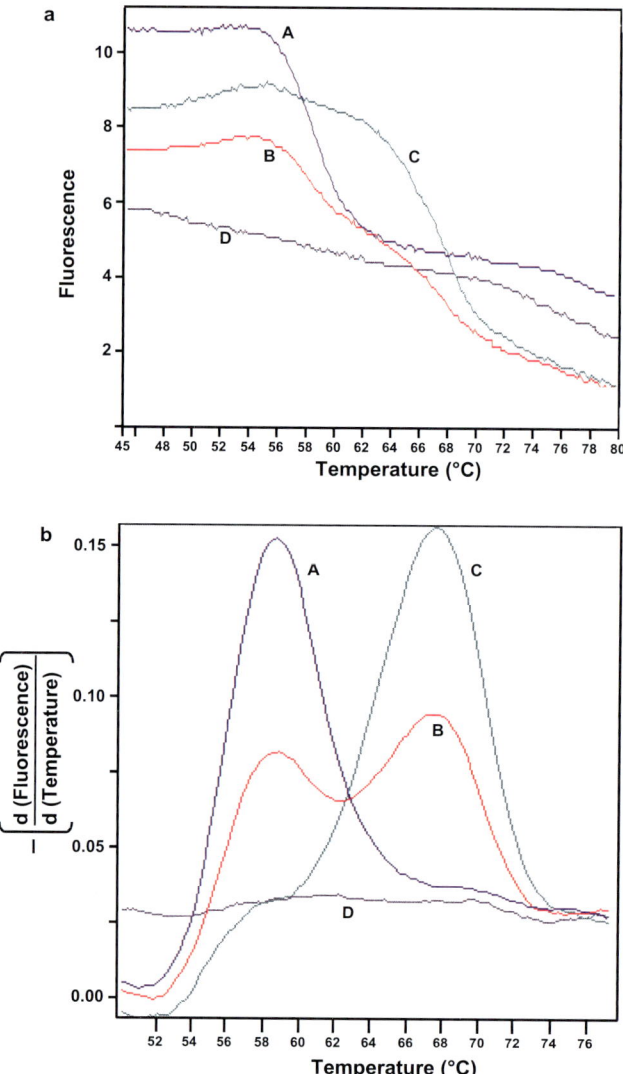

Fig. 3. Genotyping of HPA-1 with an allele-specific fluorescent probe by melting curve (a) and derivative melting curve (b) plots. Immediately after amplification, a melting analysis was performed. Data for both plots were obtained during the melting transition of the LCRed640-labeled detection probe from the amplified fragment. The temperature transition was programmed at 0.2°C/s with continuous fluorescence acquisition for each sample from 45°C to 80°C. (a) Melting curves; the fluorescence signal vs temperature is plotted for a sample homozygous for the HPA-1a allele *A*, a heterozygous sample *B*, and a sample homozygous for the HPA-1b allele *C*. Melting analysis of a no template control *D* was also performed. (b) Derivative melting-curve; the melting curve plot of fluorescence vs temperature was transformed into a derivative melting curve plot. The samples A, B, C and D correspond to the same genotypes as in Fig. 3a

Table 2. Melting temperatures of the two HPA-1 alleles

Locus	Allele	Pairing	T_m (°C) observed
HPA-1	HPA-1a	A–C mismatch	58
	HPA-1b	G–C match	68

Comments

The homogeneous protocol presented here combines a number of advantages:
- The fluorescence assay is robust and reliable, as documented by the complete concordance of 100 genoptypes determined with both the conventional allele-specific restriction fragment analysis and the LightCycler protocol.
- The processing of the samples is simple and the analysis is fast, providing rapid results as well as the possibility of high throughput genotyping.
- As this method is performed in a closed system with no post-amplification processing, potential problems with sample tracking and end-product contamination are eliminated.
- As hands-on time is shorter than in any other technique used so far and costs for reagents and consumables are not higher than for conventional assays, the fluorescence method allows HPA-1 genotyping in a very economic manner.

In summary, this homogeneous (closed tube) assay for rapid genotyping of the HPA-1 polymorphism on the LightCycler is ideally applicable to routine analysis in a clinical setting.

References

1. Newman PJ, McFarland JG, Aster RH. (1994) Alloimmune thrombocytopenias. In: A.I.Schafer JLa (ed) Thrombosis and hemorrhage. Blackwell Scientific Publications, Boston, pp 529–543
2. Newman PJ, Derbes RS, Aster RH (1989) The human platelet alloantigens, PlA1 and PlA2, are associated with a leucine33/proline33 amino acid polymorphism in membrane glycoprotein IIIa, and are distinguishable by DNA typing. J Clin Invest 83:1778–81
3. Simsek S, Faber NM, Bleeker PM, Vlekke AB, Huiskes E, Goldschmeding R, von dem Borne AE (1993) Determination of human platelet antigen frequencies in the Dutch population by immunophenotyping and DNA (allele-specific restriction enzyme) analysis. Blood 81:835–40
4. Zimrin AB, Gidwitz S, Lord S, Schwartz E, Bennett JS, White GC, Poncz, M (1990) The genomic organization of platelet glycoprotein IIIa. J Biol Chem 265:8590–5

Development and Validation of an Externally Standardised Quantitative Insulin-like Growth Factor-1 RT-PCR Using LightCycler SYBR Green I Technology

MICHAEL PFAFFL*

Introduction

The cytokine insulin-like growth factor-1 (IGF-1) is considered to mediate anabolic growth hormone actions in various tissues and species. During postnatal growth, IGF-1 stimulates protein synthesis and improves glucose utilisation [7, 3]. In addition, locally expressed IGF-1 is an important growth regulator acting in an auto- and paracrine manner [12]. To investigate local tissue-specific expression even in tissues with low abundancies, a very sensitive method is required which allows for reliable quantification of IGF-1 mRNA. Because of its high sensitivity, reverse-transcription with subsequent polymerase chain reaction (RT-PCR) is being increasingly used to quantify physiologically relevant changes in gene expression. RT-PCR has a detection limit 10–100 fold lower than other methods, e.g. protection-assay or northern-hybridisation, respectively [11]. The RT-ribonuclease PCR quantification technique of choice depends on the target sequence, the expected range of the mRNA amount present in the tissue, the degree of accuracy required, and whether quantification needs to be relative or absolute. Externally standardised RT-PCR with quantification on ethidium bromide stained gels followed by densitometry is widely used, but the degree of accuracy is limited and the quantification is more relative than absolute [10]. For an exact quantitative measurement of low abundant gene expression only a few PCR methods allow reliable mRNA quantification. At present the following RT-PCR methods are suitable for sensitive quantification:

1. Internally standardised competitive RT-PCR measured by HPLC separation and UV detection [9] or high resolution gel electrophoresis followed by densitometric analysis [8]: In a competitive RT-PCR, a reference RNA mutant is reverse transcribed and co-amplified in the same reaction tube with the native mRNA sequence of interest. Internally standardised RT-PCR is a very time-consuming and laborious technique. It is generally believed to yield the most precise results, because all parameters throughout RT-PCR act on both the analyte and reference mutant.

---

* Michael Pfaffl (✉) (e-mail: pfaffl@weihenstephan.de)
  Weihenstephaner Berg 5, 85354 Freising-Weihenstephan, Germany
  Institute of Physiology, Research Centre for Milk & Food, Technical University of Munich, Germany

2. Externally standardised RT-PCR with online-detection using LightCycler SYBR Green I technology [6, 13]: LightCycler PCR with SYBR Green I online detection produces reliable and rapid results. Because it uses an external standard curve, the amplification efficiencies for the calibration curve and the analyte must be equal for accurate quantification.
3. Externally standardised RT-PCR with online-detection using specific LightCycler hybridisation probes [14]: This detection format is based on fluorescence resonance energy transfer.

The development and validation of an IGF-1 mRNA RT-PCR assay on the LightCycler using SYBR Green I is described here.

Materials

Equipment
Eppendorf Mastercycler Gradient Thermal Cycler (Eppendorf)
ImageMaster (Pharmacia Biotech)
LightCycler Instrument (Roche Diagnostics)
LightCycler Capillaries (Roche Diagnostics)
LightCycler Software Version 3.39 (Roche Diagnostics)
Mac DNASISPro version 3.5 Primer design software (Hitachi)
SIGMA PLOT for Windows Version 4.01 (SPSS)

Reagents
RNA-Clean (AGS)
Cloning vector pCRII (Invitrogen)
dNTPs (dATP, dGTP, dCTP, and dTTP) (MBI Fermentas)
Guanidinium thiocyanate (ICN)
LightCycler-DNA Master SYBR Green I (Roche Diagnostics)
NuSieve agarose (FMC BioProducts)
Primers (Pharmacia Biotech)
Random Hexamer Primers (Pharmacia Biotech)
Rnasin RNase inhibitor (MBI Fermentas)
SP 6 polymerase (MBI Fermentas)
Superscript IIPlus RNase H$^-$ Reverse Transcriptase (Gibco Life Technologies)
Taq DNA Polymerase (Roche Diagnostics)

Procedures

Total RNA Extraction
A total of 0.5 g frozen tissue was homogenised in 4 M guanidinium thiocyanate buffer according to Chirgwin et al. [2] to destroy RNase activity. In the following steps, the RNA-Clean protocol with phenol/chloroform extraction for total RNA was used. In order to quantify the amount of RNA extracted, the optical density was determined at three different dilutions at 260 nm corrected by the 320 nm background absorption. RNA integrity was electrophoretically verified by ethidium bromide staining and by a 260/280 nm absorption ratio >1.75.

Primer Design

The primers used for the production of recombinant IGF-1 RNA and for quantitative LightCycler RT-PCR were derived from the bovine IGF-1 sequence [5] (EMBL Ac. no. X15726). They were designed to produce a 240 bp amplification product spanning two RNA splicing sites in a highly conserved region (exon 3–4) [7] of the IGF-1 sequence coding for the mature IGF-1 protein. Primer design and optimisation was done with the Mac DNASIS primer design software with >94% homology between cattle, sheep, pig, water buffalo, primates, mouse and human. Primers were additionally designed as multi-species primers with >94% homology between: cattle, sheep, pig, water buffalo, primates, mouse and human (Table 1).

Table 1. Oligonucleotides

IGF-1 (EMBL Accession # X15726)					
	Position	Exon	Length	GC (%)	T_m (°C)
TCGCATCTCTTCTATCTGGCCCTGT	88	3	25	52.0	68.9
GCAGTACATCTCCAGCCTCCTCAGA	327 R	4	25	56.0	69.0
Bovine RT-PCR product:	88–327	3–4	240		

Construction of a Recombinant IGF-1 RNA Mutant

A recombinant IGF-1 RNA was designed for validation of the LightCycler reaction and to determine test sensitivity and quantification range. An additional primer [9] was used to generate an internal deletion of 56 bases within the IGF-1 DNA target for length differentiation on gel electrophoresis. The truncated PCR product was cloned into pCRII and recombinant IGF-1 RNA was transcribed using SP 6 polymerase.

Optimization of the RT-PCR

The conditions for the RT-PCR were optimized on a gradient cycler for the annealing temperature, Taq DNA polymerase, primer, $MgCl_2$ and dNTP concentrations. RT-PCR amplification products were separated on a 4% high resolution NuSieve agarose gel and analysed with the Image Master system. Both RT and LightCycler PCR master mixes were assembled with a minimal pipetting volume of 2 µl, to minimise pipetting errors and to improve homogeneity between all reaction capillaries.

Reverse-Transcription

The RT was performed in 40 µl at the following total amounts and concentrations: 1 µg total RNA, 50 mM Tris, pH 8.3, 75 mM KCl, 3 mM $MgCl_2$, 10 mM DTT and 300 µM each dNTP (dATP, dGTP, dCTP, and dTTP) 10 µM random hexamers, 100 U of Superscript II Plus RNase H- reverse transcriptase and 12.5 U of Rnasin RNase inhibitor. The RNA, salts, DTT and dNTPs were first denaturated for 5 min at 65°C in an Eppendorf Mastercycler Gradient, followed by the addition of random hexamers, reverse transcriptase and RNase inhibitor. RT was done at 42°C for 60 min and terminated by heating for 1 min at 99°C.

LightCycler PCR Mastermix

For each LightCycler reaction a mastermix of the following components was prepared to the indicated final concentration:

	Volume [µl]	[Final]
LightCycler-DNA Master SYBR Green I	2.0	1×
$MgCl_2$ (25 mM)	2.4	4.0 mM
Primers (20 µM)	0.2 + 0.2	0.2 µM
H_2O (PCR grade)	13.2	–
Total volume	18.0	

To 18 µl of LightCycler mastermix in the LightCycler glass capillaries, a maximum of 10 ng cDNA in a 2-µl volume was added as PCR template. The capillaries were closed, centrifuged in a micro-centrifuge using the LightCycler centrifuge adapters and placed into the LightCycler rotor.

LightCycler Programs: Four-Segment IGF-1 Amplification Program and Melting Curve Analysis

The following LightCycler protocol was used for IGF-1 online detection using SYBR Green I, including a four-segment LightCycler PCR amplification program and melting curve analysis:

- Denaturation at 95°C for 30 s
- Amplification

Parameter	Value			
Cycles	50			
Type	Quantification			
	Segment 1	Segment 2	Segment 3	Segment 4
Target temperature (°C)	95	62	72	85
Incubation time (s)	1	10	20	3
Temperature transition rate (°C/s)	20	20	20	20
Acquisition mode	None	None	None	Single
Gain	F1 = 5			

- Melting Curve Analysis

Parameter	Value		
Cycles	1		
Type	Melting Curve analysis		
	Segment 1	Segment 2	Segment 3
Target temperature (°C)	95	60	95
Incubation time (s)	10	10	0
Temperature transition rate (°C/s)	20	20	0.10
Acquisition mode	None	None	Cont.
Gain	F1 = 5		

For exact length verification LightCycler PCR products were separated by gel electrophoresis. Amplified IGF-1 LightCycler PCR products were removed from the glass capillaries by reverse centrifugation into 1.5 ml reaction tubes. Samples were diluted in agarose gel loading buffer and loaded onto 4% high resolution NuSieve agarose gels in 1× TAE buffer. Gel analysis was performed with the Image Master.

Gel Electrophoresis of PCR Product

Results

Figure 1 shows specific LightCycler PCR products from a calibration curve of the synthetic template and of different species: cattle, sheep, pig [8] and primate (Callithrix jacchus*) [4], after 50 cycles. Specificity of the desired IGF-1 products was documented with melting curve analysis (LightCycler Software 3.39). The melting temperatures of the products are species dependent (Table 2). Unspecific products and primer-dimers have melting temperatures lower than 82°C (Fig. 2).

Confirmation of Primer Specificity

Table 2. Melting temperatures of IGF-1 products

Species	Observed melting temperature
Cattle (Bos taurus)	90.5°C
Sheep (Ovis aries)	89.9°C
Pig (Sus scrofa)	89.7°C
Primate (Callithrix jacchus*)	88.7°C

---

* Callithrix jacchus samples were kindly donated by Einspanier R and Einspanier A in collaboration with the German Primate Center in Göttingen.

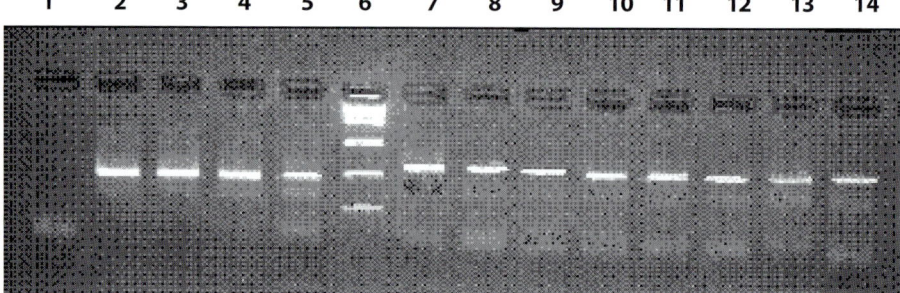

Fig. 1. IGF-1 specific LightCycler PCR products after 50 cycles: *Lane 1,* negative control; *lane 2–5,* calibration curve with synthetic 184 bp product (2×10^8 to 2×10^5 start copies of RNA); *lane 6,* 100 bp ladder; *lane 7–14*: native IGF-1 240 bp products: 2 cattle (Bos taurus), 2 sheep (Ovis aries), 2 pig (Sus scrofa) and 2 primates (Callithrix jacchus)

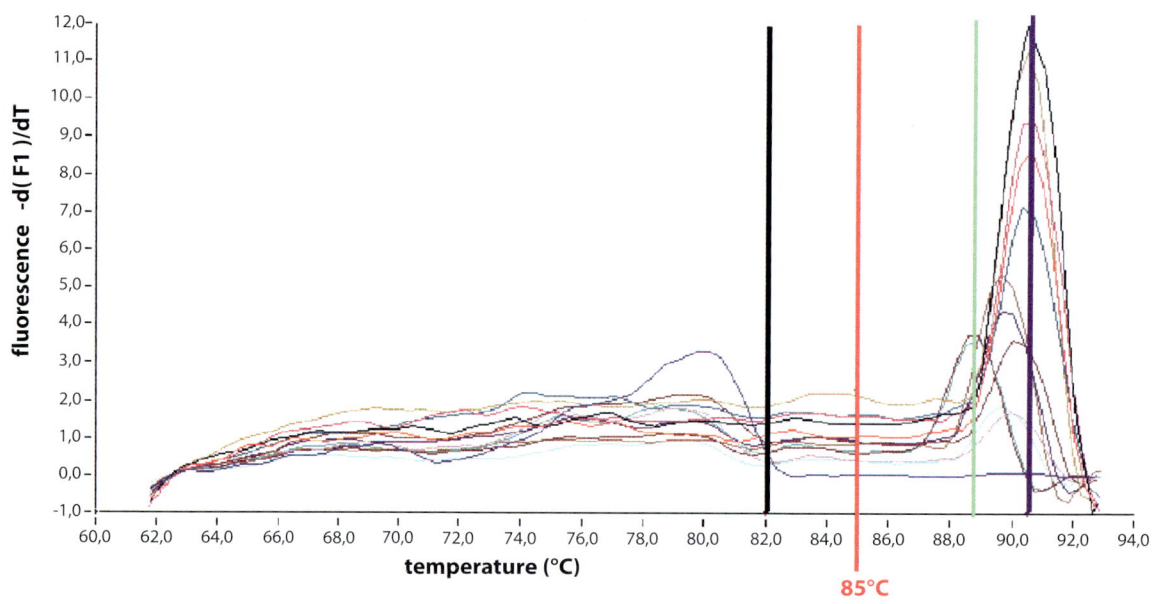

Fig. 2. Melting curves of IGF-1 LightCycler products from multiple species. Melting temperatures of IGF-1 products are between 88.7°C (Callithrix jacchus) and 90.5°C (Bos taurus) and for primer-dimers are lower than 82°C. The fourth segment during the amplification program melts the unspecific LightCycler PCR products at 85°C and eliminates any non-specific fluorescence signals

Advantage of a High Temperature Fluorescence Acquisition During Amplification

The fourth segment during the amplification program melts unspecific LightCycler PCR products at 85°C to eliminate the non-specific fluorescence signal and ensures accurate quantification of the desired IGF-1 products (Fig. 2). High temperature quantification keeps the fluorescence of the no template control around

1 unit, while the specific IGF-1 signal rises up to 40–50 fluorescence units. SYBR Green I determination at 85°C results in reliable and sensitive IGF-1 quantification with high linearity (correlation coefficient $r=0.99$) over seven orders of magnitude (10^2 to 10^9 RNA starting molecules) (Fig. 3b). In contrast, a conventional determination at 72°C results in a truncated quantification range ($r=0.99$) over only four orders of magnitude (10^5 to 10^9 RNA starting molecules) (Fig. 3a).

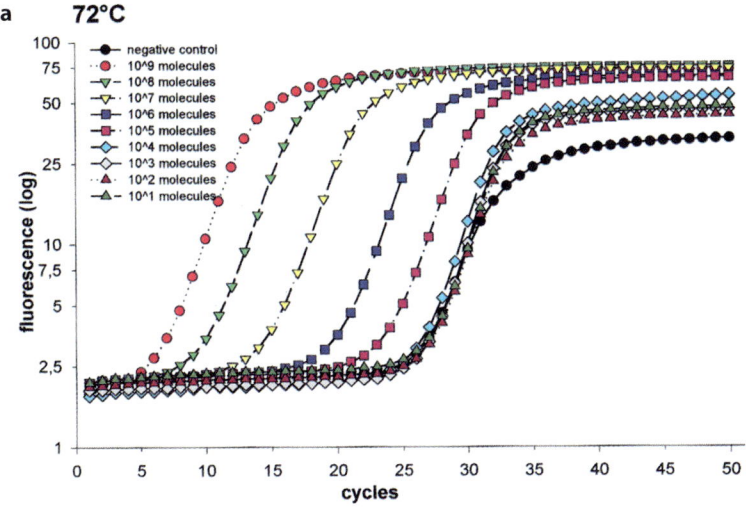

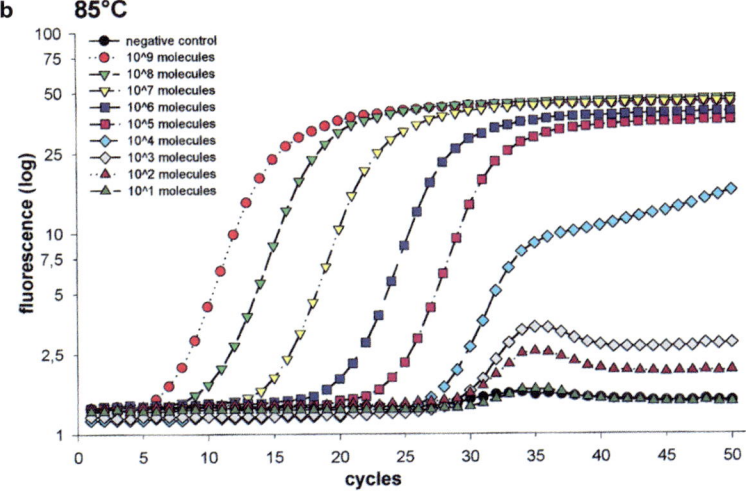

Fig. 3. The effect of SYBR Green I acquisition at 72°C in the third segment (**a**) and 85°C in the fourth segment (**b**) from 10^1 to 10^9 RNA start molecules with one negative water control. Both online quantifications were done in one LightCycler experiment with the same capillaries. Data analysis and plotting were performed with SIGMA PLOT software

Amplification Efficiencies of Recombinant IGF-1 RNA and Native IGF-1 mRNA

For reliable quantification, the amplification efficiency during PCR must be equal for recombinant RNA used in the calibration curve and native mRNA present in the investigated sample RNA. The recombinant 184 base RNA and native 240 base mRNA have previously been shown to have almost identical amplification efficiencies (E) of 66.2% and 64.7%, respectively, during the exponential phase [9]. The relationship between the initial amount A of target present and the amount Y_n of DNA produced after n PCR cycles can be expressed as $Y_n = A \times (1+E)^n$, where E is the amplification efficiency of one PCR step [1]. Figure 4 shows the log fluorescence versus cycle number during the exponential phase (cycle 23–28), inter phase (cycle 29–32) and plateau phase (cycle 35–50). In order to compare the amplification efficiencies of both targets linear regressions were calculated for the exponential-, inter- and plateau phases, using SIGMA PLOT software. The resulting efficiencies (Table 3) during the exponential phase were nearly identical with high reproducibility. In the inter phase, efficiencies were variable and approached zero during the plateau phase. Negative controls containing water and no template showed no amplification products.

Sensitivity and Linearity

The sensitivity of LightCycler PCR was evaluated using different starting amounts of IGF-1 recombinant RNA from 2.8 ag (16 RNA molecules) to 28 ng (1.6×10^{11} RNA molecules). The minimal detectable amount of IGF-1 RNA using SYBR Green I was 16 RNA molecules/capillary, with satisfactory test linearity ($r=0.985$) demonstrated from 1600 to 1.6×10^{11} RNA molecules/capillary. Using IGF-1 cDNA as template 50 molecules could be detected with high test linearity ($r=0.982$) in a range of 50 to 5×10^5 molecules/capillary.

Table 3. LightCycler PCR efficiencies (in %; mean ± std. dev.) of native IGF-1 mRNA in 5 ng liver RNA ($n=4$), of recombinant IGF-1 RNA ($n=4$) and of a negative water control ($n=1$); (r=Pearson correlation coefficient)

Template	Exponential phase (cycle 23–28)	Interphase (cycle 29–34)	Plateau phase (cycle 35–50)
Native IGF-1 mRNA	42.81±1.34 ($r=0.997$)	11.35±0.56 ($r=0.987$)	1.71±0.21 ($r=0.952$)
Recombinant IGF-1 RNA4	40.75±1.86 ($r=0.992$)	16.16±0.93 ($r=0.974$)	1.49±0.06 ($r=0.933$)
Water (no template)	0.85 ($r=0.831$)	2.59 ($r=0.952$)	[–0.43] ($r=0.789$)

Intra-assay and Inter-assay Variation

To confirm the reproducibility of LightCycler PCR even with low template copies (500 to 500,000 cDNA molecules), intra-assay variation was determined in three repeats in one LightCycler run and inter-assay variation in four experiments on 4 days using four different Master premixes (Table 4).

Table 4. Intra-assay and inter-assay variation of LightCycler IGF-1 PCR using 500–500,000 IGF-1 cDNA template molecules

cDNA template molecules	Intra-assay variation (n=3)		Inter-assay variation (n=4)	
500	13.6%	Intra-assay overall CV=11.8%	47.1%	Inter-assay overall CV=28.2%
5000	16.3%		22.5%	
50,000	11.4%		31.9%	
500,000	5.7%		11.3%	

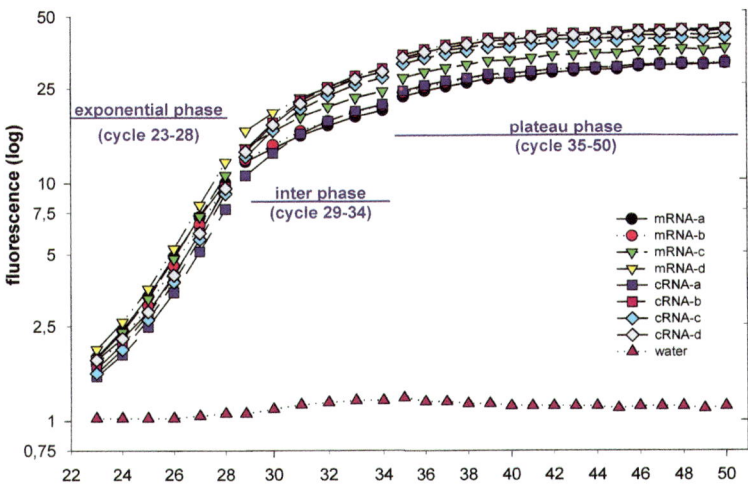

Fig. 4. Logarithmic fluorescence plot versus cycle number resulting from the amplification of liver RNA samples (5 ng) containing approximately 6.06×10^5 native IGF-1 mRNA molecules (n=4) or recombinant 6×10^5 IGF-1 RNA (n=4) and a water control containing no IGF-1 templates (n=1). Amplification efficiencies for the IGF-1 mRNA and recombinant RNA templates were recorded during the exponential phase (cycle 23–28), inter phase (cycle 29–32) and plateau phase (cycle 35–50) and calculated from fluorescence raw data

Comments

Optimal Primer Design and Influence on LightCycler PCR

Primer design is essential for successful online LightCycler PCR quantification. Even one mismatch within the primers can result in missing PCR products or reduced specificity (data not shown). Because of its high ramping rates, LightCycler PCR offers highly stringent reaction conditions for PCR.

Amplification Efficiencies in the LightCycler

The high stringency in LightCycler PCR results in lower PCR efficiencies. Prior work with conventional thermal cyclers gave efficiencies of 64.7% and 66.2% for IGF-1 mRNA and recombinant RNA, respectively, whereas amplification on the LightCycler resulted in lower values of 40.8% and 42.8% [9]. Reasons for this efficiency drop may include the higher ramping rates, and the short annealing and elongation times of LightCycler PCR. The product length determines the required elongation step duration, which is limited by the polymerase extension rate (~1000 bp per min elongation time). Efficiency is usually high with a product size of around 200–400 bp and a longer elongation step duration may enhance reaction efficiency. The starting amount of reverse transcribed total-RNA should not exceed 20 ng/capillary. Higher concentrations inhibit PCR.

Rare Transcript IGF-1 Quantification

The use of RT followed by LightCycler PCR is a simple and sensitive method of detecting low amounts of mRNA molecules and offers important insights into the local expression of transcripts present in low abundance. The reliability of the assay depends on the condition of identical amplification efficiencies for both the wild-type mRNA and the recombinant RNA. As demonstrated herein, amplification efficiencies were nearly identical. The sensitivity, linearity and reproducibility of the LightCycler PCR assay allows for the absolute and accurate quantification of IGF-1 mRNA molecules even in tissues or cells with low abundancies or when very small amounts of RNA are available. The variability of the IGF-1 test rises as the number of starting template molecules decrease. We have used this IGF-1 mRNA quantification system to compare the IGF-1 expression rates in bovine tissues (Bos taurus) [EMBL Ac. no. X15726]. The method can also be used in other species like sheep (Ovis aries) (EMBL Ac. no. M30653), pig (Sus scrofa) (EMBL Ac. no. X17492) and primates (Callithrix jacchus) (EMBL Ac. no. Z49055) with sufficiently high homologies of the amplified IGF-1 fragment.

References

1. Chelly J, Kaplan J-C, Marie P, Gautron S, Kahn A (1988) Transcription of the dystrophin gene in human muscle and non-muscle tissues. Nature 333: 856–860
2. Chirgwin JM, Przybyla EA, MacDonald J, Rutter WJ (1979) Isolation of biologically active ribonucleic acid from sources enriched in ribonucleases. J Biochem 18: 5291–5299
3. Dimitriadis G, Parry-Billings M, Leighton B, Piva T, Dunger D, Calder P, Bond J, Newsholme E (1994) Studies on the effects of growth hormone administration in vivo on the rates of glucose transport and utilization in rat skeletal muscle. Eur J Clin Invest 24: 161–165
4. Einspanier R, Gabler C, Einspanier A (1995) Fibroblast growth factor (a/bFGF), insulin-like growth factor-1 (IGF-1) and transforming growth factor-alpha (TGF-alpha) mRNAs are differentially expressed in the oviduct of the marmoset monkey. Biol Reprod 52: V 242
5. Fotsis T, Murphy C, Gannon F (1990) Nucleotide sequence of the bovine insulin-like growth factor 1 (IGF-1) and its IGF-1 A precursor. Nucleic Acids Res 18: 676
6. Morrison T, Weis JJ, Wittwer CT (1998) Quantification of low-copy transcripts by continous SYBR Green I monitoring during amplification. Biotechniques 24: 954–962
7. Ohlsen SM, Dean DM, Wong EA (1993) Characterization of multiple transcription initiation sites of the ovine insulin-like growth factor-1 gene and expression profiles of three alternatively spliced transcripts. DNA Cell Biol 12, 243–251
8. Pfaffl MW, Elsasser F, Meyer HHD, Sauerwein H (1998) Exogenous porcine growth hormone (GH) upregulates hepatic and muscle insulin-like growth factor-1 (IGF-1) mRNA expression in two different genotypes of pigs. Exp Clin Endocrinol Diabetes 106: 50, 1998
9. Pfaffl MW, Meyer HHD, Sauerwein H (1998) Quantification of the insulin-like growth factor-1 (IGF-1) mRNA: Development and validation of an internally standardised competi-

tive reverse transcription-polymerase chain reaction Exp Clin Endocrinol Diabetes 106, 502–512
10. Pfaffl MW; Rojas P; Meyer HHD, Einspanier R (1999) Comparing different RT-PCR approaches to measure specific cellular transcripts. Gene Quantification Europe Conference, 13.-14. May, Munich
11. Saiki R, Gelfand DH, Stoffel S, Sharf SJ, Higuchi R, Horn GT, Mullis KB, Erlich HA (1988) Primer-directed enzymatic amplification of DNA with thermostable DNA polymerase. Science 239: 487–491
12. Thissen JP, Ketelslegers JM, Underwood LE (1994) Nutritional regulation of the insulin-like growth factors. Endocr Rev 15: 80–101
13. Wittwer CT, Garling DJ (1991) Rapid cycle DNA amplification: time and temperature optimisation. Biotechniques 10: 76–83
14. Wittwer CT, Ririe KM, Andrew RV, David DA, Gundry RA, Balis UJ (1997) The LightCycler: a microvolume fluorimeter with rapid temperature control. Biotechniques 22: 176–181

An Application of Melting Curve Analysis to Large-Scale Genetic Analysis in Atherosclerotic Disease: Two Linked Polymorphisms of Glycoprotein Ia Gene and Myocardial Infarction in Japanese

H. Morita*, H. Kurihara, Y. Yazaki and R. Nagai

Introduction

The platelet–collagen receptor glycoprotein Ia/IIa (integrin $\alpha_2 \beta_1$) plays a fundamental role in the adhesion of platelet to fibrillar collagens followed by platelet activation and thrombus formation, and contributes to the pathogenesis of thrombotic diseases, including myocardial infarction [1, 2]. Kunicki et al. clearly demonstrated that two linked polymorphisms within the coding sequence of platelet glycoprotein Ia (integrin α_2) gene, 807C/T and 873G/A, are significantly correlated with the platelet receptor density of glycoprotein Ia/IIa [3, 4]. These polymorphisms do not alter the amino acid sequence of the translocated protein: Phe224(TT$\underline{T}$/TT$\underline{C}$) and Thr246(AC$\underline{A}$/AC$\underline{G}$), respectively; however, it appears that individuals with the 807TT (873AA) genotype have significantly higher receptor densities than those with the 807CC (873GG) genotype. Consequently, this 807C/T(873G/A) polymorphism would be a useful marker linked with the regulation of platelet function and a candidate risk factor for myocardial infarction. Indeed, one recent report showed that in the European population these polymorphisms are significantly associated with myocardial infarction [5].

In the present report, we describe a 1/1 age–sex matched, case-control study carried out to determine whether 807T and 873A variants of glycoprotein Ia gene are common in the Japanese population and whether they are associated with myocardial infarction.

As these loci have no convenient restriction sites, detection of genetic variants has conventionally required single strand conformation polymorphism (SSCP) and direct sequence analysis, which are both labor-intensive and time-consuming, especially in large-scale analysis. In this study, genotyping was accomplished using polymerase chain reaction (PCR) followed by melting curve analysis with specific fluorescent hybridization probes: a hybridization probe spanning one mismatch can still hybridize to the target sequence, but will melt off at lower temperatures than one with a perfect match. The use of melting temperature (T_m) proved to be accurate and convenient for detecting point mutations in large-scale

---

* H. Morita (✉) (e-mail: hmrt-tky@umin.ac.jp)
  Department of Cardiovascular Medicine, Graduate School of Medicine, University of Tokyo, 7-3-1 Hongo, Bunkyo-ku, Tokyo 113-8655, Japan

genetic analyses. Application of this method to association studies enabled us to assess the prevalence of glycoprotein Ia gene polymorphisms and their association with myocardial infarction.

Materials

Equipment LightCycler System (Roche Diagnostics, Mannheim, Germany)

Reagents LightCycler Red 640
Genomic DNA Extracting Columns (QIAamp blood kit, Qiagen)
Primer (Nihon Gene Research Laboratories Inc, Sendai, Japan)
Hybridization Probes (Nihon Gene Research Laboratories Inc, Sendai, Japan)

Procedure

Genomic DNA was obtained from 171 myocardial infarction patients and age- and sex-matched controls. Venous blood samples were collected in tubes containing Na_2EDTA and applied to genomic DNA extracting columns. Genotyping was accomplished, using PCR followed by melting curve analysis with specific fluorescent hybridization probes in a LightCycler System. As the large intron separating the two exons that encode each of the allelic sequences (807C/T and 873G/A) is highly polymorphic, the primers and probes used in this analysis were carefully designed (Table 1). For detection of the 807C/T genotype, a 273-bp product was amplified. Following PCR, melting curve analysis was performed, using a LightCycler Red 640-labeled anchor and a fluorescein-labeled mutation probe.

With this protocol, as temperature is slowly increased, monitored fluorescence decreases as the fluorescent mutation probe melts off and two fluorescent dyes are no longer in close proximity. When a substitution (C807→T) is present, the mismatch between the mutation probe and the target destabilizes the hybrid, leading to a lower T_m. By plotting the first negative derivative of the fluorescence (–dF/dT) as a function of the temperature, easily discriminated melting peaks are obtained (Fig. 1A). The fully homologous sequence (807C genotype) has a higher T_m than the sequence containing a mismatch with the mutation probe (807T genotype) *(compare the red and blue curves)*. The heterozygous sample displayed two smaller peaks, which are approximately one half of the maximal fluorescence, at precisely the same temperatures *(green curve)*. The partial sequences in Fig. 1B confirm that each melting curve corresponds to the respective genotype.

For the detection of the 873G/A genotype a 169-bp product was amplified. The procedures and principles of the hybridization analysis were the same as described above.

Table 1. Oligonucleotides

Integrin-2 (Genebank Accession # AF035968)				
	Position	Length	GC (%)	T_m (°C)
807C/T Primers				
AATGATTGTAGCAACATCCC-3	2749	20	40.0	54.3
TTTAACTTTCCCAGCTGCC-3′	3021R	19	47.4	56.6
Product				
807C/T Probes	2749–3021	273		
LCRed640-TGTTTGTGAGGTCCCCAC-CATATTGG-P	2801R	26	50.0	64.0
CTTGCATATTGAATTGCTCCGAAT-F	2826R	24	37.5	56.7
873G/A Primers				
GCAAGTCTTTATTTAATTTTATC-3′	6771	23	21.7	48.7
CTCAGTATATTGTCATGGTTG-3′	6939R	21	38.1	52.4
Product				
873G/A Probes	6771–6939	169		
LCRed640-AATGGTAGTTGTAACTGA-CGGTGAAT-PH-3′	6853	26	42.3	61.6
GCGACGAAGTGCTACGAAAG-F	6832	20	55.0	58.6

The following master mix was used:
- Master Mix

	Volume [µl]	[Final]
LightCycler-DNA Master Hybridization Probes	2.0	1×
MgCl$_2$ (25 mM)	0.8	2 mM
Primers (10 µM each)	1.0+1.0	0.5 µM
LightCycler Red 640-labeled anchor probe (8 µM)	1.0	0.4 µM
Fluorescein-labeled mutation probe (4 µM)	1.0	0.2 µM
H$_2$0 (PCR grade)	11.2	
Total volume	20	

- Amplification

Parameter	Value		
Cycles	40		
Type	Quantification		
	Segment 1	Segment 2	Segment 3
Target temperature [°C]	95	58	72
Incubation time [s]	3	15	11
Temperature transition rate [°C/s]	20	20	2
Acquisition mode	None	Single	None
Gains	F1=1, F2=10, F3=10		

A.

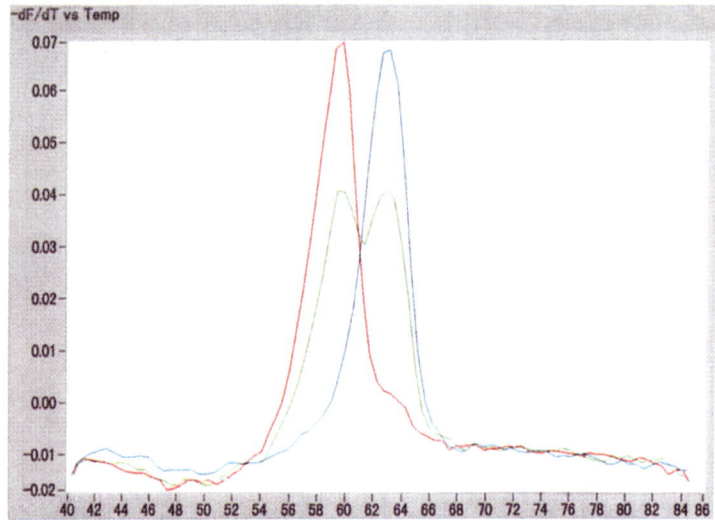

B.

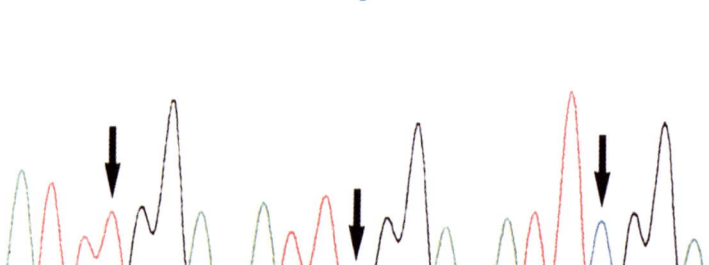

Fig. 1. A Representative melting curve analysis of platelet glycoprotein Ia 807C/T genotype. The first negative derivative of the fluorescence (−dF/dT) is plotted as a function temperature (C; T_m). The *red*, *green* and *blue* melting peaks correspond to the 807TT, 807CT and 807CC genotypes, respectively. B Partial nucleotide sequences of platelet glycoprotein Ia corresponding to red (*left*), green (*middle*) and blue (*right*) curves in A

- Melting Temperature

Parameter	Value		
Cycles	1		
Type	Melting Curve		
	Segment 1	Segment 2	Segment 3
Target temperature [°C]	95	40	85
Incubation time [s]	0	15	0
Temperature transition rate [°C/s]	20	20	0.2
Aquisition mode	None	None	Cont.
Gains	F1=1, F2=10, F3=0		

Results and Comments

In the melting curve analysis of 807C/T (Fig. 1A), the fluorescein mutation probe formed a G:T mismatch with the 807T sequence, that had a T_m that was 4°C lower than the completely complementary duplex. The observed melting temperatures of all alleles are shown in Table 2.

Distributions of Ia genotypes in control and patient groups are shown in Table 3. The 807CC, CT, and TT genotypes corresponded with the 873GG, GA, and AA genotypes, respectively, demonstrating complete linkage. Allele frequencies of the 807T (873A) variant were similar in the control and patient groups (0.357 vs. 0.351).

Table 2. Observed Melting Temperatures of Glycoprotein Ia Gene Polymorphisms

Locus	Allele	Match/Mismatch	T_m (observed)
807C/T	807C	C-G	62.2
	807T	T-G	58.5
873G/A	873G	G-C	66.2
	873A	A-C	62.1

Table 3. The distribution of 807C/T and 873G/A genotypes

	807CC (873GG)	807CT (873GA)	807TT (873AA)	807C/T (873G/A) Frequencies
Controls (n=171)	73 (42.7%)	74 (43.3%)	24 (14.0%)	0.643/0.357
Patients with MI (n=171)	70 (40.9%)	82 (48.0%)	19 (11.1%)	0.649/0.351

807CC, CT, TT are perfectly linked to 873GG, GA, AA, respectively
MI, myocardial infarction

Furthermore, the 807CC (873GG), 807CT (873GA), and 807TT (873AA) genotypes, respectively, occurred in 42.7%, 43.3%, and 14.0% of control subjects and in 40.9%, 48.0%, and 11.1% of patients. The genotype distributions of both groups were compatible with the Hardy-Weinberg equilibrium, and there was no significant difference in the distributions (χ^2 test).

Our results suggest that 807T and 873A variants of the glycoprotein Ia gene are also common in the Japanese population and in a perfect linkage with one another. Comparison of the control group in the present study with that from the earlier study by Moshfegh et al. [5] revealed that there is a difference in the distribution of the 807C/T (873G/A) allele between European and Japanese populations: the 807TT (873AA) genotype occurred in 14.0% of Japanese but in only 6.6% of Europeans. In their report, moreover, an association between 807TT (873AA) and myocardial infarction was demonstrated. However, in this study such association could not be reproduced.

Although it appears unlikely that the 807T (873A) variant is a useful marker for increased risk for myocardial infarction, it does not exclude the involvement of glycoprotein Ia/IIa in the pathogenosis of myocardial infarction. These polymorphisms could be one factor that regulates response to antiplatelet therapy. More extensive studies involving other populations will be required before a definitive conclusion can be drawn on the correlation between the 807C/T and 873G/A polymorphisms and platelet function.

Genotyping was carried out in this study, using a new protocol entailing PCR followed by melting curve analysis with specific fluorescent hybridization probes in the LightCycler System (Roche Diagnostics, Mannheim, Germany).

Recently, Bernard et al. reported the genetic analysis on hemochromatosis using this system [6]. Consistent with their report, this method enabled us to detect genetic variants quickly and precisely, eliminating the need for single strand conformation polymorphism (SSCP) and direct sequence analysis, which require multiple manual steps and are time consuming. Amplification and analysis of 32 samples required approximately 45 min and we needed no manual manipulation between amplification and genotyping. To the best of our knowledge, this is the first report of large-scale genetic analysis in cardiovascular diseases with this method. We anticipate that this method will be applicable to population screening, even larger scale genetic analyses, as a powerful tool, thereby providing an improved strategy for elucidation of genetic pathogenesis.

References

1. Stein B, Fuster V (1989) Antithrombotic therapy in acute myocardial infarction: prevention of venous, left ventricular and coronary artery thromboembolism. Am J Cardiol 64: 33B–40B
2. Saelman EU, Nieuwenhuis HK, Hese KM, de Groot PG, Heijnen HF, Sage EH, Williams S, McKeown L, Gralnick HR, Sixma JJ (1994) Platelet adhesion to collagen types I through VIII under conditions of stasis and flow is mediated by GPIa/IIa ($\alpha_2 \beta_1$-integrin). Blood 83:1244–1250

3. Kunicki TJ, Orchekowski R, Annis D, Honda Y (1993) Variability of integrin $\alpha_2\beta_1$ activity on human platelets. Blood 82:2693–2703
4. Kunicki TJ, Kritzik M, Annis DS, Nugent DJ (1997) Hereditary variation in platelet integrin $\alpha_2\beta_1$ density is associated with two silent polymorphisms in the α2 gene coding sequence. Blood 89:1939–1943
5. Moshfegh K, Wuillemin WA, Redondo M, Lammle B, Beer JH, Liechti-Gallati S, Meyer BJ (1999) Association of two silent polymorphisms of platelet glycoprotein Ia/IIa receptor with risk of myocardial infarction: a case-control study. Lancet 353:351–354
6. Bernard PS, Ajioka RS, Kushner JP, Wittwer CT (1998) Homogeneous multiplex genotyping of hemochromatosis mutations with fluorescent hybridization probes. Am J Pathol 153: 1055–1061

Infectious Organisms

Genotype-Specific Analysis of Hepatitis B Virus DNA on the LightCycler 303
Gunhild Sommer, Hans Will

Rapid and Specific Detection of *Bordetella pertussis* in Clinical Specimens
by LightCycler PCR 313
Udo Reischl, Siegfried Burggraf, Birgit Leppmeier,
Hans-Jörg Linde, Norbert Lehn

Rapid and Specific Detection of *Helicobacter pylori* by LightCycler PCR 323
Udo Reischl, Birgit Leppmeier, Markus Heep,
Daniela Beck, Norbert Lehn

Qualitative Detection of Herpes Simplex Virus DNA on the LightCycler 331
Harald H. Kessler

Quantitative Detection of *Cryptosporidium parvum* after In Vitro Excystation
by LightCycler PCR 341
Petra Krüger, Albrecht Wiedenmann, Despina Tougianidou,
Konrad Botzenhart

Quantitative Analysis of CMV in Infected Mice on the LightCycler System 349
Junichi Honda, Kotaro Oizumi

Detection and Differentiation of Equine Herpes Virus Type 1
and Type 4 on the LightCycler 359
Peter Hübert

Rapid and Quantitative Detection of *Toxoplasma gondii* by PCR
– A LightCycler Application in Prenatal Diagnosis 365
Jean-Marc Costa, Pauline Ernault, Stéphane Bretagne

Development of Quantitative PCR Tests for the Detection
of the Orthopox Virus Adsorption Protein Gene (ORF D8L) on the LightCycler ... 371
Claus-Peter Czerny, Michaela Alex, Jana Pricelius,
Christiane Zeller-Lue

Genotype-Specific Analysis of Hepatitis B Virus DNA on the LightCycler

Gunhild Sommer, Hans Will*

Introduction

There are more than 300 million chronic carriers of hepatitis B virus (HBV) worldwide. Chronicity may result in an asymptomatic carrier state, reactive fulminant hepatitis, mild and aggressive chronic hepatitis, and the development of liver cirrhosis or hepatocellurar carcinoma. Virus titers during the course of infection can vary dramatically, often being too low for detection by conventional methods. Serological analysis of viral proteins is not a reliable indicator for the HBV-DNA titers because there may be very few virions or none at all, but very high levels of subviral particles devoid of viral DNA in patients sera. Quantitative and sensitive determination of the viral DNA by PCR can provide indirect evidence for the level of viral replication, the degree of infectivity, and changes of viral DNA titers during the course of infection or in vitro studies of the viral life cycle. Moreover, monitoring of the changes in viral titers during treatment with cytokines and/or nucleoside analogues is of particular importance for evaluation of the response in terms of virus elimination, viral reactivation, or development of drug resistant viruses.

Here we present a reproducible and sensitive method for quantification of HBV-DNA from specific genotypes by real-time PCR on the LightCycler which can be performed in less than 2 hours, including sample preparation, PCR, and data evaluation. This method may help to answer open questions concerning mechanisms involved in the viral life cycle, virus infection, hepatopathogenesis, and antiviral treatment.

Materials

LightCycler Instrument (Roche Diagnostics, Mannheim, Germany) — **Equipment**

QIAamp Blood Kit (spin columns) (Qiagen) — **Reagents**
Proteinase K (Roche Diagnostics)
DNA Molecular Weight Marker VIII (Roche Diagnostics)

---

* Hans Will (✉) (e-mail: will@hpi.uni-hamburg.de)
  Heinrich-Pette-Institut für Experimentelle Virologie und Immunologie, Universität Hamburg, Martinistraße 52, 20251 Hamburg, Germany

LightCycler-DNA Master SYBR Green I (Roche Diagnostics)
LightCycler-DNA Master Hybridization Probes (Roche Diagnostics)

HBV-specific primers and fluorescent hybridization probes:

	Region	Length	GC (%)	T_m (°C)
HBV-specific primers:				
TTTTTCACCTCTGCCTAATCATC	pC	23	39	62.4
ACCCACCCAGGTAGCTAGAGTCAT	C	24	54	69.9
Product		299		
Hybridization probes:				
LCRed640-TTGGAGGCTTGAACAGT-AGGACATGA-P	Epsilon	26	46	67.4
CCAAAGCCACCCAAGGCAC-F	Epsilon	19	63	66.2

We considered previously described general principles for the successful design of hybridization probes [1].

Procedure

Isolation of DNA from the serum

Serum samples (100 µl) were incubated at 60°C for 4 h in lysis buffer containing 20 mM Tris-HCl (pH 8.0), 10 mM EDTA, 0.1% sodium dodecylsulfate (SDS), and 1 mg proteinase K per milliliter. The DNA was subsequently extracted first by phenol, then by chloroform/isoamylalcohol (24:1), and finally precipitated with two volumes of 96% ethanol containing 0.3 M sodium chloride and 10 µg tRNA as carrier. The pellet was then washed with 70% ethanol, air dried, and resuspended in 20 µl water.

Alternatively, the DNA was isolated from 100 µl serum by using a spin column method (QIAamp Blood Kit, Qiagen). The DNA was eluted from the spin column in a volume of 50 µl water. DNA obtained by both methods was stored at –20°C until use for PCR analysis.

LC-PCR

The following master mixes were used:
- SYBR Green I Master Mix

	Volume [µl]	[Final]
LightCycler-DNA Master SYBR Green I	2.0	1×
$MgCl_2$ (25 mM)	2.4	4 mM
Primers (10 µM each)	1.0	0.5 µM
Standards in appropriate dilution (1–10^5 copies)	1.0	
H_2O (PCR grade)	12.6	
Total volume	20.0	

- Hybridization Probe Master Mix

	Volume [µl]	[Final]
LightCycler-DNA Master Hybridization Probes	1.0	1×
MgCl$_2$ (25 mM)	1.2	4 mM
Primers (10 µM each)	0.5	0.5 µM
Hybridization probes (2 µM)	1.0	0.2 µM
Standards in appropriate dilution (10^5–10^8 copies)	1.0	1–10^3 pg/µl
H$_2$0 (PCR grade)	3.8	
Total volume	10.0	

For unknown samples, replace the standard DNA with 1 µl of DNA isolated from serum.

In total, 19 µl of SYBR Green I master mix and 9 µl of Hybridization Probes master mix, respectively, and 1 µl of DNA template were added in each capillary. Sealed capillaries were centrifuged in a microcentrifuge and placed into the LightCycler rotor.

The following PCR protocol was used for the SYBR Green detection:
- Denaturation for 30 s at 95°C
- Melting Curve Analysis

Parameter	Value		
Cycles	40		
Type	Melting curve		
	Segment 1	Segment 2	Segment 3
Target temperature [°C]	95	55	72
Incubation time [s]	0	10	15
Temperature transition rate [°C/s]	20	20	20
Acquisition mode	None	None	Single
Gain		F1=5	

Melting Curve Analysis and Cooling according to the instruction of the packaging insert of the LightCycler-DNA Master SYBR Green I (Ver.1, 07/1998).

For the hybridization probes the amplification program was modified as follows:
- Amplification

Parameter	Value		
Cycles	40		
Type	Quantification		
	Segment 1	Segment 2	Segment 3
Target temperature [°C]	95	55	72
Incubation time [s]	0	15	15
Temperature transition rate [°C/s]	20	20	20
Acquisition mode	None	Single	None
Gain		F1=1 ; F2=15	

Gel Electrophoretic Analysis of PCR Products

PCR products were recovered from the capillaries by reverse centrifugation into Eppendorf tubes. Samples were diluted in loading buffer and loaded on 2% agarose gels (w/v) in 1×TAE buffer. After staining in ethidium bromide solution (1 µg/ml) gels were analyzed under UV-light.

Results

Selection of Primers, Probes, and PCR Conditions for Quantitative Detection of HBV-DNA in Sera

To quantify HBV-DNA isolated from sera, PCR-primer and hybridization probes should anneal to a conserved region of the viral genome. Furthermore, 300 bp is an optimal amplicon length for an efficient PCR. Therefore, we chose the PCR-Primer P1 and P2 to obtain an amplicon of 299 bp in length (Fig. 1). The hybridization probes HP1 and HP2 anneal to the DNA region called epsilon, a rather conserved region of the viral genome which serves as encapsidation signal on the viral pregenome [2].

The amplification product was analyzed by gel electrophoresis (Fig. 2a) and its melting behaviour was determined (Fig. 2b). As a standard we used HBV-DNA from a cloned HBV genome of known sequence (subtype ayw) [3]. The chosen PCR conditions resulted in a single amplification product of the expected length (299 bp), indicating that only the predicted priming had occured (Fig. 2a). Moreover, the PCR-product specific melting curve, apparent as a single peak with a T_m of 85.3°C (peak 2) (Fig. 2b), corroborates this conclusion and also indicates that no misprimed products were amplified. The first peak in Fig. 2b represents primer-dimers which have a significant lower T_m than the HBV-DNA PCR products. They were formed only in samples with ten or fewer than ten viral genomes in the reaction mixture. Since primer-dimers often dramatically reduce the sensitivity of a PCR, the low amount of primer-dimer formation only under extreme reaction conditions

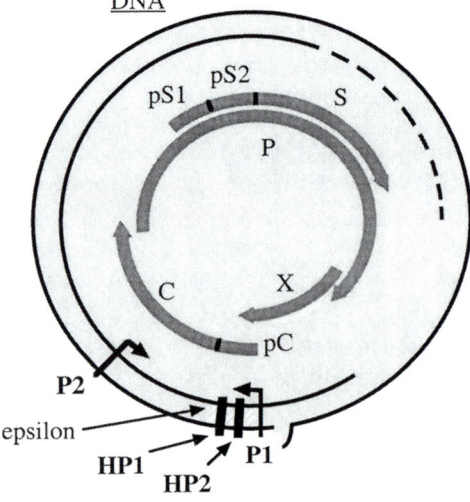

Fig. 1. Structure and gene organisation of the partially double-stranded circular HBV-DNA genome. *P1* and *P2* stand for PCR-primer 1 and 2. The hybridization probes HP1 and HP2 anneal to a conserved DNA region called epsilon which is marked by hatching. The abbreviations *pS1, pS2, S, P, X, pC,* and *C* stand for the viral proteins

with borderline HBV-DNA template concentrations in our assay underscores the optimal choice of amplification primers and PCR conditions.

For quantitative and specific detection of the amplified HBV-DNA the Hybridization Probe format was used. Gel electrophoretic analysis of the PCR products obtained under these conditions revealed a single amplification product of the

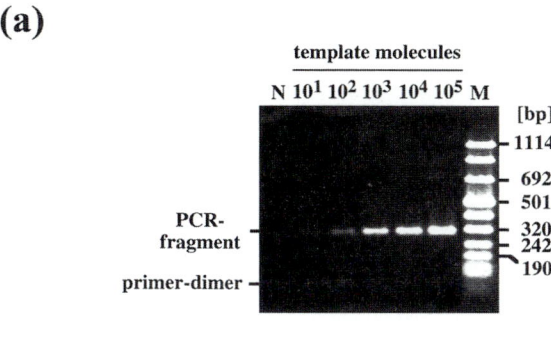

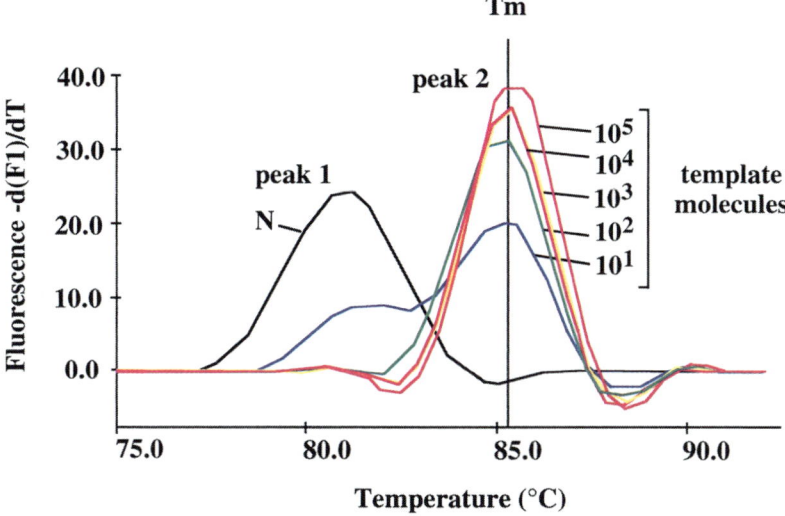

Fig. 2. Gel electrophoretic and melting curve analysis of amplified HBV-DNA. Defined amounts of cloned HBV-DNA (subtype ayw) [3], were used as templates for amplification using the SYBR Green I format. (a) The 299 bp PCR products were resolved on an agarose gel (2%) and stained with ethidium bromide. Lane M, DNA Molecular Weight Marker VIII; lane N, mock PCR without HBV-DNA as template (negative control). (b) The melting curve analysis shows two peaks: peak 1 represents primer-dimers, and peak 2 the amplified HBV-DNA with a melting temperature (T_m) of 85.3°C

expected length (lane 1 and 2) (Fig. 3a). Using sequential dilutions of the standard cloned HBV-DNA, a calibration curve perfectly linear in the range of 10^4–10^8 copies of template was obtained. In the Hybridization Probe format, all amplification curves had the same slope. This indicates that cloned HBV-DNA and viral DNA isolated from sera were amplified with similar efficacy and accuracy (Fig. 3b).

(a)

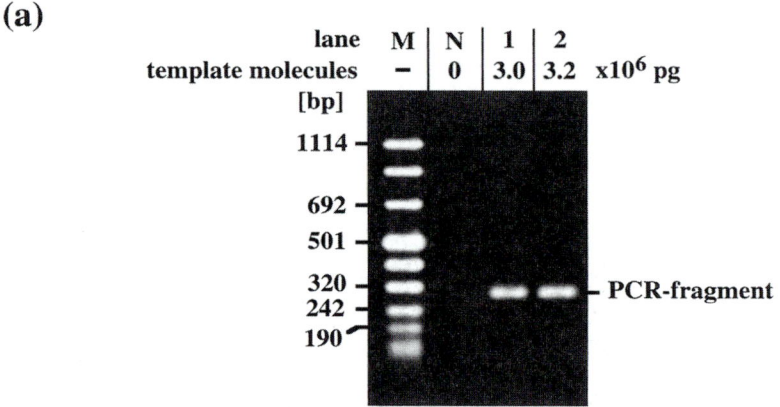

(b)

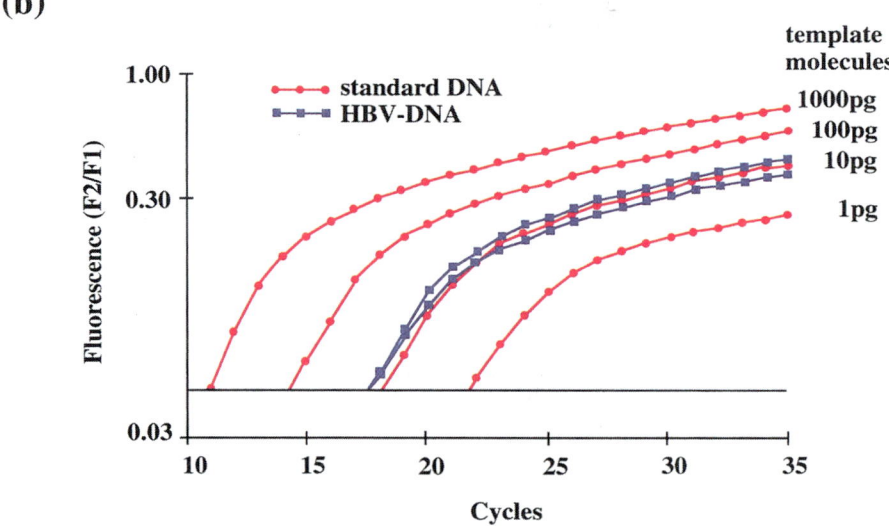

Fig. 3. Gel electrophoretic analysis of PCR products after quantitative detection of HBV-DNA isolated from serum. (**a**) HBV-DNA isolated from sera was amplified in duplicates for 35 cycles (lanes 1 and 2) by using the Hybridization Probe format. After separation on a 2%-agarose gel the PCR products were stained with ethidium bromide. Lane M and lane N, length marker and negative control as described in *Fig. 2*. (**b**) For quantitative analysis of viral DNA from the serum defined amounts of cloned HBV-DNA (subtype ayw) [3] were used as templates to establish a calibration curve

Quantification of HBV-DNA Isolated from Sera of Different Patients

For application of this method we quantified the HBV-DNA titer in the sera of different patients after extraction of the DNA by the spin column method. The HBV-DNA from ten different sera were thus analyzed on the LightCycler (Fig. 4a). Each quantitative PCR was carried out in duplicates. On average, a very low intra-assay variation factor of only 1.24 was found (range: 1.00–1.72). The inter-assay variation factor was only slightly higher (1.54 in average, range: 1.01–2.37). In addition, very similar quantitative HBV-DNA data were obtained when using HBV-DNA extracted independently twice by the spin column method (serum 1) (Fig. 4a).

(a)

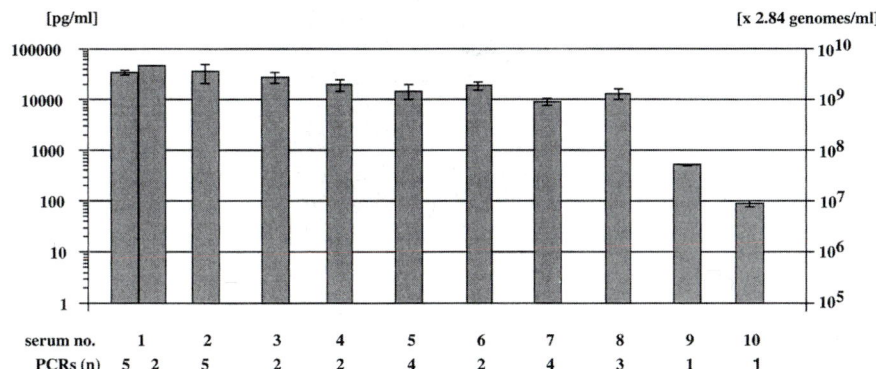

(b)

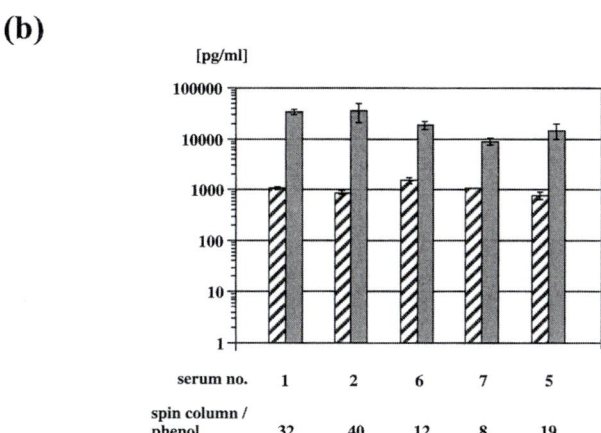

Fig. 4. Quantitative analysis of HBV DNA isolated from sera of chronically infected patients. (a) Quantitative PCR was always carried out in duplicates and in addition, where indicated, also in independent repeat experiments. (b) Comparison of quantitative DNA data obtained with template DNA prepared from five human sera either by the conventional phenol extraction method (*hatched column*) or the spin column method (*colored column*). The ratio of DNA titers measured with DNA obtained by either extraction methods is given

Taken together, these results imply that our method allows highly reliable, reproducible, and sensitive quantitative determination of HBV-DNA in sera of patients. Whether all genotypes of HBV are amplified with equal efficacy remains to be established.

Phenol-extracted HBV-DNA from sera is most commonly used in many laboratories for PCR. Therefore, we also isolated HBV-DNA by using this method, and subjected it to quantitative PCR analysis. On average, about 20-fold lower titers (range: 8–40) of HBV-DNA were obtained than by using HBV-DNA extracted with the spin column method (Fig. 4b). This may be due to traces of phenol known to inhibit the Taq polymerase or other inhibiting factors not removable by the phenol extraction method. These results indicate that the spin column method is superior to the phenol extraction method for quantitative and sensitive real-time PCR determination of serum derived HBV-DNA.

Comments

Further improvement of our method may be needed for monitoring viral DNA in very low viremic sera. Hot start techniques are known to improve sensitivity, specificity and yield of PCR by decreasing primer dimer formation. Furthermore, a mixture of Taq polymerase and a thermostable polymerase with proofreading activity, such as the Pwo polymerase, can increase the sensitivity of the PCR at least tenfold [4].

For quantification of HBV-DNA we established a method based on the HBV genotype D (subtype ayw) sequence (Fig. 5) [3]. However, there are six known HBV genotypes which diverge by more than 8% on the DNA sequence level (reviewed in [5]). Genotype A, D, and E are most common in Europe, North and Central America and Africa, genotype B and C are most prevalent in the Far East, and genotype F occurs almost exclusively in South America. Genotype-specific

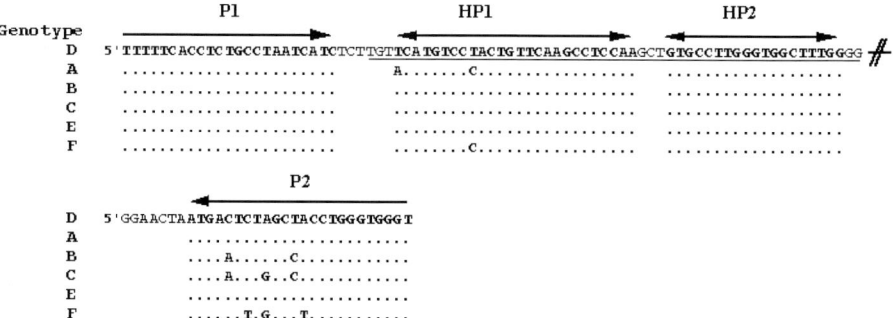

Fig. 5. HBV genotype-specific sequences in the regions to which PCR-primers (P1 and P2) and hybridization probes (HP1 and HP2) bind. Sequence identities with the primer and probe sequences are shown as *dots*; substitutions are indicated in *bold* letters. Sequences of the encapsidation signal are *underlined*

substitutions can influence the sensitivity of the real-time PCR, especially if they occur in PCR-primer binding sites. This is consistent with approximately five times less efficient amplification of genotype C HBV-DNA when using our PCR-primers matching perfectly with genotype A, D, and E (data not shown). Taken together, accurate quantification of HBV-DNA for all HBV genotypes should be possible, although our method is less sensitive for genotypes more common in the Far East and South America. It is necessary to evaluate the efficacy of quantitative real-time PCR determination of HBV-DNA, with additional mutations in the primer binding region which may exist in HBV variant genomes isolated from patients' sera.

References

1. Caplin BE, Rasmussen RP, Bernard PS, Wittwer CT (1999). LightCycler Hybridization Probes. Biochemica 1: 5–8
2. Nassal M, Schaller H (1996). Hepatitis B virus replication – an update. J Viral Hepat 3: 217–226
3. Galibert F, Mandart E, Fitoussi F, Tiollais P, Charnay P (1979). Nucleotide sequence of the hepatitis B virus genome (subtype ayw) cloned in E.coli. Nature 281: 645–650
4. Günther S, Sommer G, von Breunig F, Iwanska A, Kalinina T, Sterneck M, Will H (1998). Amplification of full-length hepatitis B virus genomes from samples from patients with low levels of viremia: frequency and functional consequences of PCR-introduced mutations. J Clin Microbiol 36: 531–538
5. Pult I, Günther S, Burda M, Dandri M, Iwanska A, Kalinina T, Possehl C, Rang A, Sommer G, Wollersheim M, Will H (1998). Hepatitis B virus variants and their role in hepatopathogenesis. Hepatologia Polska 5: 57–70.

Rapid and Specific Detection of *Bordetella pertussis* in Clinical Specimens by LightCycler PCR

Udo Reischl*, Siegfried Burggraf, Birgit Leppmeier, Hans-Jörg Linde, Norbert Lehn

Introduction

The bacterium *Bordetella pertussis* is the causative agent of whooping cough, which is an infectious disease occurring worldwide with a high incidence among young, unvaccinated infants. Related species which may also cause pertussis syndrome are *B. parapertussis* (milder respiratory infection) and *B. bronchiseptica* (primarily infecting animals). The striking and unique presentation of classical pertussis does not usually present a clinical diagnostic dilemma. Atypical pertussis, however, which may occur with mild or absent symptoms in adults or previously vaccinated children, offers a greater diagnostic challenge to the clinician. It has been shown that atypical illness in adults is common, endemic, and usually unrecognized [1]. The epidemiological implications of unrecognized pertussis are that exposure of unimmunized infants to individuals with pertussis places them at high risk and that pertussis remains endemic in society. The recent increase in the incidence of pertussis in the United States [2] and the fact that other agents (such as adenovirus, cytomegalovirus, or chlamydiae) are thought to cause pertussis syndrome underscore the need for rapid and accurate diagnostic methods to guide therapeutic and preventive interventions.

Although culture of *B. pertussis* is considered to be the "gold standard" for the diagnosis of pertussis due to its high specificity, its use is possible only in the initial phase of infection and colonies have to be verified by agglutination and biochemical tests. Moreover, successful culture requires special media and incubation periods up to 7 days and is highly dependent upon specimen collection and laboratory techniques. Diagnostic sensitivities below 60% are observed when nasopharyngeal secretions are obtained outside the early catarrhal stage of the illness, from older or vaccinated persons, from persons treated with certain antibiotics, or in the case of prolonged sample transport.

Serology is considered highly sensitive and can provide rapid diagnosis for a patient with classical pertussis, but there are two major drawbacks: it cannot be used in the acute phase of the disease, and it can be difficult to differentiate between vaccine effects and a pertussis infection. Moreover, serological testing is

---

* Udo Reischl (✉) (e-mail: Udo.Reischl@klinik.uni-regensburg.de)
  Institute of Medical Microbiology and Hygiene, University of Regensburg,
  Franz-Josef-Strauß-Allee 11, 93053 Regensburg, Germany

considered less reliable in infants and, due to its poor specificity, it should not be used to detect atypical disease. Thus there is a need for more rapid diagnostic methods with high degrees of specificity and sensitivity, preferably for use early in infection.

Similar to other fastidious organisms that are present in the asymptomatic individual, nucleic acid amplification techniques, such as PCR, are the methods of choice for direct detection of *B. pertussis* in clinical specimens [3–7]. Potential *B. pertussis*-specific target regions include insertion sequences, repeat elements, the pertussis toxin (PT) promoter region, the adenylate cyclase gene, and the porin gene. Amplification of target regions within repetitive elements, which are present in 50–100 copies in *B. pertussis* chromosomal DNA, are more sensitive than amplifying single-copy targets.

Here we describe the development of a sensitive and specific hybridization probe-based LightCycler assay and its validation for diagnosis of *B. pertussis* infections in clinical specimens. Thirty-five well-characterized nasopharyngeal specimens and cultured strains of *B. pertussis*, and 80 isolates of gram-negative and gram-positive bacterial species other than *Bordetella*, were tested by PCR and compared to clinical findings and the results of conventional microbiological testing.

Materials

Equipment LightCycler Instrument (Roche Diagnostics, Mannheim, Germany)
Oligo Primer analysis software (Molecular Biology Insights, Inc., Cascade, CO)

Reagents Amplification Primers (Metabion, Munich, Germany)
Hybridization Probes (TIB MOLBIOL, Berlin, Germany)
  The following reagents were purchased from Roche Diagnostics:
High Pure PCR Template Purification Kit
LightCycler-DNA Master Hybridization Probes
LightCycler Control Kit DNA

Procedure

Preparation of Template DNA Bacterial DNA was prepared, using the High Pure PCR Template Preparation Kit according to the manufacturer's instructions. Starting with nasopharyngeal specimens (throat swabs, nasal swabs, nasopharyngeal aspirates or nasopharyngeal swabs), aliquots with a volume of approx. 100 µl, or directly the cotton wools of the swabs, were suspended in 200 µl of tissue lysis buffer and 40 µl of proteinase K solution (20 mg/ml) and incubated at 55°C for at least 30 min. After complete disintegration of the tissue, which can be examined visually, 200 µl of binding buffer were added and a further incubation was performed at 70°C for 10 min. Subsequently, 100 µl of isopropanol were added and the mixture was applied to the High Pure spin column.

Starting with cultured bacteria, a 150 μl aliquot was suspended in 200 μl of PBS buffer, 15 μl of a lysozyme solution (10 mg/ml in Tris-HCl, pH 8.0) was added and incubated at 37°C for 10 min. After adding 200 μl of binding buffer and 40 μl of a proteinase K solution (20 mg/ml) with a further incubation at 70°C for 10 min, 100 μl of isopropanol were added and the mixture was applied to the High Pure spin column.

Following the centrifugation and wash steps, bacterial DNA was eluted with 200 μl of elution buffer and a 2 μl aliquot was directly transferred to PCR. The remainder was stored at −20°C for further experiments.

Primer and Probe Design

A previously published primer pair [4] (Table 1) was used for amplifying a segment within the *B. pertussis* repetitive insertion sequence IS481 (GenBank Accession #M28220), which is species specific and, due to its high copy number per cell, a very sensitive target. Based on the primer annealing sites, an alignment was performed with all of the different *B. pertussis* sequence entries deposited in the GenBank and EMBL databases. The short length of the 181-bp amplicon restricted the selection of candidate oligonucleotide sequences serving as LightCycler hybridization probes for the sequence-specific detection of amplicons. The selection of appropriate oligonucleotides was aided by the Oligo software in order to obtain comparable T_m values, GC contents within the range of 35–60%, and to avoid stable secondary structures, regions of significant self-complementarity, stretches of palindromic sequences and close proximity between the primers and the probes.

Inhibition Control

The LightCycler Control Kit DNA was used for the detection of Taq DNA polymerase inhibitors, possibly present in DNA preparations from clinical samples. The latter kit provides control DNA template, primers and hybridization probes for amplification and specific detection of the human β-globin gene. To identify even weak inhibition events, 300 pg of human genomic DNA was amplified in each case.

Table 1. Oligonucleotides

Bordetella pertussis Repetitive Insertion Sequence (Genebank Accession # M 22031)				
	Position	Length	GC (%)	T_m (°C)
Primers				
GATTCAATAGGTTGTATGCATGGTT	12	25	36.0	61.1
TTCAGGCACACAAACTTGATGGGCG	192 R	24	50.0	67.6
Product	12–192	181		
Probes				
GGGTGTTCATCCGGCCGGGCTC-F	112 R	22	72.7	71.7
LCRed640-TGAGTGAACTGGGGGG-TTGGCGATT-P	88 R	25	56.0	69.8

LightCycler PCR   The following master mix was used for amplification and hybridization probe-based detection of the *B. pertussis*-specific amplicon:

	Volume [µl]	[Final]
LightCycler-DNA Master Hybridization Probes	2	1×
MgCl$_2$ stock solution	1.6	3 mM
Primers (10 µM each)	1.0	0.5 µM
Hybridization Probes (2 µM)	2+2	0.2 µM
H$_2$O (PCR grade)	9.4	–
Total volume	18	

To complete the amplification mixtures, 18 µl of master mix and 2 µl of the corresponding template DNA preparation were added to each capillary. After a short centrifugation, the sealed capillaries were placed into the LightCycler rotor.

The inclusion of negative as well as positive controls in each set of experiments is considered to be obligatory in the field of diagnostic PCR. The negative control sample was prepared by replacing the DNA template with PCR grade water. The positive control sample was prepared by adding 2 µl of *B. pertussis* genomic DNA (approx. 1 ng) to the master mix.

The following PCR protocol was used used for amplification and hybridization probe-based detection of the *B. pertussis*-specific amplicons:

- Denaturation for 2 min at 95°C
- Amplification

Parameter	Value		
Cycles	50		
Type	Quantification		
	Segment 1	Segment 2	Segment 3
Target temperature (°C)	95	50	72
Incubation time (s)	0	10	20
Temperature transition rate (°C/s)	20	20	20
Acquisition mode	None	Single	None
Gains	F1=1; F2=15; F3=30		

- Cooling 2 min at 40°C

Results

Based on an established PCR protocol for the detection of *B. pertussis* directly on nasopharyngeal specimens, the LightCycler system was examined to determine if it could simplify the diagnostic laboratory workflow by automation and by reducing the possibility of product contaminations frequently associated with post-

PCR amplicon manipulation. As with every modified or new PCR protocol in diagnostic microbiology, the sensitivity and specificity of any assay has to be carefully examined before introducing it into routine testing.

Sensitivity

Starting with a proven pair of PCR primers [4], a set of hybridization probes was selected, targeting the amplified segment within the *B. pertussis*-specific repetitive element IS481. PCR experiments were performed on serial dilutions of genomic DNA prepared from cultured *B. pertussis* organisms in order to determine the assay's lower limit of detection. A detection limit of 1 pg of template DNA was determined (Fig. 1). The diagnostic sensitivity of the assay was further examined with a dilution series performed on a typical nasopharyngeal swab specimen obtained from a *B. pertussis*-positive patient. After 35 cycles of amplification, positive PCR results were observed, even with template DNA prepared from a 1:100 dilution of the specimen (Fig. 2).

Specificity

The PCR assay was evaluated with a collection of 25 well-characterized nasopharyngeal specimens (originating from 12 *B. pertussis*-positive and 13 *B. pertussis*-negative patients), 10 cultured strains of *B. pertussis*, and 80 isolates of gram-neg-

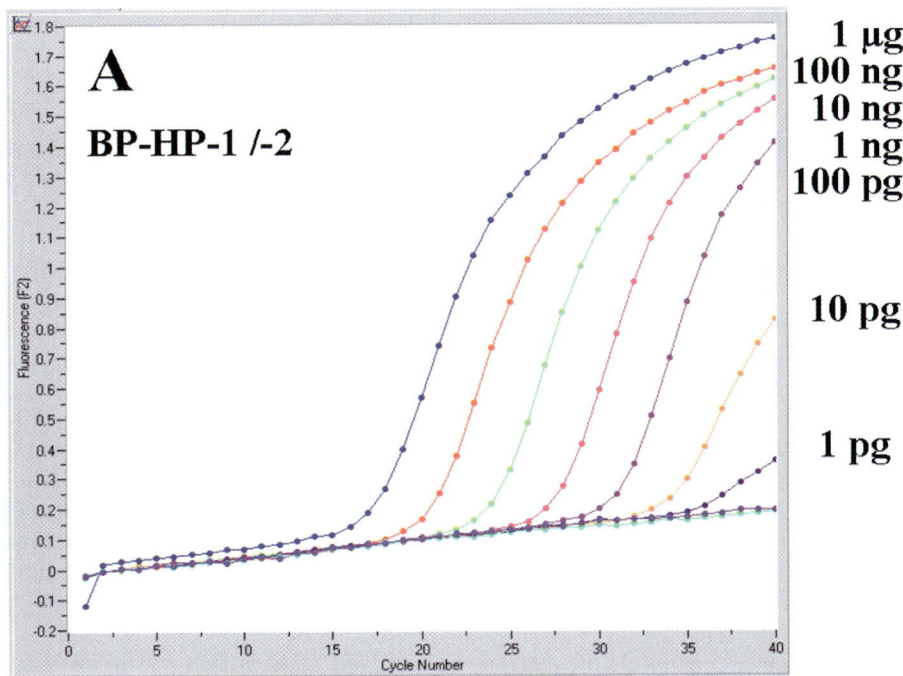

Fig. 1. Sensitivity of the PCR assay for the detection of the specific inverted repeat element IS481 determined with serial dilutions of *B. pertussis* genomic DNA. For each dilution, the corresponding amount of *B. pertussis* genomic DNA present in the reaction mixture is given next to the amplicon curve

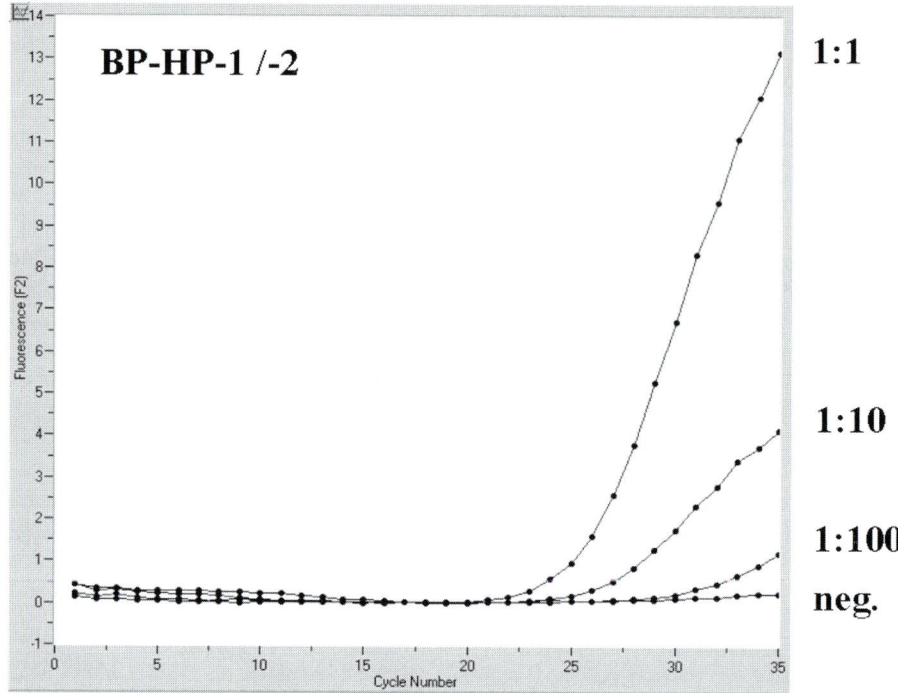

Fig. 2. Sensitivity of the *B. pertussis*-specific PCR assay determined with serial dilution of a typical nasopharyngeal swab specimen obtained from a *B. pertussis*-positive patient. The dilutions are indicated next to the corresponding LC curves

ative and gram-positive bacterial species other than *Bordetella*. All DNA preparations from *B. pertussis* strains and from nasopharyngeal specimens originating from *B. pertussis*-positive patients showed amplification by fluorescence, whereas none of the DNA preparations from the other nasopharyngeal specimens or cultured bacterial species examined were recognized.

Experimental results on cultured bacteria and clinical specimens are shown in Figure 3. Even after as few as 35 cycles of amplification, the curves generated by the LightCycler software allowed discrimination between *B. pertussis*-positive and -negative samples. One specimen, however, which originated from a pertussis patient who had already received antibiotic treatment, gave a very weak signal after 35 cycles of amplification. Therefore, a total of 40 amplification cycles should be performed if the sample is from an antibiotic-treated patient.

Inhibition Events To investigate PCR inhibition, a separate set of PCR experiments was performed on DNA preparations from the nasopharyngeal specimens and 5 cultured isolates (spiked with 300 pg of human genomic DNA). Here the primers and probes of the LightCycler Control Kit were used for amlification of the human ß-globin gene. No inhibition events were observed with any of the DNA preparations tested.

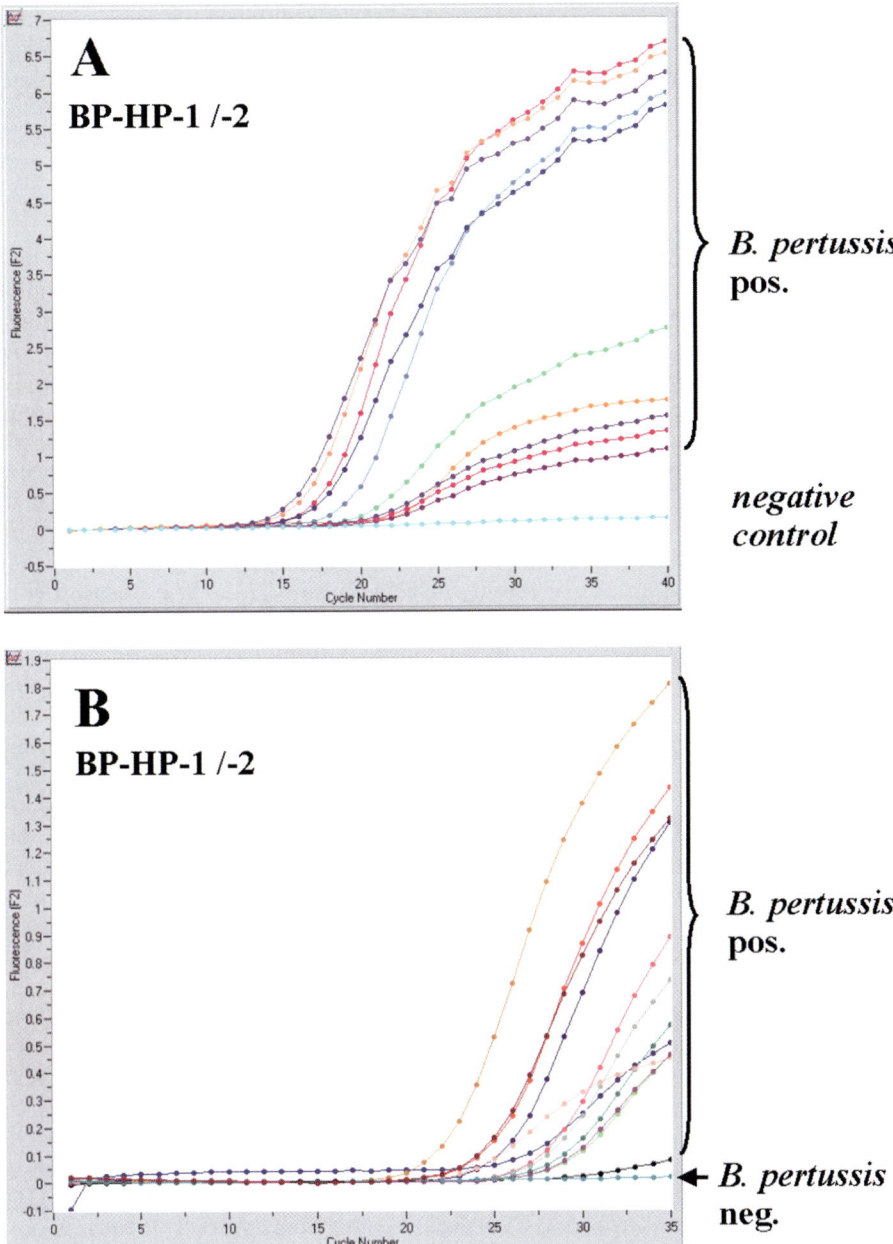

Fig. 3. Evaluation of the *B. pertussis*-specific PCR assay with a representative set of 10 cultured strains of *B. pertussis* (A) and 12 well-characterized nasopharyngeal swab specimens originating from *B. pertussis*-positive patients (B). The curves representing *B. pertussis*-positive specimens are indicated by *brackets*

In contrast to direct testing of clinical specimens known to be prone to inhibition, such as blood or stool samples, there was no apparent inhibition with nasopharyngeal swabs or cultured bacteria processed with the HighPure PCR Template Preparation Kit.

Comments

LightCycler PCR

Previous methods for diagnostic PCR have employed a variety of time-consuming and laborious procedures (such as Southern blot, DNA sequencing or solid-phase capturing) for sequence-specific characterization of the amplicons generated with clinical specimens. Enhancing the reliability of the results by sequence-specific probes and simplifying the PCR workflow by a completely automated amplification and online detection procedure, the LightCycler system proved itself as a valuable tool for rapid and sensitive identification of *B. pertussis* in the environment of an academic microbiological laboratory. Once DNA is extracted from suitable specimens (like throat swabs, nasal swabs, nasopharyngeal aspirates or nasopharyngeal swabs) and reaction mixtures are completed, the results of specific and sensitive PCR are available within a period of 60 minutes. Current limitations of the LightCycler system are a total reaction volume of 20 µl (which limits the input of template DNA), individual sample handling of the reaction cuvettes and the lack of an automated device for template DNA preparation.

Comparison of PCR with Direct Fluorescent-Antibody (DFA) Testing and Culture

Previous studies suggest that PCR-based assays are comparable or, in certain cases, even superior to culture with respect to sensitivity and specificity [3–7]. Recently, the performance of culture, DFA and PCR testing were compared on 319 consecutive specimens [8]. Only 14% of the positive PCR results were confirmed by culture. However, after comparison of the performance of culture, DFA, and PCR against an expanded gold standard defining infection (multiple positive tests or a single positive test other than culture combined with symptoms meeting the CDC clinical case defintion [9]), it became apparent that the difference was due largely to false-negative culture results rather than to false-positive PCR results. Indeed, the resolved sensitivities of culture and PCR were determined to be 15.2 and 93.5%, respectively. The sensitivity of the DFA was 52.2%. Provided that the method is executed and applied properly, these data support the suitability of PCR as the primary diagnostic test for direct detection of *B. pertussis* in clinical specimens.

Differentiation Between B. pertussis and B. parapertussis

From the clinical point of view, a rapid and reliable method for the differentiation between *B. pertussis* and *B. parapertussis* organisms is highly desirable [10. 11]. The use of the *B. parapertussis* repetitive insertion sequence IS1001 as target and the selection of a second specific probe labelled with LCRed705 may offer an attractive approach for the simultaneous detection of both *Bordetella* species. Amplification could be performed in a multiplex assay format with detection of each amplicon using the dual color option of the LightCycler system.

References

1. Deville JG, Cherry JD, Christenson PD, Pineda E, Leach CT, Kuhls TL, Viker S (1995) Frequency of unrecognized Bordetella pertussis infection in adults. Clin Infect Dis 21:639–642
2. Centers for Disease Control and Prevention (1993) Resurgence of pertussis: United States 1993. Morbid Mortal Weekly Rep 42:952–960
3. Wadowsky RM, Michaels RH, Libert T, Kingsley LA, Ehrlich GD (1996) Multiplex PCR-based assay for detection of Bordetella pertussis in nasopharyngeal specimens. J Clin Microbiol 34:2645–2649
4. Glare EM, Paton JC, Premier R, Lawrence AJ, Nisbet LT (1990) Analysis of a repetitive DNA sequence from Bordetella pertussis and its application to the diagnosis of pertussis using the polymerase chain reaction. J Clin Microbiol 28:1982–1987
5. Backman A, Johansson B, Olsen P (1994) Nested PCR optimized for detection of Bordetella pertussis in clinical nasopharyngeal samples. J Clin Microbiol 32:2544–2548
6. Schlapfer G, Cherry JD, Heininger U, Uberall M, Schmitt-Grohe S, Laussucq S, Just M, Sterhr K (1995) Polymerase chain reaction identification of Bordetella pertussis infection in vaccinees and family members in a pertussis vaccine efficacy trial in Germany. Pediatr Infect Dis J. 14:209–214
7. Van der Zee A, Agterberg C, Peeters M, Schellekens J, Mooi FR (1993) Polymerase chain reaction assay for pertussis: simultaneous detection and discrimination of Bordetella pertussis and Bordetella parapertussis. J Clin Microbiol 31:2134–2140
8. Loeffelholz MJ, Thompson CJ, Long KS, Gilchrist MJR (1999) Comparison of PCR, culture, and direct fluorescent -antibody testing for detection of Bordetella pertussis. J Clin Microbiol 37:2872–2876
9. Centers for Disease Control and Prevention. 1997. Case definitions for infectious conditions under public health surveillance. Morbid Mortal Weekly Rep 46:25
10. Lind-Brandberg L, Welinder-Olsson C, Lagergard T, Taranger J, Trollfors B, Zackrisson G (1998) Evaluation of PCR for diagnosis of Bordetella pertussis and Bordetella parapetussis infections. J Clin Microbiol 36:679–683
11. Farell DJ, Daggard G, Mukkur TKS (1999) Nested duplex PCR to detect Bordetella pertussis and Bordetella parapertussis and its application in dioagnosis of pertussis in nonmetropolitan Southeast Queensland, Australia. J. Clin Microbiol 37:606–610

Rapid and Specific Detection of *Helicobacter pylori* by LightCycler PCR

Udo Reischl*, Birgit Leppmeier, Markus Heep,
Daniela Beck, Norbert Lehn

Introduction

Helicobacter pylori, a spiral-shaped, microaerophilic bacterium that colonizes the stomach in humans, is an important cause of many gastrointestinal disorders, ranging from chronic gastritis, peptic ulcer disease (PUD), mucosa-associated lymphoid tissue lymphoma, to adenocarcinoma. It is most widely known as the cause of duodenal and gastric ulcers [1, 2]. Infection occurs mainly in childhood and, once established, it may reside in the gastric mucosa for years, possibly for the entire lifetime of the host [3, 4]. Up to 40% of the population may be infected with *H. pylori*, and approximately 1 in 6 infected individuals will develop a peptic ulcer.

Treatment regimes with short courses of antibiotics are successful in eradicating the infection in the majority of cases. Consequently, the recurrence rate of ulcers has fallen from a range of 50–100% to less than 5% in *H. pylori*-negative patients. Nevertheless, due to the emerging cases of drug resistant organisms, medication side-effects and costs, the presence of *H. pylori* has to be verified in each case and empiric treatment should be avoided [5].

Since confirmation of the organism's role in gastric ulcer disease, a number of tests have been developed to confirm the presence of *H. pylori*. These fall into 2 categories: those that rely on non-invasive methods such as serology, ^{14}C-labeled urea breath testing, stool antigen testing, and those requiring endoscopy. Endoscopic examination is used to locate the ulcers and obtain tissue samples for examination. Methods such as histology, culture, the rapid urease test, or PCR testing are then used to detect the organisms in the biopsy specimen.

Diagnostic culture, which is the gold standard for many infectious diseases, is rarely practised for primary diagnosis of *H. pylori*. Successful growth of the fastidious organisms depends mainly on laboratory expertise, timely specimen handling, appropriate media, and the incubation environment. Since only the detection of *H. pylori* from biopsy specimens gives absolute specificity, direct PCR-testing of biopsy specimens has proved to be a valuable tool in the rapid and specific detection of *H. pylori* infections, thereby complementing the results of indirect screening methods.

---

* Udo Reischl (✉) (e-mail: Udo.Reischl@klinik.uni-regensburg.de)
 Institute of Medical Microbiology and Hygiene, University of Regensburg,
 Franz-Josef-Strauß-Allee 11, 93053 Regensburg, Germany

Intense urease activity is a key pathogenic factor of *H. pylori*. The urease gene of *H. pylori* was first sequenced in 1990 [6] and these genes are highly conserved among the so-called *Helicobacter*-like organisms. Due to the absence of urease activity (and the corresponding gene) in most other bacterial species, conserved regions within the urease gene were selected as suitable targets for specific PCR assays.

The present study describes the development and evaluation of a rapid PCR assay for the detection of *H. pylori* directly on biopsy material. Since amplification and sequence-specific detection of the amplicons was performed on a LightCycler with ultrarapid thermal cycling, a total assay time of less than 60 minutes was achieved. The assay was evaluated with a collection of 40 cultured strains of *H. pylori* and other *Helicobacter* species, as well as 40 gram-negative and 40 gram-positive bacterial isolates. A good overall correlation of the PCR results was observed when compared to clinical findings and the results of conventional microbiological testing.

Materials

Equipment
Micro pistil (Eppendorf, Hamburg, Germany)
LightCycler Instrument (Roche Diagnostics, Mannheim, Germany)
Oligo Primer analysis software (Molecular Biology Insights, Inc., Cascade, CO)

Reagents
Amplification primers (Metabion, Munich, Germany)
Hybridization probes (TIB MOLBIOL, Berlin, Germany)
The following reagents were purchased from Roche Diagnostics:
High Pure PCR Template Purification Kit
LightCycler-DNA Master Hybridization Probes
LightCycler-Control Kit DNA

Procedure

Preparation of Template DNA

Bacterial DNA was prepared using the High Pure PCR Template Preparation Kit according to the manufacturer's instructions. Starting with cultured bacteria, a loopful of bacteria was harvested from the plate using a sterile disposable plastic loop. The bacteria were suspended in 200 µl of PBS buffer, and 15 µl of a lysozyme solution (10 mg/ml in Tris-HCl, pH 8.0) was added and incubated at 37°C for 10 min. After adding 200 µl of binding buffer, 40 µl of a proteinase K solution (20 mg/ml) and a further incubation at 70°C for 10 min, 100 µl of isopropanol were added and the mixture was applied to the High Pure spin column.

Following the centrifugation and wash steps, bacterial DNA was eluted with 200 µl of elution buffer and a 2-µl aliquot was directly transferred to PCR. The remainder was stored at –20°C for further experiments.

Primer and HybProbe Design

A previously published primer pair [7] was used for amplifying a specific segment within the gene coding for the large subunit of the *H. pylori* urease (Table 1). Based on the primer annealing sites, an alignment was performed with all of the different *H. pylori* sequence entries deposited in the GenBank and EMBL databases. Defined sequence variations observed at certain nucleotide positions within the 411-bp amplicon restricted the selection of candidate oligonucleotide sequences serving as LightCycler hybridization probes for the sequence–specific detection of the amplicons. The selection of appropriate hybridization probe oligonucleotides was further aided by the Oligo software in order to obtain comparable Tm values and GC contents within the range of 35–60% and to avoid stable secondary structures, regions of significant self-complementarity or stretches of palindromic sequences.

Inhibition control

The LightCycler Control Kit DNA was used for the detection of *Taq* DNA polymerase inhibitors, possibly present in DNA preparations from clinical samples. This kit provides control DNA template, primers and hybridization probes for amplifying and specific detection of the human ß-globin gene. To identify even weak inhibition events, 300 pg of human genomic DNA was amplified in each case.

Table 1. Oligonucleotides

Helicobacter pylori Urease Gene (Genebank Accession # X 17079)				
	Position	Length	GC (%)	T_m (°C)
Primers				
GCCAATGGTAAATTAGTT	360	18	33.3	51.8
CTCCTTAATTGTTTTTAC	770 R	18	27.8	47.5
Product	360–770	411		
Hybridization Probes				
GACATTGCGAGCGGGACAGCGGT-F	555	23	65.2	71.5
LCRed640-AGGTTTGAGCCT-GGCGAAGAAAAATC-P	579	26	46.2	65.8

LightCycler PCR

The following master mix was used for amplification and hybridization probe-based detection of the *H. pylori*-specific amplicons:

	Volume (µl)	Final
LightCycler-DNA Master	2	1×
MgCl$_2$ stock solution (25 mM)	1.6	3 mM
Primers (10 µM each)	1.0	0.5 µM
Hybridization Probes (2 µM)	2+2	0.2 µM
H$_2$O (PCR grade)	9.4	–
Total volume	18	

To complete the amplification mixtures, 18 μl of master mix and 2 μl of the corresponding template DNA preparation were added to each capillary. After a short centrifugation of the sealed capillaries, they were placed into the LightCycler rotor.

The inclusion of negative as well as positive controls in each set of experiments is considered to be obligatory in the field of diagnostic PCR. The negative control sample was prepared by replacing the DNA template with PCR grade water. The positive control sample was prepared by adding 2 μl of *H. pylori* genomic DNA (approx. 1 ng) to the master mix.

The following PCR protocol was used for amplification and hybridization probe-based detection of the *H. pylori*-specific amplicons:

- Denaturation for 2 min at 95°C
- Amplification

Parameter	Value		
Cycles	50		
Type	Quantification		
	Segment 1	Segment 2	Segment 3
Target temperature (°C)	95	50	72
Incubation time (s)	0	10	20
Temperature transition rate (°C/s)	20	20	20
Acquisition mode	None	Single	None
Gains	F1=1, F2=15, F3=30		

- Cooling 2 min at 40°C

Results

Based on an established PCR protocol for the sensitive and specific detection of *H. pylori*, the LightCycler system was examined to determine if it could simplify the diagnostic laboratory workflow by automation and by reducing the possibility of product contaminations frequently associated with post-PCR amplicon manipulation. As with every modified or new PCR protocol in diagnostic microbiology, the sensitivity and specificity of any assay has to be carefully examined before introduction into routine testing.

Sensitivity Starting with a proven pair of PCR primers [7], a set of hybridization probes was selected for targeting an amplified fragment of the *H. pylori* urease gene. PCR experiments were performed on serial dilutions of genomic DNA prepared from cultured *H. pylori* organisms, in order to determine the assay's lower limit of detection. Using the probe set HPU-HP-3 and HPU-HP-4, a detection limit of 1 pg of template DNA was determined, corresponding to approx. 5 CFU of *H. pylori* (Fig. 1).

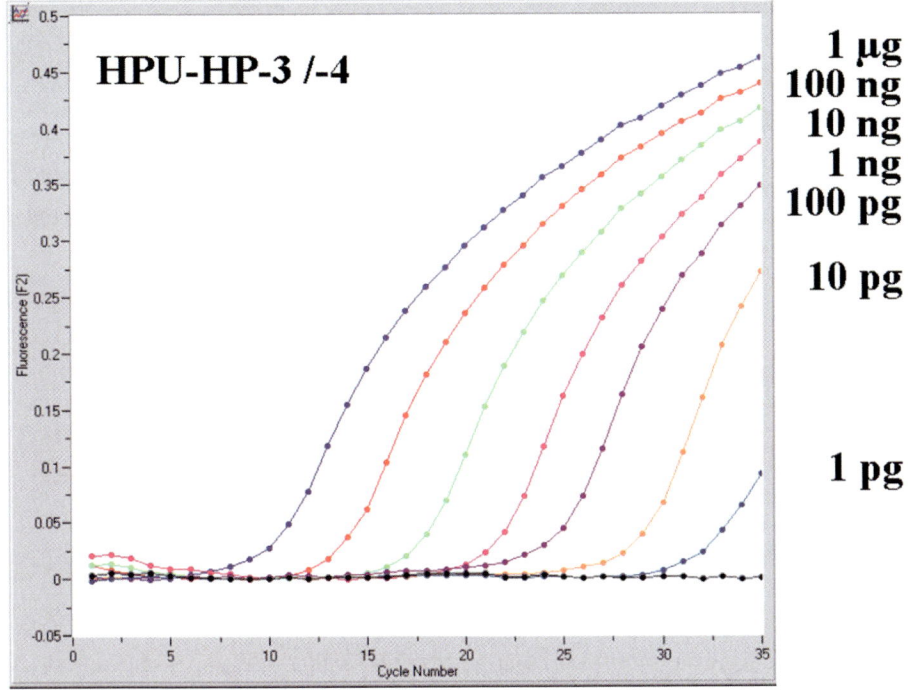

Fig. 1. Sensitivity of the PCR assay for the detection of the *H. pylori* urease gene determined with serial dilutions of *H. pylori* genomic DNA. For each dilution, the corresponding amount of *H. pylori* genomic DNA present in the reaction mixture is given next to the amplicon curve

Specificity

The PCR assay was evaluated with a collection of 40 cultured strains of *H. pylori* and other *Helicobacter* species, as well as 40 gram-negative and 40 gram-positive bacterial isolates other than *Helicobacter*. All DNA preparations from *H. pylori* strains originating from *H. pylori*-positive patients were amplified and detected, whereas none of the DNA preparations from the other bacterial species examined were recognized. Experimental results on representative panels of cultured bacteria are shown in Fig. 2 A. Even after as few as 35 cycles of amplification, the curves generated by the LightCycler software allowed for a clear discrimination between *H. pylori*-positive and -negative samples.

Inhibition Events

To investigate PCR inhibition, DNA preparations from the cultured isolates (spiked with 300 pg of human genomic DNA) were subjected to PCR. Amplification and detection were performed with the primers and probes of the LightCycler Control Kit, allowing for specific detection of the human ß-globin gene. No inhibition events were observed with any of the DNA preparations tested. In contrast to direct testing of clinical specimens known to be prone to inhibition, such as blood or stool samples, there was no apparent inhibition with cultured bacteria processed with the High Pure PCR Template Preparation Kit.

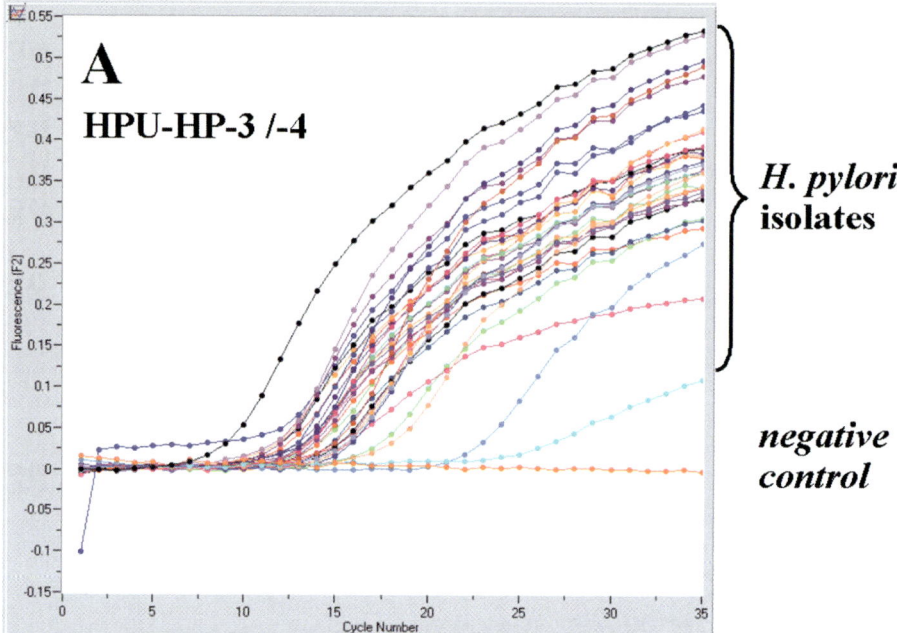

Fig. 2 A,B. Evaluation of the *H. pylori*-specific PCR assay with a representative set of 30 well-characterized cultured strains of *H. pylori* (**A**) and 30 gastric biopsies originating from patients suspected to have an infection with *H. pylori* (**B**). The depicted curves, representing *H. pylori*-positive results, are indicated by *brackets*

Comments

LightCycler PCR Previous methods for diagnostic PCR have employed a variety of time-consuming and laborious procedures (such as Southern-blot, DNA sequencing or solid-phase capturing) for sequence-specific characterization of the amplicons generated with clinical specimens. Enhancing the reliability of the results by sequence-specific probes and simplifying the PCR workflow by a completely automated amplification and online detection procedure, the LightCycler system proved itself as a valuable tool for rapid identification of *H. pylori* in the environment of an academic microbiological laboratory. Once DNA is extracted from the gastric biopsies and reaction mixtures are prepared, results are available within 60 min. Current limitations of the LightCycler system are a total reaction volume of 20 µl (which limits the input of template DNA), individual sample handling in the reaction cuvettes, and the lack of an automated device for template DNA preparation.

PCR Testing of Gastric Biopsies Direct detection of *H. pylori* organisms in gastric biopsies obtained in the course of endoscopic examination would be an ideal complement to the results of histology and rapid urease testing. In order to evaluate the sensitivity and specifici-

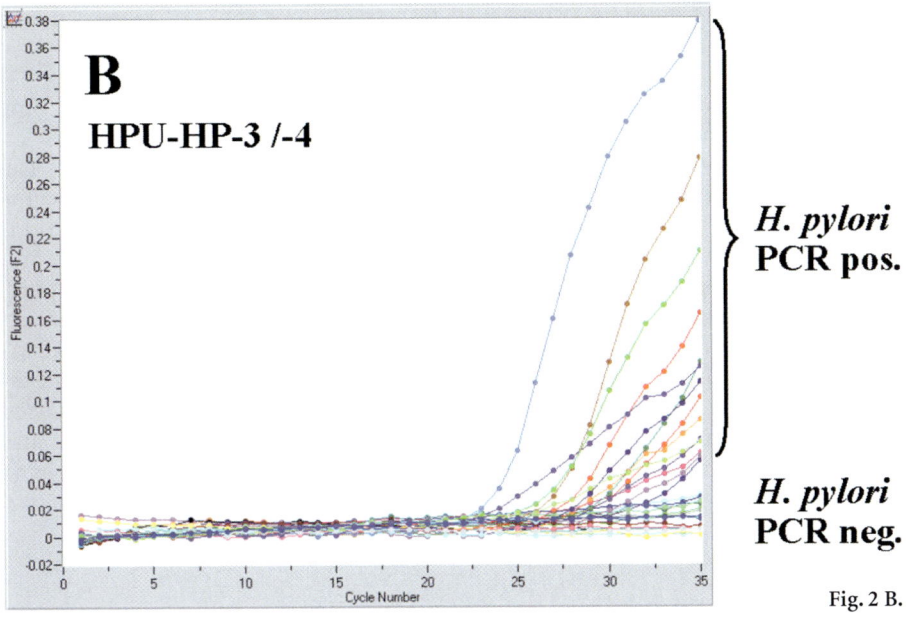

Fig. 2 B.

ty of PCR testing, we performed a randomized trial on 30 gastric biopsies from patients suspected to be infected with *H. pylori*. Template DNA from gastric biopsies was prepared with the High Pure PCR Template Preparation Kit. Initial PCR results, which are shown in Fig. 2B, document the feasibility of this approach. After comparing the performance of PCR testing against culture, 11 of the culture-positive specimens were found positive with PCR and 10 of the culture-negative specimens were found negative with PCR. However, discrepant results were observed with nine specimens, where five culture-negative specimens were found positive with PCR and four of the culture-positive specimens were found negative with PCR. The significance of these discrepancies will be further investigated with the help of the patient's clinical records and supplementary diagnostic testing.

PCR Testing of Stool Samples

There is evidence that infected individuals excrete *H. pylori* in feces, since the pathogen can be detected in stool specimens by PCR, antigen enzyme immunoassays, or even culture [8]. Although PCR is a powerful technique for the detection of target DNA in various clinical specimens, its application to stool specimens has been limited due to the presence of inhibitory substances. With the application of solid-phase extraction and purification procedures for template DNA (e.g., High Pure PCR Template Preparation Kit) the inhibition problem may be overcome in a significant portion of stool specimens. Having an appropriate inhibition control available (e.g., primers and hybridization probes of the LightCycler Control Kit), PCR may turn out to be a suitable tool for the detection of *H. pylori* infections with stool samples. However, *H. pylori* can persist in the stool of patients after successful antibiotic treatment [8], so PCR testing of stool may not be useful as a therapeutic monitor.

PCR Testing for Macrolide Resistance

Resistance to antimicrobial agents is an important factor for the clinical outcome of anti-*Helicobacter* treatment. Next to the metronidazol-resistant strains, an increasing prevalence of strains resistant to macrolides has been observed. The major cause for macrolide-resistance in *H. pylori* is the lack of binding of the macrolides to the components of the bacterial ribosome due to point mutations in the peptidyltransferase region of the 23 S rRNA. *H. pylori* contains two copies of the 23 S rDNA gene, and at least six distinct point mutations have been reported that are associated with macrolide resistance [9]. In contrast to the time-consuming and laborious antibiotic susceptibility testing of cultured strains, DNA-based methods offer a rapid and reliable alternative approach for macrolide susceptibility testing. Amplification of the *H. pylori*-specific urease gene can be used to identify infection. In addition, the 23 S rDNA region could be amplified and macrolide resistance mutations identified with real-time hybridization probes or DNA sequencing.

References

1. Blaser MJ (1987) Gastric Campylobacter-like organisms, gastritis and peptic ulcer disease. Gastroenterology 93:371–383
2. Hopkins RJ, Morris JG (1994) Helicobacter pylori: the missing link in perspective. Am J Med 97:265–277
3. Banatvala N, Mayo K, Megraud F, Jennings R, Deeks JJ, Feldmann RA (1993) The cohort effect and Helicobacter pylori. J Infect Dis 168:219–221
4. Mataly HM, Graham DY (1994) Importance of childhood socioeconomic status on the current prevalence of Helicobacter pylori infection. Gut 35:742–745
5. NIH. (1994) Consensus development panel on Helicobacter pylori in peptic ulcer disease. JAMA 272:65–69
6. Clayton CL, Pallen MJ, Kleanthous H, Wren BW, Tabaqchali S (1990) Nucleotide sequence of two genes from Helicobacter pylori encoding for urease subunits. Nucl Acids Res 18:362
7. Clayton CL, Kleanthous H, Coates PJ, Morgan DD, Tabaqchali S (1992) Sensitive detection of Helicobacter pylori by using polymerase chain reaction. J Clin Microbiol 34:2778–2730
8. Makristathis A, Pasching E, Shütze K, Wimmer M, Rotter ML, Hirschl AM (1998) Detection of Helicobacter pylori in stool specimens by PCR and antigen enzyme immunoassay. J Clin Microbiol 36:2772–2774
9. Doorn L-J, Debets-Ossenkopp YJ, Marais A, Sanna R, Megraud F, Kusters JG, Quint WGV (1999) Rapid detection, by PCR and reverse hybridization, of mutations in the Helicobacter pylori 23 S rRNA gene, associated with macrolide resistance. J Clin Microbiol 43:1779–1782

Qualitative Detection of Herpes Simplex Virus DNA on the LightCycler

Harald H. Kessler*

Introduction

Herpes simplex virus (HSV) is the most important agent of sporadic encephalitis in industrialized countries. HSV type 1 (HSV-1) is commonly associated with oropharyngeal infections, keratoconjunctivitis, and infections of the central nervous system, whereas HSV-2 commonly produces genital infections that are considered one of the most important sexually transmitted diseases. Neonatal HSV infection following exposure to the virus at delivery can produce severe disseminated infection and death. Powerful therapeutic management exists today, however, antiviral drugs must be administered early. On the other hand, symptoms mimicking HSV infection could lead to the unnecessary application of drugs. Therefore, a rapid and safe method for the detection of HSV appears to be of paramount importance for decreasing the lethality as well as the sequelae of HSV infection.

Viral culture is considered the gold standard in the diagnosis of HSV, but cultures may require up to 1 week and demonstration of HSV by viral culture from cerebrospinal fluid is seldom successful. Therefore, culture is not commonly used in the clinical diagnostic laboratory. Serologic diagnosis, which is based on increasing specific IgM antibodies is difficult to interpret, especially in immunocompromised patients and in neonates. Furthermore, in many patients with recurrent HSV infections, serology is complicated by the difficulty in detecting significant changes in serum antibodies during the course of the disease. IgG levels may not change significantly, and IgM antibodies may be present or absent. Polymerase chain reaction (PCR) combined with nucleic acid hybridization results in the highest sensitivity and specificity. Molecular assays can be performed within 1 working day. In earlier studies, hybridization techniques usually involved radioactive agents. Nonradioactive and more automated methods have recently been introduced [1]. With the development of LightCycler technology, PCR amplification and the detection of amplification products occur simultaneously. The entire process is finished in 15 to 30 minutes. Protocols for the qualitative detection of HSV DNA using LightCycler technology are presented.

---

* Harald H. Kessler (✉) (e-mail: harald.kessler@kfunigraz.ac.at)
  Institute of Hygiene, Medical Faculty, Karl-Franzens-University Graz, Universitaetsplatz 4, 8010 Graz, Austria

Materials

Equipment LightCycler Instrument (Roche Diagnostics, Mannheim, Germany)
LightCycler Capillaries (Roche Diagnostics)
LightCycler Centrifuge Adapters (Roche Diagnostics)

Kits LightCycler-DNA Master SYBR Green I (Roche Diagnostics)
LightCycler-DNA Master Hybridization Probes (Roche Diagnostics)

Reagents Analytical grade Chelex resin, 100–200 mesh, sodium form (Bio-Rad Laboratories, Richmond, Calif.)
Amplification Primers (MWG-Biotech GmbH, Ebersberg, Germany)
TaqMan Probe (TIB MOLBIOL, Berlin, Germany)
TaqStart Antibody (Clontech, Palo Alto, Calif.)

Procedure

Study Design In a first step, a plasmid (pS4), kindly provided by K. W. Knopf, German Cancer Research Center, Heidelberg, Germany, containing a single copy of a SalI restriction fragment of the HSV polymerase gene from the HSV-1 strain Angelotti, served as a standard for the determination of the detection limit. Tenfold dilutions of the plasmid were prepared and tested with each of the following assays five times on different days: SYBR Green I without "hot start," SYBR Green I with "hot start," TaqMan Probe without "hot start," and TaqMan Probe with "hot start."

In a second step, the First European Union Concerted Action HSV Proficiency Panel, which contained different concentrations of HSV type 1, strain MacIntyre (American Type Culture Collection), HSV type 2, strain MS (American Type Culture Collection), VZV, and negative samples, was used (Table 1). Samples were tested five times on different days using the TaqMan Probe with a "hot start".

In a third step, a total of 58 clinical specimens was investigated. Twenty-five cerebrospinal fluid (CSF) samples had been collected from patients who were admitted to the Department of Pediatrics, University Hospital Graz, Austria. All of them had clinical presentations compatible with HSV encephalitis. Thirty-three vaginal swabs had been collected from pregnant women at the outpatient clinics of the Department of Obstetrics and Gynecology, University Hospital, Graz. All of them had clinical presentations compatible with genital HSV infection.

Sample Preparation DNA was rapidly extracted in a 1.5-ml tube. A total of 50 µl of sample were added to 150 µl of a suspension consisting of 20% (wt/vol) Chelex resin in 10 mM Tris-HCl (pH 8.0)-0.1 mM EDTA-0.1% sodium azide. After vortexing for 10 s, incubation at 100°C for 10 min and vortexing for another 10 s, the tube was allowed to cool to room temperature. Following complete settlement of the resin, the supernatant was directly used for amplification.

Primer Design

Oligonucleotides deduced from the published sequence of the DNA polymerase gene-coding region from HSV were used [2, 3]. This set of primers, which was chosen within a highly conserved region of the DNA polymerase gene from the herpesvirus group, allows amplification of a 92-bp fragment of the HSV-1 and HSV-2 DNA polymerase genes in clinical samples [4, 5]. The primer and probe sequences and characteristics are shown in Table 2.

Table 1. First European Union Concerted Action Herpes Simplex Proficiency Panel

Vial number	Virus	Genome equivalents (GE) per ml	Positive results at six reference labs
1	Negative	Not tested	0/7
2	HSV-1	$2–5 \times 10^4$	7/7
3	HSV-2	$2–5 \times 10^3$	1/7
4	HSV-1	$2–5 \times 10^2$	3/7
5	HSV-1	$2–5 \times 10^5$	7/7
6	HSV-2	$2–5 \times 10^6$	7/7
7	HSV-1	$2–5 \times 10^3$	7/7
8	Negative	Not tested	0/7
9	HSV-2	$2–5 \times 10^5$	6/7
10	HSV-2	$2–5 \times 10^4$	6/7
11	VZV	Not tested	0/7
12	HSV-1	$0.7–1.7 \times 10^3$	5/7

Panel tested by 6 reference laboratories using a total of 7 different molecular methods before distribution.

Table 2. Oligonucleotides

HSV Type 1 Complete Genome (Genebank Accession Number X14112); HSV Type 2 Complete Genome (Genebank Accession Number Z86099)					
	Position	Length	GC (%)	T_m (°C)	
CATCACCGACCCGGAGAGGGAC	HSV-1, 65866 HSV-2; 66339	22	68.2	71.7	
GGGCCAGGCGCTTGTTGGTGTA	HSV-1, 65957 R HSV-2, 66430 R	22	63.6	73.3	
Product	HSV-1, 65866-65957 HSV-2, 66339-66430	92			
TaqMan Probe FAM-CCGCCGAACTGAGCAGACA-CCCGCGC-TAMRA	HSV-1, 65907 HSV-2, 66380	26	73.1	79.1	

LC-PCR SYBR Green I Master Mix for each 20 µl reaction:

	Volume [µl]	[Final]
LightCycler-DNA Master SYBR Green I	2	1×
MgCl$_2$ (25 mM)	2.4	4 mM
Primers (50 µM each)	0.2 each	0.5 µM
H$_2$O (PCR grade)	10.2	
Total volume	15	

TaqMan Master Mix for each 20 µl reaction:

	Volume [µl]	[Final]
LightCycler-DNA Master Hybridization Probes	2	1×
MgCl$_2$ (25 mM)	2.4	4 mM
Primers (50 µM each)	0.2 each	0.5 µM
TaqMan Probe (7.4 µM)	0.5	0.185 µM
H$_2$O (PCR grade)	9.7	
Total volume	15	

When the hot start technique was used, 0.16 µl TaqStart Antibody was added directly to each 2 µl of 10 × DNA Master solution and incubated at room temperature for 5 min. Then, MgCl$_2$, primers, TaqMan Probe, and water (to make 15 µl) were added.

A total of 15 µl of master mix and 5 µl of DNA template were added in each capillary. Sealed capillaries were centrifuged in a microcentrifuge and placed into the LightCycler rotor.

The following PCR protocol was used for SYBR Green I detection:

- Denaturation for 2 min at 95°C
- Amplification

Parameter	Value		
Cycles	40		
Type	Quantification		
	Segment 1	Segment 2	Segment 3
Target temperature (°C)	95	60	72
Incubation time [s]	0	5	5
Temperature transition rate [°C/s]	20	20	5
Acquisition mode	None	None	Single
Gains	F1=5		

- Melting Curve Analysis

Parameter	Value		
Cycles	1		
Type	Melting curve		
	Segment 1	Segment 2	Segment 3
Target temperature (°C)	95	55	95
Incubation time [s]	0	30	0
Temperature transition rate [°C/s]	20	20	0.1
Acquisition mode	None	None	Step
Gains	F1=5		

For TaqMan monitoring, the amplification program was modified as follows:
- Amplification

Parameter	Value	
Cycles	55	
Type	Quantification	
	Segment 1	Segment 2
Target temperature (°C)	95	60
Incubation time [s]	2	20
Temperature transition rate [°C/s]	20	10
Acquisition mode	None	Single
Gains	F1=1	

Results

Plasmid

Results obtained with all four techniques (SYBR Green I or TaqMan, with or without a hot start) were comparable: When tenfold dilutions of plasmid pS4 were tested, the detection limit was found to be 10^4 copies per ml, i.e., 12.5 copies per LC-PCR run. With the dilution containing 10^3 copies per ml, i.e., approx. 1 copy per LC-PCR run, all assays were inconsistent and produced either negative or positive results.

When SYBR Green I detection without "hot start" was used, a significant number of primer-dimers accumulated (Fig. 1). Formation of primer-dimers can reduce the sensitivity of PCR, especially for targets with low copy number and give false-positive results if melting curve analysis is not performed. This can be inhibited by using a "hot start" technique (Fig. 2). Results obtained with the TaqMan Probe and a "hot start" are shown in Fig. 3.

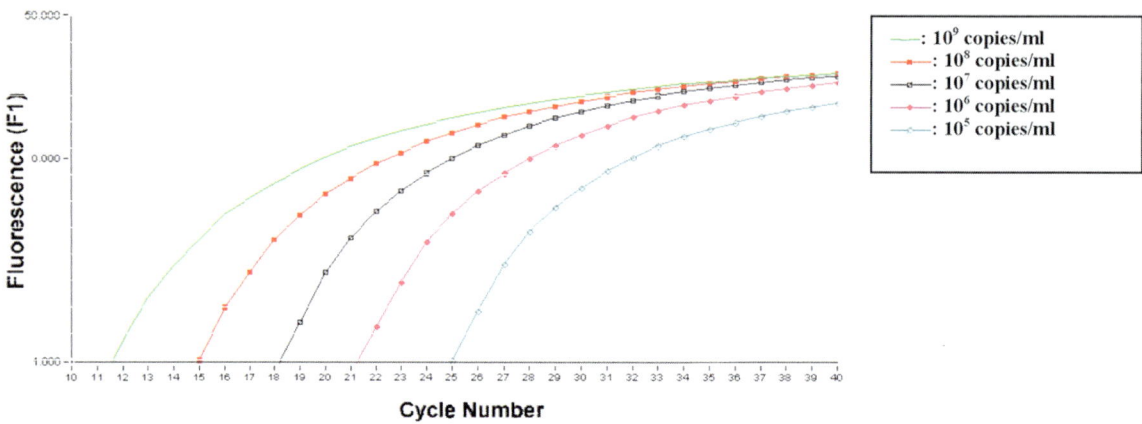

Fig. 1 a. LC-PCR of tenfold dilutions of plasmid pS4. SYBR Green I detection without TaqStart™ antibody, fluorescence curves

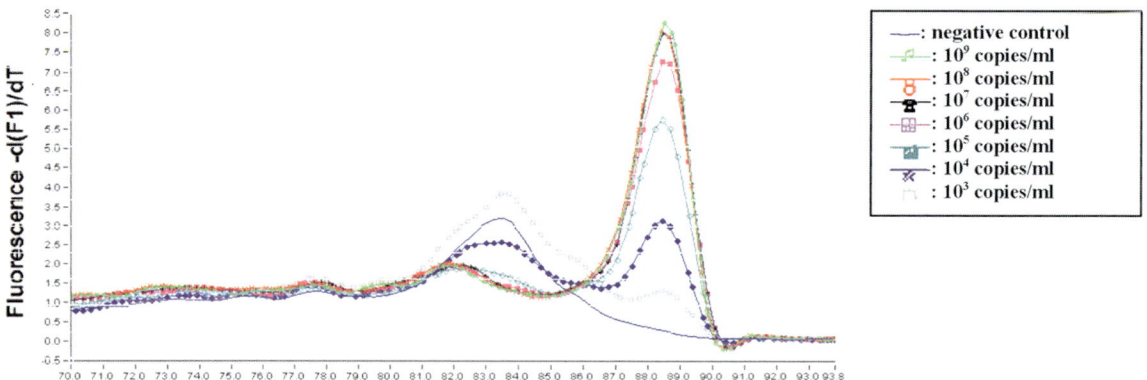

Fig. 1 b. LC-PCR of tenfold dilutions of plasmid pS4. SYBR Green I detection without TaqStart™ antibody, melting curves

First European Union Concerted Action Herpes Simplex Virus Proficiency Panel

When samples of the First European Union Concerted Action Herpes Simplex Virus Proficiency Panel were tested with the TaqMan Probe and the hot start technique, $2–5 \times 10^3$ HSV-1 genome equivalents (GE) per ml, i.e., 2.5–6.3 GE per LC-PCR run, could consistently be detected (Fig. 4). With the dilution containing $0.7–1.7 \times 10^3$ HSV-1 genome equivalents (GE), i.e., 1–2 GE per LC-PCR run, inconsistent results (either positive or negative) were obtained. When HSV-2 samples from the same panel were tested, $2–5 \times 10^4$ GE per ml, i.e., 25–62.5 GE per LC-PCR run, could consistently be detected, whereas $2–5 \times 10^3$ HSV-2 GE per ml, i.e., 3–6 GE per LC-PCR run, were never detected (Fig. 4).

Qualitative Detection of Herpes Simplex Virus DNA on the LightCycler

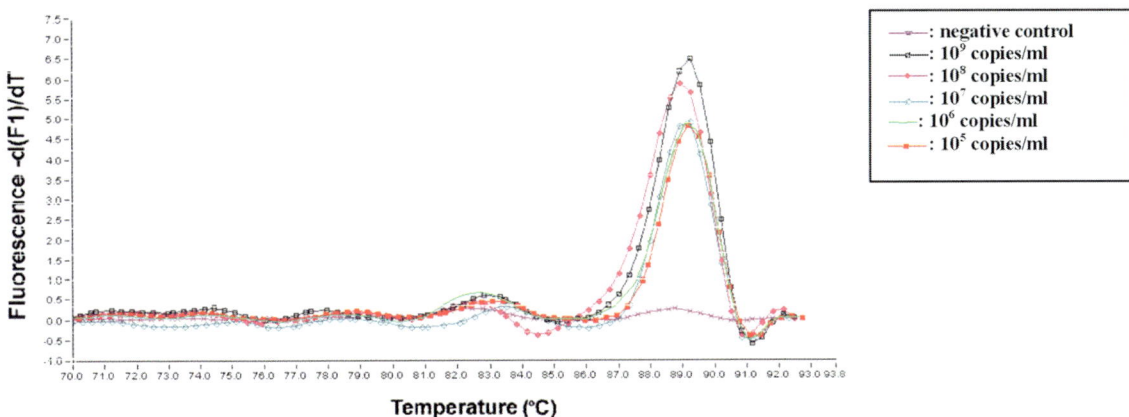

Fig. 2. LC-PCR of tenfold dilutions of plasmid pS4. SYBR Green I detection with "hot start" technique, melting curves

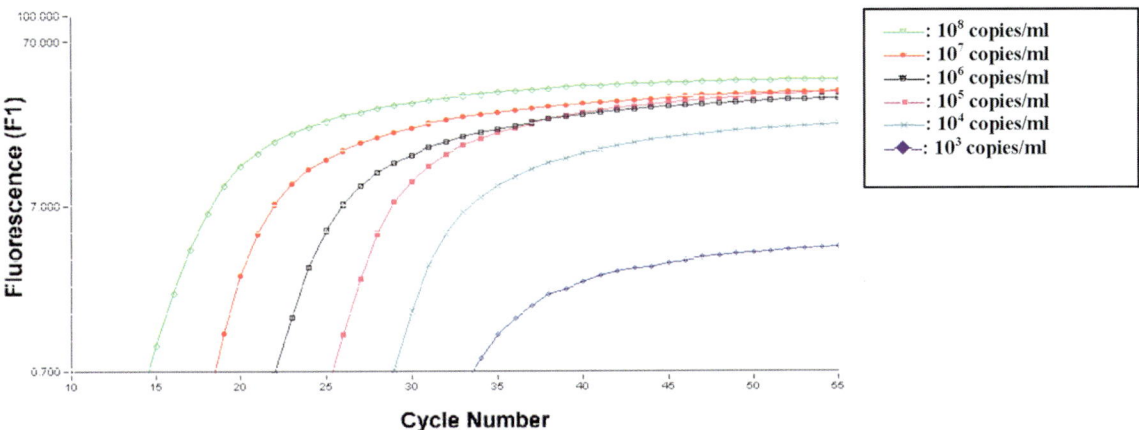

Fig. 3 a. LC-PCR of tenfold dilutions of plasmid pS4. TaqMan™ Probe detection with "hot start" technique, fluorescence curves

Infectious Organisms

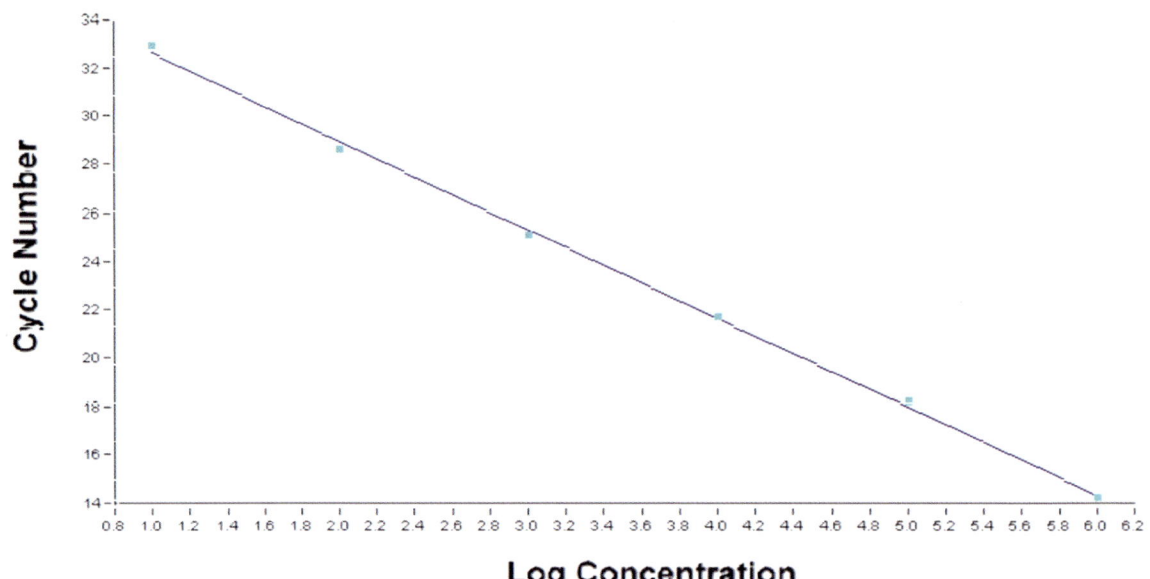

Fig. 3 b. Standard curve from values of Fig. 3a

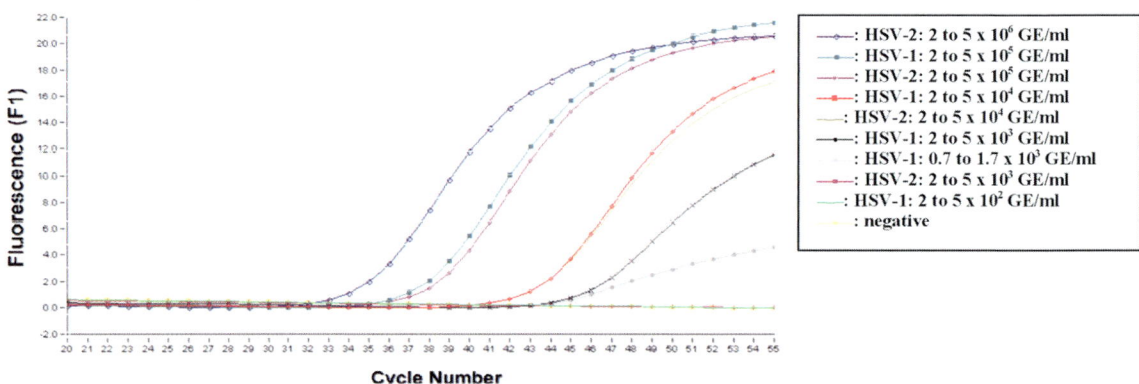

Fig. 4. LC-PCR of samples of the First European Union Concerted Action Herpes Simplex Virus Proficiency Panel. TaqMan™ Probe detection with "hot start" technique, fluorescence curves

When clinical specimens were tested, 1 of 25 CSF samples was found positive, whereas 6 of 33 vaginal swabs gave a positive result. Results of clinical samples (vaginal swabs) are shown in Fig. 5.

Clinical Specimens

Comments

Sample preparation is one of the most critical aspects of molecular assays because the probability of false-positive results from contamination increases as the number of manipulations involved increases, and results should be available as rapidly as possible. Use of the cation-exchanger Chelex 100 requires only a few manipulation steps and allows DNA extraction within less than half an hour. In comparison with classic nucleic acid extraction protocols, it has been shown to increase sensitivity of molecular assays for the detection of CMV in cultures and clinical samples [6], for the detection of HIV-1 proviral DNA in blood [7], and for the detection of *Legionella pneumophila* in bronchoalveolar lavage fluids [8]. However, Chelex 100 itself inhibits PCR, mainly by reducing the optimum magnesium ion concentration. Therefore, Chelex 100 must settle completely before the extracted DNA is used, to avoid the transfer of resin with the sample.

Sample Preparation

LightCycler-PCR proved to be suitable for the routine diagnostic laboratory. This novel method was found to be very quick, easy to use, and sensitive. Because nonspecific amplification products must be excluded in routine diagnostic molecular assays, specific probes appear to be most suitable. The current "hot start" technique requires an additional pipetting step and may increase the danger of contamination. Any real-time technique may be used for the quantitative estimation of HSV DNA.

Use of LC-PCR in the Routine Diagnostic Laboratory

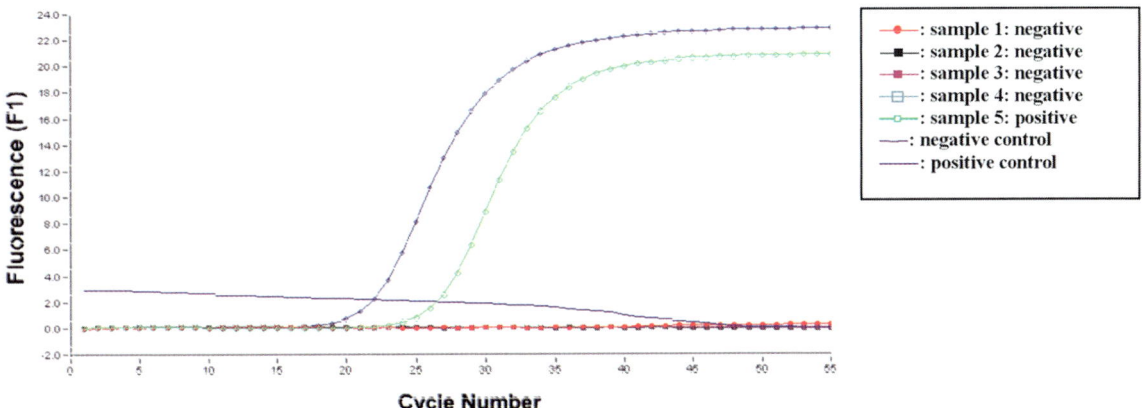

Fig. 5. LC-PCR of clinical samples (vaginal swabs). TaqMan™ Probe detection with "hot start" technique, fluorescence curves

References

1. Kessler HH, Pierer K, Weber B, Sakrauski A, Santner B, Stuenzner D, Gergely E, Marth E (1994) Detection of herpes simplex virus DNA from cerebrospinal fluid by PCR and a rapid, nonradioactive hybridization technique. J Clin Microbiol 32: 1881–1886
2. Larder BA, Kemp SD, Darby G (1987) Related functional domains in virus DNA polymerases. EMBO J 6: 169–175
3. Tsurumi T, Maeno K, Nishiyama Y (1987) Nucleotide sequence of the DNA polymerase gene of herpes simplex virus type 2 and comparison with the type 1 counterpart. Gene 52: 129–137
4. Cao M, Xiao X, Egbert T, Darragh TM, Yen TSB (1989) Rapid detection of cutaneous herpes simplex virus infection with the polymerase chain reaction. J Invest Dermatol 92: 391–392
5. Brice SL, Krzemien D, Weston WL, Huff JC (1989) Detection of herpes simplex virus DNA in cutaneous lesions of erythema multiforme. J Invest Dermatol 93: 183–187
6. Zandotti C, de Lamballrie X, Guignole-Vignoli C, Bollet C, de Micco P (1993) A rapid DNA extraction method from culture and clinical samples. Suitable for the detection of human cytomegalovirus by the polymerase chain reaction. Acta Virol 37:106–108
7. Essary LR, Kinard SJ, Butcher A, Wang H, Laycock KA, Donegan E, McCreedy B, Connell S, Batchelor J, Harris J, Spadoro J, Pepose JS (1996) Screening potential corneal donors for HIV-1 by polymerase chain reaction and a colorimetric microwell hybridization assay. Am J Ophthalmol 122: 526–534
8. Jaulhac B, Reyrolle M, Sodahlon YK, Jarraud S, Kubina M, Monteil H, Piemont Y, Etienne J (1998) Comparison of sample preparation methods for detection of *Legionella pneumophila* in culture-positive bronchoalveolar lavage fluids by PCR. J Clin Microbiol 36:2120–2122

Quantitative Detection of *Cryptosporidium parvum* after In Vitro Excystation by LightCycler PCR

Petra Krüger, Albrecht Wiedenmann*, Despina Tougianidou, Konrad Botzenhart

Introduction

Cryptosporidium parvum is a protozoan parasite (protist) which can infect a wide variety of vertebrates including man. It can cause a severe but usually self-limiting diarrhoea. It is transmitted by the fecal–oral route. Human infections and epidemics have been linked to the consumption of contaminated drinking water, swimming pool water, recreational water, milk, cider, and berries [1]. The infection starts with the oral uptake of infective sporulated oocysts. Inside each oocyst are four sporozoites which actively penetrate the oocyst wall when certain physiologic triggers such as temperature, changes in pH, and the presence of bile salts and pancreatic enzymes are present. This biological process can be simulated in vitro and is described as "in vitro excystation". In vitro excystation has been widely used as a surrogate marker for viability and infectivity of *Cryptosporidium parvum* oocysts [2–8]. Usually the quantification of in vitro excystation is achieved by a microscopic evaluation of the excystation rate. On the other hand, various PCR protocols, including a TaqMan protocol, have been described in combination with in vitro excystation; they lack, however, a quantification of the PCR results [9–17].

Outline

The method described here consists of an initial isolation and purification step of *Cryptosporidium parvum* oocysts by immunomagnetic separation (IMS), a washing step to remove any free DNA which might be present in the sample material, an in vitro excystation step, the isolation and purification of the DNA of excysted sporozoites, and the quantitative detection of this DNA by LightCycler PCR with hybridization probes.

---

* Albrecht Wiedenmann (✉) (e-mail: albrecht.wiedenmann@uni-tuebingen.de)
  Eberhard-Karls-University of Tübingen, Hygiene Institute, Environmental Hygiene Department, Wilhelmstr. 31, 72074 Tübingen, Germany

Materials

Equipment Black cellulose acetate membrane filters, $\varnothing$ 13 mm, porosity 0.8 µm (Millipore)
Spin-X Centrifuge Tube Filters, 0.45 µm (Corning Costar Germany, Bodenheim, Germany)

Reagents Cryptosporidium parvum oocysts, live in water (Waterborne Inc., LA, USA), 1 month old at use
Cryptosporidium immunofluorescence labelling kit (Medac, Hamburg, Germany/Cell Labs, Sydney, Australia)
Dynabeads anti-Cryptosporidium (Dynal, Norway)
HCl pH 2.6 or Hanks balanced salt solution (HBSS) acidified to pH 2.6
Excystation medium: 1.5% (wt/vol) tauroglycholic acid sodium salt (Merck), 0.5% (wt/vol) bovine trypsin (2.2 U/mg [BAEE, pH 7.6, 25°C]; Biochrom KG, Berlin, Germany), in 0.05 M Tris, pH 7.8
Sterile pyrogen free water (Aqua ad iniectabilia, Ampuwa, Fresenius AG, Bad Homburg, Germany)
High-Pure PCR Template isolation kit (Roche Diagnostics, Mannheim, Germany)
Amplification Primer (Interactiva, Ulm, Germany)
Hybridization Probes (TIB MOLBIOL, Berlin, Germany)
LightCycler-DNA Master Hybridization Probes (Roche Diagnostics, Mannheim, Germany)

Primers and Hybridization Probes

	GeneBank	Length	GC (%)	T_m (°C)
Primers				
CCT TTT GTA GCT CCT CAT ATG CCT TA	M59419	26	42	67.0
ACT TCA CGT GTG TTT GCC AAT G	M59419	22	45	67.3
Hybridization Probes				
GGT AAA AAG TAT AGA AAG CTC TCA TTA TTG ATC C-F	M59419	34	32	65.3
LCRed640-CCC TGA TAA GAC AAG TCA TGA AAA GGC TAG -P	M59419	30	43	68.0

Procedure

Production of External Standards

Enumeration of Oocyst Suspensions

Currently no well defined quantitative external standards consisting of *Cryptosporidium* DNA or exactly quantified oocysts are commercially available. To produce external standards for quantification of PCR results 5×10^6 oocysts of *Cryptosporidium parvum*, live in water, were purchased from Waterborne Inc., USA, and stored at 5 ± 2°C for no longer than 3 months. Standard operation procedures for enumerating oocysts in stock suspensions and for the preparation of dilution series have recently been set up by the US Environmental Protection

Agency [18] and the British Drinking Water Inspectorate [19]. For the experiment described here, oocyst concentrations were evaluated by placing 5 aliquots of a dilution, providing approximately between 50 and 100 oocysts each, on black cellulose acetate membrane filters. The filters were placed on microscope slides coated with a thin film of white Vaseline. Then, 15 µl of fluorescein labelled monoclonal antibodies (Cellabs) from a commercially available labelling kit, was applied on the membrane. The Vaseline on the slide prevents floating of the filters and keeps the staining solution on the membrane. The membranes were incubated at 37°C in a humid chamber for 30 min, and a cover glass was placed on top without removing the staining solution. The membranes were screened under an epifluorescence microscope using 200× magnification. The arithmetic mean of the five counts was used to calculate the concentration of the oocyst stock suspension, and working dilutions with the intended numbers of oocysts were prepared accordingly.

Microscopic Quantification of In Vitro Excystation Ability

The ability of sporozoites to excyst in an in vitro excystation assay is called the "excystation ability". The number of oocysts, which have the ability to excyst, divided by the total number of oocysts in the oocyst suspension is called the "excystation rate". If the in vitro excystation ability is to be quantified in absolute numbers of oocysts, the figures for the external standards in the LightCycler PCR assay have to be corrected by multiplication with the microscopically verified excystation rate. This is not necessary in other cases, e.g., in experiments where only the relative reduction of the excystation ability is of interest. Protocols for the microscopic determination of the excystation rate can be adopted from literature [3].

Sample Preparation

Isolation and Purification of Oocysts from Environmental or Clinical Samples

Oocysts from environmental or clinical samples must be carefully isolated and purified in order to prevent any substances which are inhibitory to in vitro excystation or PCR. Immunomagnetic separation (e.g., Dynabeads anti-*Cryptosporidium*; Dynal, Norway) has been shown to be a suitable procedure for this purpose [9]. For small sample volumes (1 ml) Dynal IMS can be performed according to the manufacturer's instructions with the following modifications: the capture reaction is performed in 1.5 ml Eppendorf cups instead of Dynal L10 tubes (flat sided Leighton tubes), and the bead detachment step may be omitted, as low numbers of beads do not interfere with the subsequent in vitro excystation procedure or the isolation of DNA in Spin-X centrifuge tube filters. If larger sample volumes of up to 10 ml are used for IMS, the beads must be detached prior to in vitro excystation. In this case the detachment process as described by the manufacturer should be modified as follows: It should be performed at pH 2.6 for 1 h at 37°C without the subsequent neutralization. The acidification step of the in vitro excystation protocol is thereby replaced in one operation.

In Vitro Excystation and DNA Extraction

In vitro excystation and DNA extraction can be performed with external standards and sample oocysts in the same manner. The oocyst suspensions or the suspensions with the Dynabeads-oocyst-complex are placed onto the membrane of the filter unit of a Spin-X centrifuge tube filter (Corning Costar). The tube is centrifuged for 1 min at 8000 ×g. During centrifugation the oocysts remain on the surface of the filter membrane while excess fluid and any free DNA from already disintegrated oocysts, which might be present in the oocyst suspensions, is passing through the membrane and retained in the collection tube. The oocysts on the filter membrane are washed with 100 µl sterile water (1 min, 8000×g). For in vitro excystation 100 µl of HCl, pH 2.6 (or acidified HBSS) are applied onto the membrane, and the tube is incubated at 37°C for 30 min. After incubation the tube is centrifuged again to remove the HCl. The membrane is washed once with 100 µl excystation medium (1 min at 8000×g), 80 µl of excystation medium is added again, and the tube is incubated at 37°C for 1 h. After incubation, the filter unit is placed in a second collection tube (Eppendorf cup) and centrifuged for 1 min at 12,000×g. To enhance the release of DNA from excysted sporozoites even further, the membrane is washed once with 50 µl AMPUWA (1 min at 12,000×g). Another 50 µl is added, and the tube is incubated for 5 min at 90°C in a heating block. After centrifugation (1 min at 12,000×g) the membrane is washed again with 50 µl of AMPUWA. DNA is extracted from the 230 µl collected fluid (80 µl excystation medium and 150 µl AMPUWA) using a PCR template preparation kit (High-Pure, Roche Diagnostics, Mannheim) according to the manufacturer's instructions with the following modification: Prior to the elution of DNA, the spin column is dried at 72°C for 5 min, and the volume of the elution buffer can be decreased to a minimum of 25 µl.

LightCycler PCR   In the experiment, which is used for illustration, the master mix was prepared according to the following recipe:
- Hybridization Probe Mastermix

	Volume [µl]	[Final]
LightCycler DNA Master Hybridization Probes	2	1×
MgCl$_2$ (25 mM)	3	3.75 mM
Primers (10 µM each)	1	0.5 µM each
Hybridization probes (2 µM each)	2	0.2 µM each
H$_2$O (PCR grade)	2	

A total of 10 µl of DNA template and 10 µl of master mix were loaded into each of the capillaries. The capillaries were sealed, centrifuged (10 s at 1000×g) and placed into the LightCycler rotor.

The program for PCR in the LightCycler instrument was set up as follows:
- Denaturation for 2 min at 95°C
- Amplification

Parameter	Value		
Cycles	50		
Type	Quantification		
	Segment 1	Segment 2	Segment 3
Target temperature [°C]	95	52	72
Incubation time [s]	0	15	5
Temperature transition rate [°C/s]	20	20	20
Acquisition mode	None	Single	None
Gains	FL1=1; FL2=15		

- Melting Curve Analysis

Parameter	Value		
Cycles	1		
Type	Melting Curves		
	Segment 1	Segment 2	Segment 3
Target temperature [°C]	95	50	95
Incubation time [s]	0	10	0
Temperatire transition rate [°C/s]	20	20	0.2
Acquisition mode	None	None	Continuous
Gains	FL1=1; FL2=15		

- Cooling for 1 min at 40°C

Results

Using the protocol described, *Cryptosporidium parvum* could be quantified over 4–5 logs of concentration. An example with 25,000, 2,500, 250, 25 and 2.5 oocysts after in vitro excystation (sample type="standard"), and equivalent numbers of oocysts without in vitro excystation (sample type="unknown"), is given in Fig. 1. A high correlation ($r=1.00$) between the microscopically determined number of oocysts and the LightCycler PCR results could be achieved. In this experiment, only the lowest concentration, with theoretically 2.5 oocysts after in vitro excystation, was not detectable. Equivalent numbers of oocysts were not detectable at all when the in vitro excystation step was omitted. The microscopically verified excystation rate of the oocysts used in this experiment was 85.6% (SD 2,99; $n=5\times100$ oocysts).

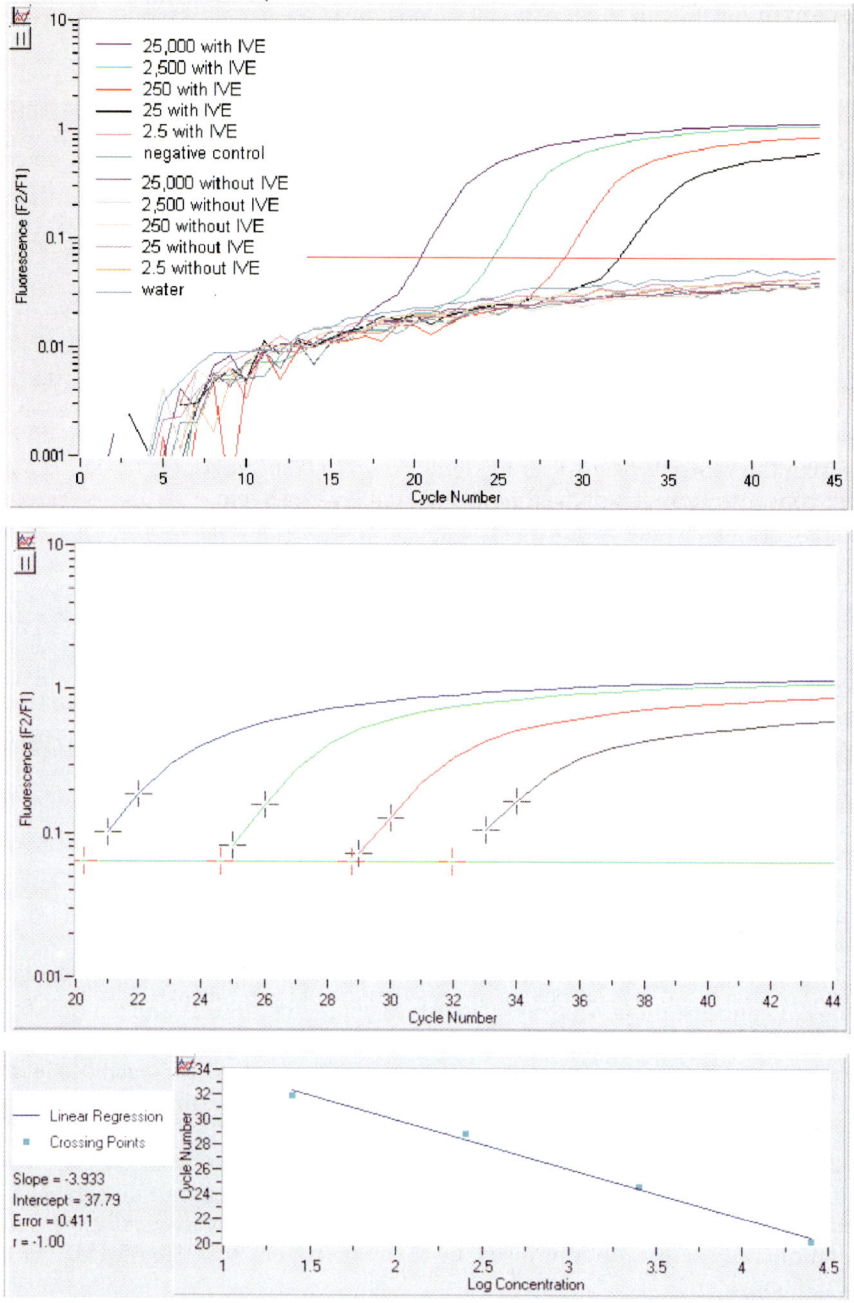

Fig. 1. Quantification of DNA derived from 25,000–2.5 *Cryptosporidium parvum* oocysts with and without in vitro excystation (IVE): 25,000–25 oocysts with IVE are positive; all results without IVE are negative

Comments

The sample volume which can be used in one LightCycler reaction is very small (10–12 µl). However, the volume for an efficient elution of DNA from the spin columns of the DNA purification kit is usually much higher (200 µl is recommended). This may represent a problem when the original volume of a sample has to be examined in total. According to our own experience the absolute minimum for elution is 25 µl. This means that the total elution volume has to be divided and distributed between several capillaries. However, splitting of the total volume into aliquots may lead to a decrease in the total amount of DNA per reaction below the detection limit. It is therefore recommended to initially sample a multiple of the volume equivalent, which can be examined in a single LightCycler reaction.

When the method is used to quantify the amount of excystable oocysts in samples with unknown concentrations, the oocyst concentrations, which are defined as standards, should be adjusted by the microscopically determined excystation rate. Example: number of oocysts in standard 1=25,000; excystation rate=0.856; known concentration of standard 1=25,000×0.856=21,400 etc.

Applications

Possible applications include all methods which utilize in vitro excystation as a surrogate marker for viability or infectivity of *Cryptosporidium parvum*, such as detection methods for *Cryptosporidium* oocysts in water and food, disinfection experiments, studies of the survival of oocysts under various conditions, and the screening of antiprotozoal drugs.

References

1. Casemore DP (1990) Epidemiological aspects of human cryptosporidiosis. Epidemiol Infect 104: 1–28
2. Korich DG, Mead JR, Madore MS, Sinclair NA, Sterling CR (1990) Effects of ozone, chlorine dioxide, chlorine, and monochloramine on *Cryptosporidium parvum* oocyst viability. Appl Environ Microbiol 56: 1423–1428
3. Campbell AT, Robertson LJ, Smith HV (1992) Viability of *Cryptosporidium parvum* oocysts: correlation of in vitro excystation with inclusion or exclusion of fluorogenic vital dyes. Appl Environ Microbiol 58: 3488–3493
4. Robertson LJ, Campbell AT, Smith HV (1992) Survival of *Cryptosporidium parvum* oocysts under various environmental pressures. Appl Environ Microbiol 58: 3494–3500
5. Campbell AT, Robertson LJ, Smith HV (1993) Effects of preservatives on viability of *Cryptosporidium parvum* oocysts. Appl Environ Microbiol 59: 4361–4362
6. Finch GR, Black EK, Gyürek L, Belosevic M (1993) Ozone inactivation of *Cryptosporidium parvum* in demand-free phosphate buffer determined by in vitro excystation and animal infectivity. Appl Environ Microbiol 59: 4203–4210
7. Campbell AT, Robertson LJ, Snowball MR, Smith HV (1995) Inactivation of oocysts of *Cryptosporidium parvum* by ultraviolet irradiation. Wat Res 29: 2583–2586
8. Rennecker JL, Marinas BJ, Owens JH, Rice EW (1999) Inactivation of *Cryptosporidium parvum* oocysts with ozone. Wat Res 33: 2481–2488

9. Wiedenmann A, Krüger P, Botzenhart K (1998) PCR detection of *Cryptosporidium parvum* in environmental samples – a review of published protocols and current developments. J Indust Microbiol Biotechnol 21: 150–166
10. Filkorn R, Wiedenmann A, Botzenhart K (1994) Selective detection of viable *Cryptosporidium* oocysts by PCR. Zentralbl Hyg Umweltmed 195: 489–494
11. Wagner-Wiening C, Kimmig P (1995) Detection of viable *Cryptosporidium parvum* oocysts by PCR. Appl Environ Microbiol 61: 4514–4516
12. Rochelle PA, Ferguson DM, Handojo TJ, De Leon R, Stewart MH, Wolfe RL (1996) Development of a rapid detection procedure for *Cryptosporidium*, using in vitro cell culture combined with PCR. J Eukaryot Microbiol 43: 72 S
13. Wiedenmann A, Krüger P, Filkorn R, Botzenhart K (1996) Selective detection of viable *Cryptosporidium* oocysts by PCR after free DNA digestion and in vitro excystation. In: Persing DH (ed) PCR protocols for emerging infectious diseases, chapter p.11: 163–168
14. Deng MQ, Cliver DO, Mariam TW (1997) Immunomagnetic capture PCR to detect viable *Cryptosporidium parvum* oocysts from environmental samples. Appl Environ Microbiol 63: 3134–3138
15. Rochelle PA, Ferguson DM, Handojo TJ, De Leon R, Stewart MH, Wolfe RL (1997) An assay combining cell culture with reverse transcriptase PCR to detect and determine the infectivity of waterborne *Cryptosporidium parvum*. Appl Environ Microbiol 63: 2029–2037
16. Krüger P, Wiedenmann A, Botzenhart K (1999) Detection of *Cryptosporidium* oocysts in water: comparison of the conventional immunofluorescent method with PCR and TaqMan PCR. In: EAWAG/OECD: Proceedings of the workshop of the Organization for Economic Co-operation and Development (OECD) on 'Molccular Technologies for Safe Drinking Water' at Interlaken, Switzerland, 1998, http://www.eawag.ch/pulications_e/proceedings/oecd.html
17. Rochelle PA, De Leon R, Johnson A, Stewart MH, Wolfe RL (1999) Evaluation of immuno-magnetic separation for recovery of infectious *Cryptosporidium parvum* oocysts from environmental samples. Appl Environ Microbiol 65: 841–845
18. United States Environmental Protection Agency (1999) Method 1622: *Cryptosporidium* in water by filtration/IMS/FA http://www.epa.gov/nerlcwww/1622ja99.pdf
19. Drinking Water Inspectorate (1999) Standard operating protocol for the monitoring of *Cryptosporidium* oocysts in treated water supplies to satisfy water supply (water quality) (amendment) regulations 1999, SI No 1524, Part 2 – Laboratory and analytical procedures – *Cryptosporidium* regulations. http://www.dwi.detr.gov.uk/prot2.pdf

Quantitative Analysis of CMV in Infected Mice on the LightCycler System

Junichi Honda*, Kotaro Oizumi

Introduction

Cytomegalovirus (CMV) induced interstitial pneumonia sometimes leads to death in patients with severe immunodificiency. Drugs such as Ganciclovir and Foscarnet are available for treatment of CMV infection. However, early diagnosis and early treatment are essential. Progress in molecular biology has led to the development of the polymerase chain reaction (PCR) which can detect CMV sequences with high sensitivity. Moreover, monitoring the expression of CMV antigen in peripheral blood leukocytes, so-called CMV antigenaemia [1] has proved to be a sensitive, specific and rapid technique for the diagnosis of CMV infection. Based on the importance of detecting CMV activation and being able to initiate treatment for CMV infection before the on set of disease, we developed a new method for CMV-DNA detection using a quantitative PCR method.

Recently, a novel technique for "real time" PCR which measures product accumulation during the exponential phase of PCR has been developed [2, 3]. This new method is easy, reproducible and rapid and appears to be useful in monitoring activation of CMV. In this paper, we measured mouse-CMV DNA levels in salivary glands of infected mice by using LightCycler system [4], and compared with plaque assays.

Materials

LightCycler Instrument **Equipment**

Amplification Primers (Greiner Japan Co,Tokyo) **Reagents**
Hybridization Probes (Nihon Gene Research Lab., Sendai)
QIAamp DNA Mini Kit (QIAGEN Inc., Tokyo)
TA Cloning Kit (Invitrogen Co. San Diego, CA)
Wizard® Plus Minipreps DNA Purification System (Promega Co. Madison, WI)
The following reagents were purchased from Roche Molecular Biochemicals, Mannheim:
LightCycler-DNA Master SYBRGreen I
LightCycler-DNA Master Hybridization Probes

---

* Junichi Honda (✉) (e-mail: junbmlc@med.kurume-u.ac.jp)
   First Department Internal Medicine, Kurume University School of Medicine, Asahi-machi 67, Kurume Fukoka 830-011, Japan

Mice	Six-to eight-week-old female BALB/c mice
Virus	The Smith strain of mouse CMV (MuCMV; American Type Culture Collection. Manassas, VA;ATCC#VR-1399)

Procedure

MuCMV infection Mice were exposed intranasally to 4×10^3 plaque forming units (PFU) of MuCMV. Mice were given cyclophosphamide (CTX) (0.2 mg/g) intraperitoneally a day after Virus challenge.

Sample Preparation Five, eight, eleven, fifteen, and eighteen days after virus inoculation, mice were anesthetized and killed for getting the salivary glands. For virus quantitation, tissues (0.2 g) were prepared as 10% (wt/vol.) homogenates with MEM containing 10% FCS and 10% dimethylsulfoxide and stored at –70 °C until the viral plaque assay was performed [5]. Genomic DNA was isolated from the salivary gland tissues using the QIAamp DNA kit. The DNA was resuspended in distilled water (final concentration: 40 µg/ml) and stored at –20 °C until PCR analysis.

Olygonucleotides For the detection of murine cytomegalovirus DNA, the primers for amplification of the 296 bases in the Smith HindIII J fragment (Genebank Accession #U10326) were designed.

Olygonudeotides

(Genebank Accession #U10326)				
	Position	Length	GC (%)	T_m (°C)
Primers				
TCCAGGCTTAATAGCAGGCG	1026	20	55	58.4
ACGAAAGATCCGATCGAGGC	1321R	20	55	58.4
product	1026-1321	296	62.4	84.7
Hybridization probes				
TCAGGCATGGACGCGCCATCAG-FL	1149	22	63.6	64.1
Red640-CGTACTTGTTGTCCGCTTCAGCA	1174	24	50.0	61.7

LC-PCR • SYBR Green Hot Start for each 20 µl reaction:

	Volume [µl]	[Final]
LightCycler DNA Master SYBR®Green I	2	1×
MgCl$_2$ stock solution (25 mM)	0.8	2 mM
Primer stock (l5 mM each)	0.8	0.6 µM
TagStart antibody	0.8	
PCR grade H$_2$O	13.2	
Total volume	18	

- Hybridization Probes Hot Start for n samples

	Volume [µl]	[Final]
LightCycler DNA Master Hybridization Probes	2	1×
MgCl$_2$ stock solution (25 mM)	2.4	4 mM
Primer stock (15 µM each)	0.8	0.6 µM
Probes (2 µM each)	2+2	0.2 µM
TagStart antibody	0.4	
PCR grade H$_2$O	8.4	
Total volume	18	

To each capillary, 18 µl of master mix and 2 µl of DNA template were added in each capillary. Sealed capillaries were centrifuged in a microcentrifuge and placed into the LightCycler rotor.

The following PCR protocol was used for the SYBR Green detection:
- Denaturation for 2 min at 95 °C.
- Amplification

Parameter	Value		
Cycles	40		
Type	Quantification		
	Segment 1	Segment 2	Segment 3
Target temperature (°C)	95	55	72
Incubation time [s]	2	10	10
Temperature transition rate [°C/s]	20	5	20
Acquisition mode	None	None	Single
Gains	F1=5; F2=10; F3=30		

- Melting Curve Analysis:

Parameter	Value		
Cycles	40		
Type	Quantification		
	Segment 1	Segment 2	Segment 3
Target temperature (°C)	95	65	95
Incubation time [s]	0	10	0
Temperature transition rate [°C/s]	20	20	0.1
Secondary Target Temperature [°C/s]	0	0	0
Step Size	0	0	0
Step Delay	0	0	0
Acquisition mode	None	None	Cont.

For the hybridization probes the amplification program was modified the following way:
- Denaturation for 2 min at 95°C
- Amplification

Parameter	Value		
Cycles	45		
Type	Quantification		
	Segment 1	Segment 2	Segment 3
Target temperature (°C)	95	55	72
Incubation time [s]	2	10	12
Temperature transition rate [°C/s]	20	5	20
Acquisition mode	None	Single	None
Gains	F1=1; F2=10; F3=30		

Standard   Standards of mouse CMV-DNA were prepared by using TA CloningKit. The PCR product was ligated into pCR™II and transformed into One Shot competent cells. After the cloning, the competent cells were incubated to select recombinant plasmids were collected by using Wizard *Plus* Minipreps DNA Purification System. These recombinant plasmids with inserted MuCMV-DNA were used as standards.

Gel analysis of PCR product   PCR products amplified in the SYBRGreen format were recovered from the capillaries by reverse centrifugation into microtiter V-bottomed plates. The amplifications were subjected to electrophoresis in 2% agarose gels, and the presence of bands was analyzed by visualization with ultraviolet light following ethidium bromide staining.

Results

PFU assay   The viral titer of the mice salivary glands revealed a rather sharply rising curve to a plateau at 8 days following intranasal exposure, with substantial quantities of MuCMV still being present at day 18 (Figure 1). The titers of MuCMV in salivary glands at 5, 8, 11, 15, and 18 day after virus challenge were 2.5, 4, 1, 0.5, and 0.2×10^4 PFU/ml, respectively.

Quantitative LC-PCR   Figure 2a shows the amplification of standard DNA and sample DNA from the salivary glands using the SYBRGreen Hot Start protocol. For analyzing quantification data, the noise band was set to 3.075 on the analysis screen. We could get a best-fit crossing line (error = 0.008, correlation coefficient value = –1) of standard samples automatically. The calculated concentrations of samples are shown in Table 1. Figure 2b shows the results of melting curve analysis. The resulting melting curves allow discrimination between primer dimers and spe-

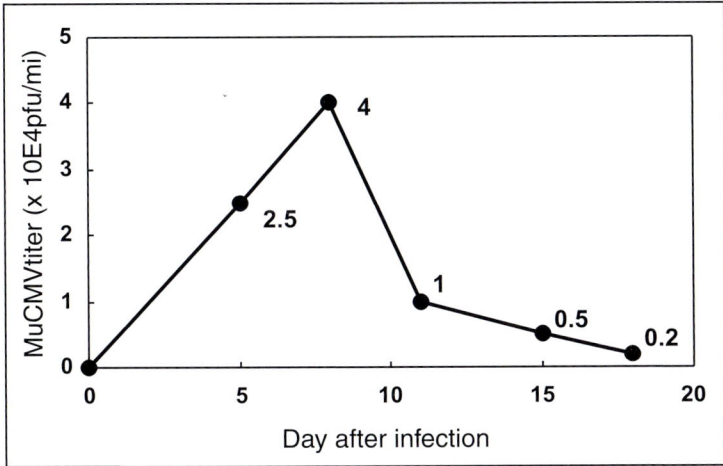

Fig. 1. MuCMV replication in salivary glands of BALB/c mice at various times after intranasal inoculation with 4×10^3 pfu of MuCMV.

Table 1. The calculated concentrations of samples (by SYBRGreen Hot Start protocol)

	Days After Infection	Concentrations (pg)	Crossing Point
sample 1	5	49.78	15.38
sample 2	8	107.6	14.24
sample 3	11	18.83	16.83
sample 4	15	9.617	17.83
samples	18	1.092	21.06

cific product. As shown on figure 2b and gel analysis (data not shown), the formation of these primer dimers could not be detected in the samples.

Figure 3 and table 2 shows the quantitative LC-PCR using the Hybridization Probes Hot Start protocol.

We compared the MuCMV titers and the results of quantitative LC-PCR by the SYBR Green Hot Start protocol in the salivary glands 5, 8, 11, 15, and 18 days after virus challenge (Figure 4a). The quantitative LC-PCR was well correlated to the results of PFU assay. Moreover, we compared the MuCMV titers and the results of quantitative LC-PCR by Hybridization Probes Hot Start protocol (Figure 4b).

Relation of the MuCMV titer (PFU assay) and the result of quantitative LC-PCR

Comments

MuCMV-DNA levels in salivary glands of MuCMV infected mice were analyzed by using a "real time" PCR method, and compared with the plaque assay. The quantitative LC-PCR was well correlated to the results of PFU assay. However,

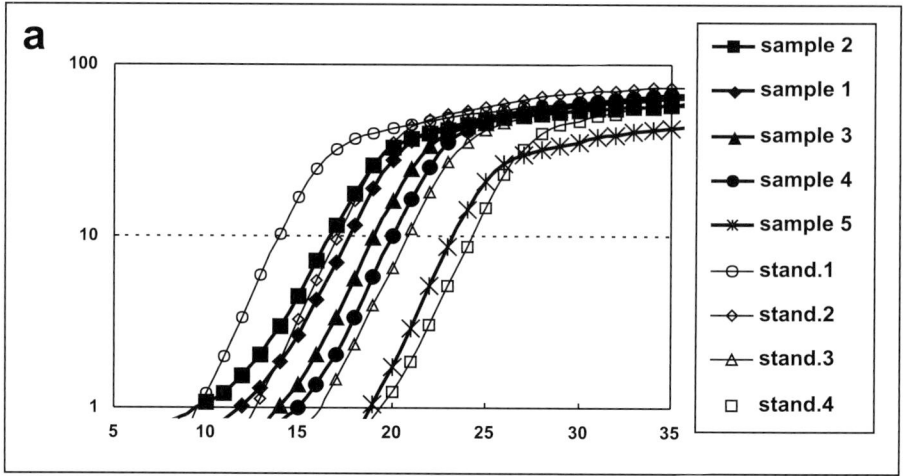

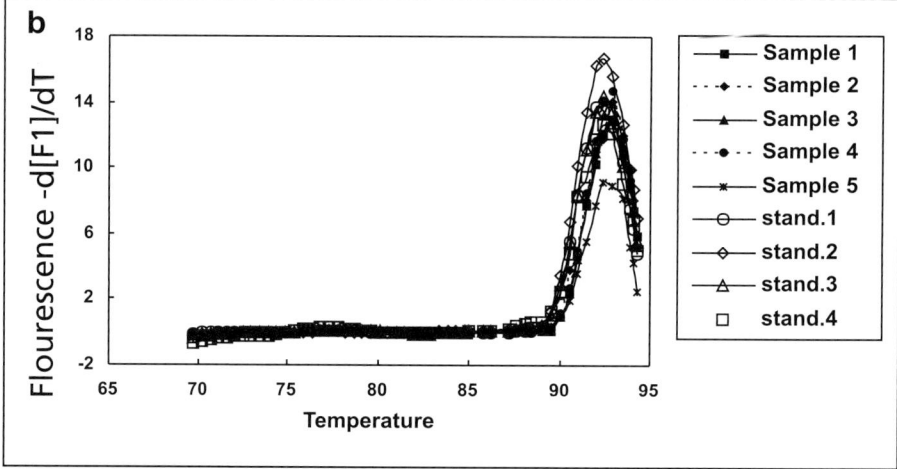

Fig. 2. Quantitative analysis in LC-PCR by using SYBRGreen Hot Start protocol. (a) The Light-Cycler software V.3.0 construct amplification plots from the extension phase fluorescent emission data collected during the PCR amplification. Overlay of amplification plots of salivary glands DNA samples amplified with MuCMV primers. *Y-axis*: fluorescence (log), *x-axis*: cycles. (b) Melting peak analysis of MuCMV primers

the SYBRGreen Hot Start method was more sensitive and better correlated with the results of PFU assay than the Hybridization Probes Hot Start method. We found it difficult to optimize the Hybridization Probes method. Because the effects of probe length, melting temperature and concentration, distance between hybridization probes, distance to primer, temperature profiles have not been systematically optimized. In our case, we tested a wide range of $MgCl_2$ concentrations, annealing temperatures, and enzyme concentrations. However,

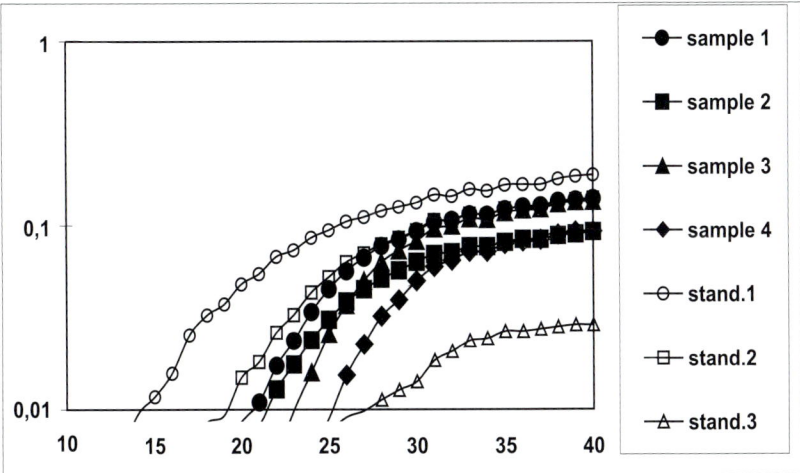

Fig. 3. Quantitative analysis in LC-PCR by using Hybridization Probes Hot Start protocol

Table 2. The calculated concentrations of samples (by Hybridization Probes Hot Start protocol)

	Days After Infection	Concentrations (pg)	Crossing Point
sample 1	5	152	22.2
sample 2	11	112	22.9
sample 3	15	67	24
sample 4	18	23	26.4

we could not get an optimal condition. We suggest that hybridization probe and primer designs are one of the most important and factors for the hybridization probes method.

Our SYBRGreen Hot Start protocol for detecting MuCMV-DNA was optimal condition. The formation of primer-dimers often results in false positive values using the SYBRGreen method. Therefore, we performed a hot start technique using an antibody to Taq polymerase. In our results, we could not detect the primer-dimers on the melting curve analysis and gel analysis. We recommend using the hot start techniques in every protocol. Moreover, $MgCl_2$ concentration is absolutely critical for the successful development of quantitative LC-PCR. In our SYBRGreen Hot Start protocol for detecting MuCMV-DNA, various $MgCl_2$ concentrations were tested. As shown in figure 5, 2 mM of the final concentration was most sensitive and specific condition. The magnesium concentration may affect all of the following: primer annealing, product specificity, and enzyme activity. We recommend testing the magnesium concentration before quantitative LC-PCR analysis. Until now, the plaque assay was a common method for detecting CMV activation. However, the plaque assay needs a lot of time for obtaining results and complicated methods. In this report, we demonstrated that

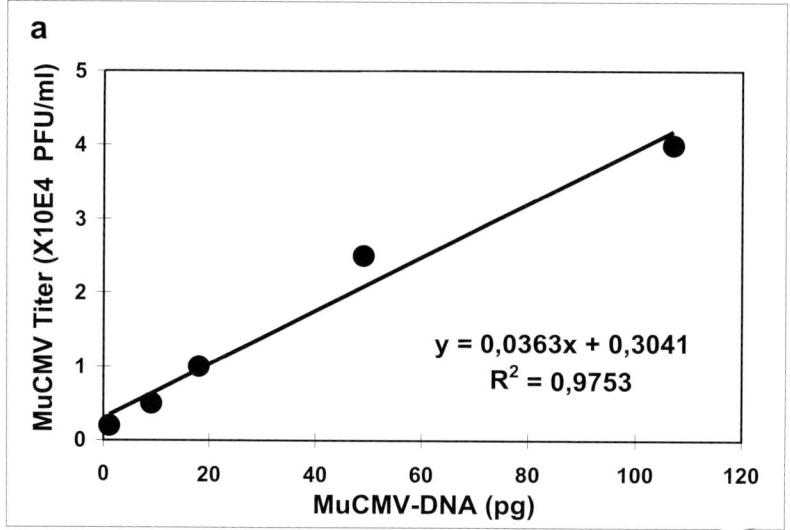

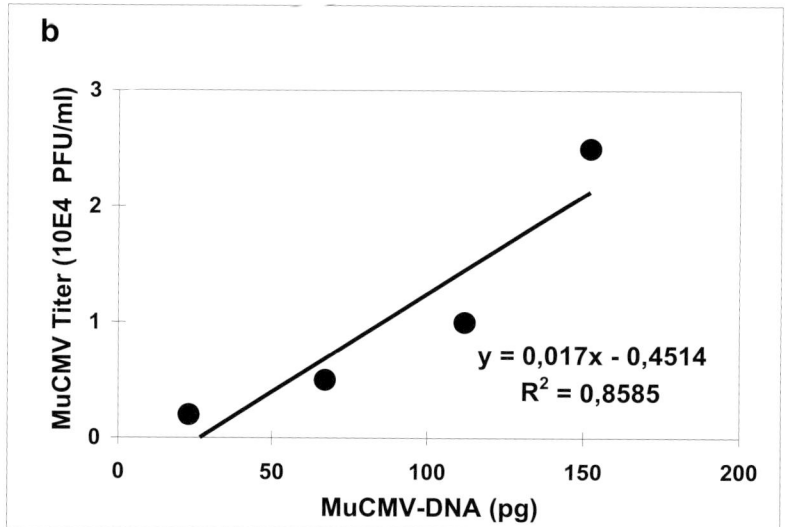

Fig. 4. The relation of the MuCMV titers and the results of quantitative LC-PCR by using (a) SYBR Green Hot Start protocol and (b) Hybridization Probes Hot Start protocol

the quantitative LC-PCR was well correlated with the results of PFU assay. We suggest that LC-PCR method is very useful in monitoring activation of CMV and can contribute to analyzing CMV activation more precisely.

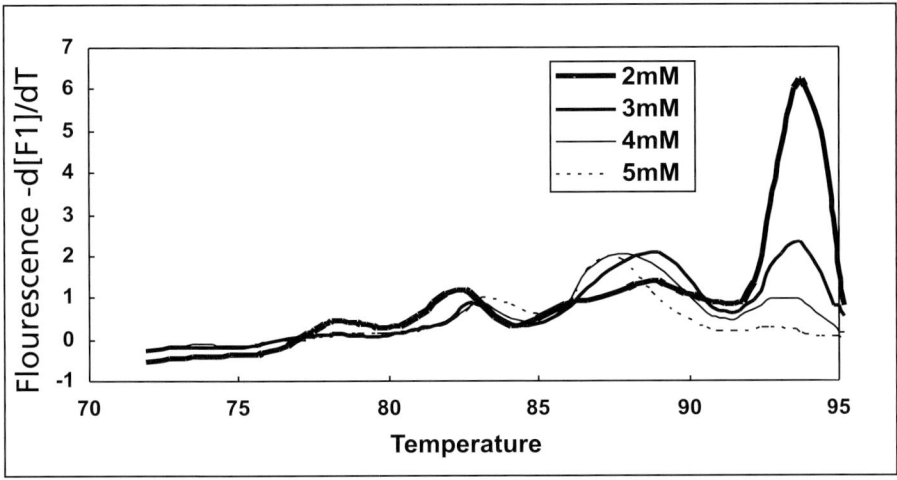

Fig. 5. Effect of MgCl$_2$ concentration to PCR efficiency. Same sample and same protocol without MgCl$_2$ concentration used for LC-PCR by using SYBR Green Hot Start and analyzed melting curves

References

1. Bacigalugo A, Tedone E, Isaza A, Soracco M, Van Lint MT, Sanna A, Frassoni F, Occhini D, Gualandi F, Lamparelli T, Fifari O, Faffo MR, Marmont AM (1995) CMV-antigenemia after allogenic bone marrow transplantaion: correlation of CMV-antigen positive cell numbers with transplant-related mortality. *Bone Marrow Transplantation* 16:155–161.
2. Preudhomme C, Revillion F, Merlat A, Hornez L, Roumier C, Duflos-Grardel N, Jouet JP, Cosson A, Peyrat JP, Fenaux P (1999) Detection of BCR-ABL transcription in chronic leukemia (CMV) using a "real time" quantitative RT-PCR assay. *Leukemia* 13:957–964.
3. Aiuti A, Turchetto L, Cota M, Cipponi A, Brambilla A, Arcelloni C, Paroni R, Vicenzi E, Brodignon C, Poli G (1999) Human CD34+ cells express CXCR4 and ligand stromal cell-derived factor-1. Implications and infection by T-cell tropic human immunodificiency virus. *Blood* 94:62–73.
4. Wittwer CT, Ririe KM, Andrew RV, David DA, Gundry RA, Balis UJ (1997) The LightCycler™: A microvolume multisample fluorimeter with rapid temperature control. *BioTechniques* 22:176–181.
5. Shanley JD (1987) The modification of acute murine cytomegalovirus adrenal gland infection by adoptive spleen cell transfer. *J Virol* 61:23–8.

Detection and Differentiation of Equine Herpes Virus Type 1 and Type 4 on the LightCycler

Peter Hübert*

Introduction

The control of many herpes virus infections is complicated by a latent state following the infection. Once infected, an individual may carry this virus for life. The latent infection can be reactivated by various factors such as stress or other infections [1]. Reactivated herpes virus can lead to typical clinical symptoms, such as the well known labial or genital herpes. Various herpes viruses are associated with various patterns of disease. The typical clinical symptoms of equine herpes virus type 1 (EHV1) infections are abortion (synonym: equine abortion virus), rhinopneumonitis (synonym: equine rhinopneumonitis virus), and encephalitis. However, two distinct viruses have seemed to be involved. One of them was almost exclusively recovered in cases of rhinopneumonitis, while the other was associated with cases of abortion or neurological disorders [2–4]. The differentiation by serology of these two viruses was difficult because the two viruses have a common set of antigens [5]. However, because their genomic homology is less than 20% [6] they have been taxonomically separated into two distinct types: EHV1 and EHV4 [7]. We have designed a LightCycler polymerase chain reaction (PCR) application that differentiates between EHV1 and EHV4 and avoids the potential for contamination during nested processes and makes restriction enzyme analysis unnecessary. A 834-bp fragment from a highly conserved region within the glycoprotein B (gB) gene was amplified and the viral types analyzed by melting curves.

Materials

Oligo Primer analysis software **Equipment**
LightCycler instrument

Amplification primers (Life Technologies) **Reagents**
  The following reagents were purchased from Roche Diagnostics, Mannheim, Germany:
dNTP mix for PCR
LightCycler-DNA Master SYBR Green I

---

* Peter Hübert (✉) (e-mail: Peter.Huebert@lvua-sh.de)
  Food and Veterinary Diagnostic Institute of Schleswig-Holstein, 24517 Neumünster, Germany

Procedure

Sample Preparation

Organ tissue samples (lung, spleen, and liver) and cell fractions eluted from nasal swabs were chopped with sea sand in cell culture medium (MEM with Earle's salts). The supernatant of these preparations or infected cell cultures were purified by a modified proteinase K lysis procedure [8].

Proteinase K lysis

100 µl supernatant
100 µl proteinase K (100 µg/ml in 1xPCR buffer : 10 mM Tris-HCl pH 9.0, 25 mM KCl, 0.1% Triton X-100; Promega, Mannheim, Germany) with 0.5% Tween 20
Mix thoroughly
Incubate for 45 min. at 65°C
Centrifuge 1 min. at 1.000xg
Incubate supernatant for 10 min. at 95°C
Centrifuge 1 min. at 1.000×g
Add 2 µl supernatant to PCR assay

Primer Design

Sequence data were derived from published EHV1 and EHV4 gB sequences [9, 10]. The amplification primer selection was performed using the Oligo computer program. The theoretical T_m of the products was calculated from sequence data assuming 1×SSC and 0% formamide [11]. Annealing temperatures and $MgCl_2$ concentrations were optimized experimentally (Table 1).

Table 1. Oligonucleotides

(Equine herpes virus, Genebank accession # HSE1GBA)				
	Position	Length	GC (%)	T_m (°C)
ACGACGACGAGGACGAGGTG	1345	20	65	57.3
GGTTGGCTGAGGTTCTGGTA	2198	20	55	51.1
Product EHV1		853	52.2	89.1
Product EHV4		848	46.3	86.7

LC-PCR

The following master mixes were used:

	Volume [µl]	[Final]
LightCycler-DNA Master SYBR Green I	2	1x
$MgCl_2$ (25 mM)	1.5	2.9 mM
Primers (20 µM each)	1+1	1 µM each
PCR grade H_2O	12.5	

For hot start PCR 0.4 µl Platinum Taq Antibody (5 U/µl, Life Technologies, Karlsruhe, Germany) per reaction were added to the master mix. Then 18 µl of master mix and 2 µl of DNA template were filled into each capillary.

The following LC PCR protocol was used:
- Denaturation for 2 min at 95°C
- Amplification

Parameter	Value		
Cycle type	40 Quantification		
	Segment 1	Segment 2	Segment 3
Target temperature (°C)	94	65	95
Incubation time (s)	3	10	34
Temperature transition rate (°C/s)	20	20	0.2
Acquisition mode	None	None	Single
Gain	F1=5		

- Melting curve analysis

Parameter	Value		
Cycle type	1 Melting Curves		
	Segment 1	Segment 2	Segment 3
Target temperature (°C)	95	65	95
Incubation time (s)	0	10	0
Temperature transition rate (°C/s)	20	20	0.2
Acquisition mode	None	None	Cont.
Gain	F1=5		

Results

Figure 1 shows the results of melting curve analysis after amplification of DNA isolated from, EHV4, EHV1, and EHV1 substrain Rac H viral standards, and organ tissue extract (lung and spleen combined) and nasal swab obtained from a foal, which suddenly died on its first day of life. Low melting products can be detected in the negative control, interpreted as primer dimer formation. This peak at 82°C is also detectable in the nasal swab specimen, where the amount of specific template DNA is very low. The melting temperatures of the specific products are clearly distinct: 88°C for EHV4 and 91°C for EHV1. When a hotstart is performed (Fig. 2), no primer dimer formation is observed in the negative control. The melting temperatures of several EHV1 or EHV4 isolates segregate into 2 T_m peaks. However, at 87°C and 90°C, these melting peaks differ by one degree from the values obtained in Fig. 1. These results are in good agreement with the predicted T_m values, where a salt concentration of 1×SSC is assumed. In addition, both peaks at 88°C and 91°C were detected in the samples obtained from the foal

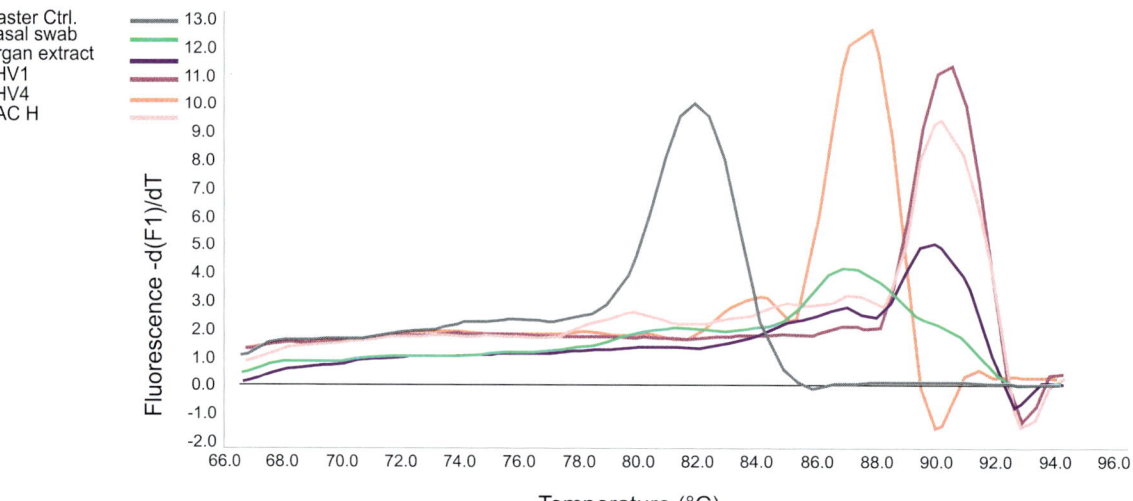

Fig. 1. Melting curve analysis of LC-PCR products. Analysis of purified preparations of reference DNA of EHV1 strain Rac H, EHV4, EHV1, an organ extract of lung and spleen tissue, and a nasal swab extract. A control for primer dimer formation contains all components except template DNA

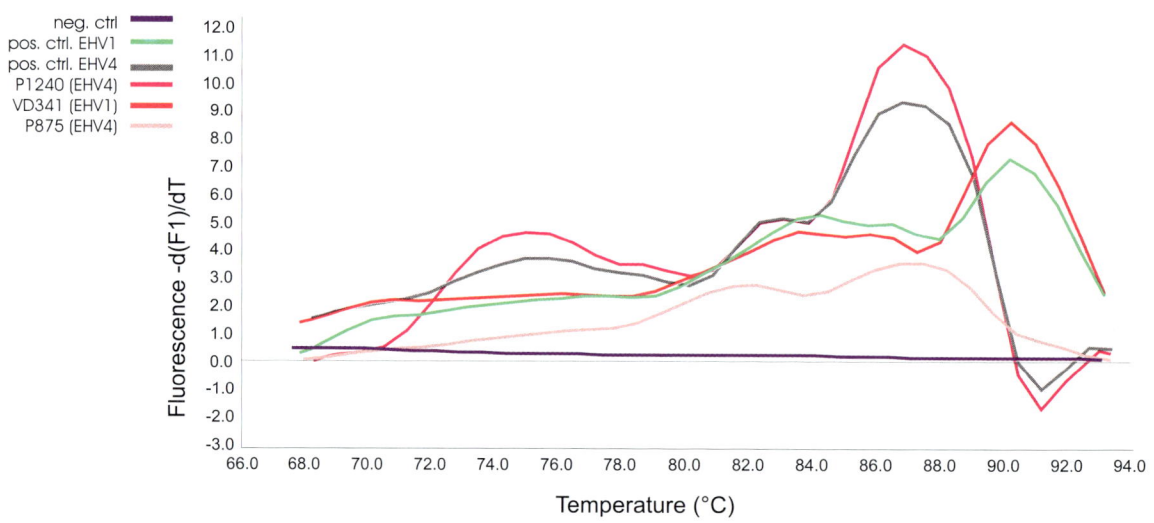

Different EHV1 and EHV4 isolates with Hotstart

Fig. 2. Melting curve analysis of hot start LC-PCR products. Analysis of purified preparations of reference DNA of EHV1, EHV4, and different field isolates determined as EHV1 or EHV4 previously. A control for primer dimer formation contains all components except template DNA. Addition of hot start antibody reduces primer dimer formation but has no effect on the reproducibility of melting curves

specimen (Fig. 1). Whereas the EHV4 specific peak is dominant over the EHV1 specific peak in the nasal swab PCR product, the opposite is the case in the organ specific material.

Figure 3 shows the restriction enzyme pattern of the products shown in Fig. 1 produced by conventional nested PCR. The results clearly differentiate between EHV1 and EHV4 as well as between EHV1 and EHV1 strain Rac H. The pattern of the nasal swab and of the tissue sample PCR products look like EHV1, demonstrating that the viral genome can be detected in clinical cases of herpes virus infections caused by EHV1.

Residual subgenomic bands are present in the product obtained from the nasal swab and organ extract. These bands may result from partial digestion, or coexistence of EHV1 and EHV4 in the specimen. Comparison of this result with the observation of both peaks in the melting curve analysis of this specimen in Fig. 1 suggests that simultaneous infection of this animal with both equine herpes viruses occurred. In conclusion, LC-PCR provides an easy, fast, and convenient method to detect and differentiate EHV1 and EHV4.

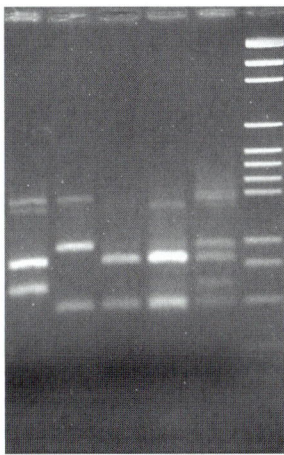

Fig. 3. Restriction enzyme pattern of conventional nested PCR. *Lane 1,* Rac H; *lane 2,* EHV4; *lane 3,* EHV1; *lane 4,* organ extract; *lane 5,* nasal swab; *lane 6,* DNA marker D-15 (Novex). The template for this PCR is the product of the LC-PCR presented here. The product of the nested PCR is 336 bp long and is suitable to differentiate EHV1, and EHV4 and Rac H by a double digest of Hinc II/Sac II restriction enzymes. Digestion with Hinc II generates two fragments of 215 and 121 bp, which is typical for EHV4. The PCR product obtained from EHV1 strain Rac H is only cut by Sac II into 148 and 188 bp fragments. An EHV1 field virus double digestion will cut the 148 bp *Sac* II fragment into 121 bp and 27 bp Hinc II/Sac II fragments

References

1. Gibson JS, Slater JD, Awan AR, Field HJ (1992) Pathogenesis of equine herpesvirus-1 in specific pathogen-free foals: primary and secondary infections and reactivation. Arch Virol 123:351–366
2. Allen GP, Bryans JT (1986) Molecular epizootiology, pathogenesis and prophylaxis of equine herpesvirus-1 infections. Prog Vet Microbiol Immun Basel: Karger
3. Meyer H, Thein P, Hübert P (1987) Characterization of two equine herpesvirus (EHV) isolates associated with neurological disorders in horses. J Vet Med B 34:545–548
4. Studdert MJ, Crabb BS, Ficorilli N (1992) The molecular epidemiology of equine herpesvirus 1 (equine abortion virus) in Australasia 1975 to 1989. Aust Vet J 69:104–11
5. Crabb BS, Studdert MJ (1990) Comparative studies of the proteins of equine herpesviruses 4 and 1 and asinine herpesvirus 3: antibody response of the natural hosts. J Gen Virol 71:2033–41.
6. Allen GP, Turtinen LW (1982) Assessment of the base sequence homology between the two subtypes of equine herpesvirus 1. J Virol 44:249–255
7. Studdert MJ, Fitzpatrick DR, Horner GW, Westbury HA, Gleeson LJ (1984) Molecular epidemiology and pathogenesis of some equine herpesvirus type 1 (equine abortion virus) and type 4 (equine rhinopneumonitis virus) isolates. Austr Vet J 61:345–348
8. Meyer H, Hübert PH, Schwend C, Eichhorn W (1992) Rapid identification and differentiation of the vaccine strain Rac H from EHV-1 field isolates using a non-radioactive probe. Vet Microbiol 30:13–20
9. Riggio MP, Cullinane AA, Onions DE (1989) Identification and Nucleotide Sequence of the Glycoprotein-gB Gene of Equine Herpesvirus 4. J Virol 63:1123–1133
10. Whalley JM, Robertson GR, Scott NA, Hudson GC, Bell CW, Woodworth LM (1989) Identification and nucleotide sequence of a gene in equine herpesvirus 1 analogous to the herpes simplex virus gene encoding the major envelope glycoprotein gB [published erratum appears in J Gen Virol (1989) 70:3513]. J Gen Virol 70:383–94
11. Baldino FJ, Chesselet M-F, Lewis ME (1989) High resolution in situ hybridization histochemistry. Methods Enzymol 168:761–777

Rapid and Quantitative Detection of *Toxoplasma gondii* by PCR – A LightCycler Application in Prenatal Diagnosis

Jean-Marc Costa*, Pauline Ernault, Stéphane Bretagne

Introduction

Toxoplasmosis is an infectious disease that is distributed worldwide and caused by the protozoan parasite *Toxoplasma gondii (T. gondii)*. The infection is usually asymptomatic and harmless in immunocompetent patients but can be life-threatening or responsible for severe sequelae in immunocompromised individuals such as the fetus and HIV and transplant patients. In these situations, early treatment significantly reduces the extent of the damage [1]. However, the classical diagnosis of toxoplasmosis based on serological tests is inefficient and inadequate in these patients. Therefore, the diagnosis is based on the direct demonstration of the parasite in tissues or biological fluids. This can be achieved by using tissue-culturing or mouse inoculation. However, tissue-culturing has a low sensitivity and inoculation into mice requires waiting more than 3 weeks for the results [2]. PCR overcomes these shortcomings [3]. Indeed, the PCR is sensitive and the diagnosis can be given in 1 day. The use of PCR has been particularly useful in facilitating the greatly improved and simplified prenatal diagnosis of toxoplasmosis, by making it faster and more sensitive [4]. However, the main risk is that of false positive results due to contamination with previously amplified products. The recent development of a real-time, on-line PCR procedure gives the opportunity to control this risk, because it uses a closed tube format for amplification and detection. We have, therefore, tested the feasibility of this technique in a routine laboratory and compared the results to PCR with ELISA detection of biotin-labeled probes (conventional PCR).

Materials

Equipment

LightCycler Instrument
ThermalCycler 480 (PE Biosystem)
Microtiterplate washer and reader

Reagents

Amplification Primers, Biotin-Labeled Primers and Digoxigenin-Labeled Probes (Genset)

---

* Jean-Marc Costa (✉) (e-mail: jean-marc.costa@ahp-paris.com)
  American Hospital of Paris, Molecular Biology Laboratory, 63 bd Victor Hugo, 92200 Neuilly-sur-Seine, France

Hybridization Probes (Tib Molbiol)
Nucleotides (Pharmacia)
AmpliTaq Gold (Perkin-Elmer)
TaqStart Antibody (Clontech)
Uracil-DNA Glycosylase (Biolabs)

The following reagents were purchased from Roche Diagnostics, Mannheim:
High Pure PCR Template Preparation Kit
LightCycler DNA Master Hybridization Probes
Heat-labile UDG
PCR Elisa DIG Detection Kit

Procedure

Samples

T. gondii DNA (strain RH) was prepared with the use of a conventional phenol-chloroform procedure from purified parasite obtained from ascite fluids of inoculated mice. The equivalent amount of parasite was estimated by spectrophotometry (one parasite ≈0.1 pg of DNA). Ref 0.1 pg.

Nucleic acids from amniotic fluids were extracted using the High Pure PCR Template Preparation kit according to the manufacturer's recommendations. A total of 5–20 µl of the eluate were submitted to PCR amplification.

Primers and Probes

The primers were designed to amplify specifically a 126-bp fragment in a well-conserved region of the tandemly repeated B1 gene of *T. gondii* [5]. For the conventional procedure, the forward primer was 5' biotin-labeled to allow detection and specificity checking of PCR products using a digoxigenin-labeled probe (see below). The same set of primers was used in combination with specific hybridization probes to detect the PCR products, using fluorescence resonance energy transfer on the LightCycler (Table 1).

Conventional PCR

The samples (20 µl) were amplified in a 50-µl reaction mixture containing 2.5 mM $MgCl_2$; 50 mM KCl; 10 mM Tris-HCl pH 8.3; 0.2 mM each dATP, dGTP and dCTP; 0.4 mM dUTP; 20 pmoles each of *T. gondii* primers; 0.5 units of Uracil-DNA glycosylase (UDG); and 1.25 units of AmpliTaq Gold. The samples were initially incubated for 5 min at 50°C to allow the action of UNG. This incubation was followed by a 10-min step at 95°C to denature the DNA and to activate the AmpliTaq Gold. The temperature cycling (42 cycles at 95°C and 60°C for 30 s each, and 72°C for 1 min.) was performed in a 48-well thermal cycler. Amplification products were detected by PCR ELISA using biotin-labelled primers and the PCR ELISA DIG Detection Kit.

Real-Time Detection

PCR amplification was performed with the LightCycler DNA Master Hybridization Probes Kit in a standard PCR reaction containing 0.5 µM of each primer and 0.25 µM of each probe with 5 µl of sample. A hot-start procedure was systematically used by introducing an anti-Taq DNA polymerase to the amplification reaction mixture. Prevention of carry-over was systematically achieved by using the

Table 1. Primers and fluorescent hybridization probes used for Toxoplasma gondii detection

	Length	GC (%)	T_m (°C)
Primers:			
GGAGGACTGGCAACCTGGTGTCG	23	65.2	63.5
TTGTTTCACCCGGACCGTTTAGCAG	25	52.0	63.8
Product	126		
Hybridization probes:			
LCRed640-ACGGGCGAGTAGCACCTGAGGAGAT-P	25	60.0	64.1
CGGAAATAGAAAGCCATGAGGCACTCC-F	27	51.8	64.6

heat-labile Uracil-DNA glycosylase (UDG). The amplification was carried out in a LightCycler as follows: The reaction mixture was initially incubated for 5 min at room temperature to allow the action of UDG. This incubation was followed by a 2-min step at 95°C to denature the DNA and to inactivate both the UDG and the anti-Taq DNA polymerase. Amplification was performed according to the following protocol:

- Amplification

Parameter	Value		
Cycles	50		
Type	Quantification		
	Segment 1	Segment 2	Segment 3
Target temperature [°C]	95	60	72
Incubation time [s]	5	10	15
Temperature transition rate [°C/s]	20	20	2
Acquisition mode	None	Single	None
Gains	F1=1, F2=10		

- Hybridization Probe Master Mix

	Volume [µl]	[Final]
LightCycler-DNA Master Hybridization Probes	2	1 ×
TaqStart antibody (5 min incubation at room temperature)	0.2	
$MgCl_2$ (25 mM)	3.2	5 mM
Primers (20 µmM each)	0.5	0.5 µM
Hybridization probes (20 µmM each)	0.25	0.25 µM
Uracil-DNA glycosylase (1 U/µml)	0.5	0.5 U/tube
Standard DNA in appropriate dilution (0.75–0.75 10^{+6} tachyzoites)	5	
H_2O (PCR grade)	7.6	
Total volume	20	
For unknown samples, replace the standard DNA with amniotic DNA extract		

Results

Using purified DNA diluted in a Tris-HCl buffer, the same amount of DNA can be detected with real-time PCR and with conventional PCR. The amount of DNA equivalent to one tachyzoite was systematically detected by both methods (Fig. 1). Using both PCR techniques, the results obtained with 22 frozen amniotic fluids (11 negative and 11 positive) were similar (Fig. 2).

For quantitative analysis of the parasite DNA, the results obtained with the LightCycler were compared with the standard curve of *T. gondii* DNA (Fig. 3). The parasite burden was extremely variable, from $1.62\ 10^2$/ml to $0.98\ 10^{-1}$/ml, although most of the parasite burden was weak (see Table 2). Nevertheless, even the weak parasite burden must be detected, as the consequences on the fetus can be damaging.

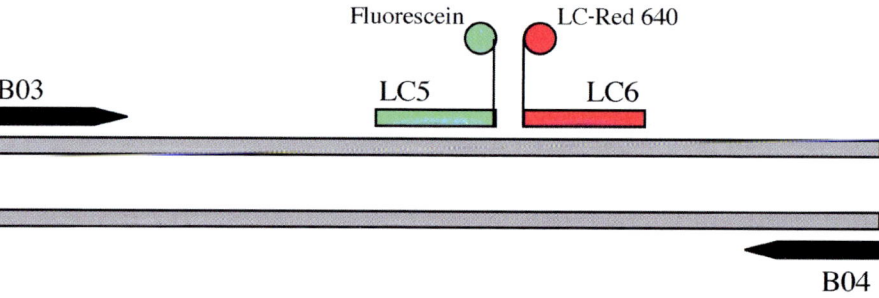

Fig. 1. Hybridization Probes procedure. Fluorescence is emitted by LC-Red 640 when both Hybridization Probes are annealed to one strand of the amplification product that is generated during PCR. The PCR fragment size is 126 basepairs

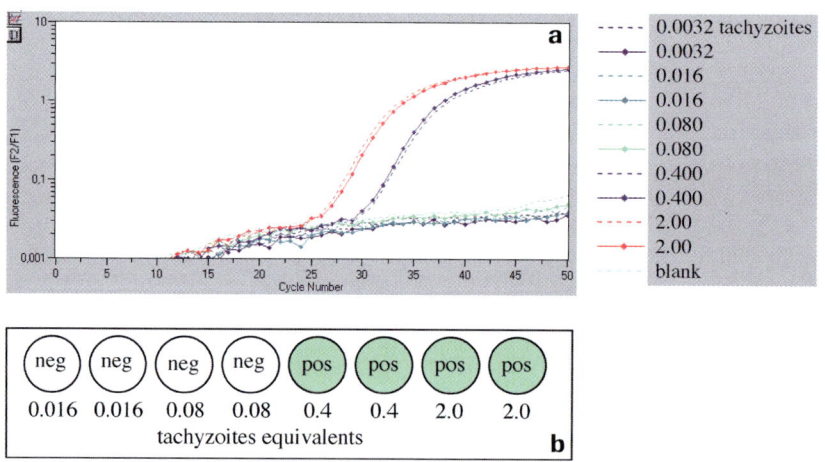

Fig. 2. Comparative sensitivity between real-time PCR and the convertional procedure. **2a.** Screenshot of real-time detection, showing the development of fluorescent signal during PCR. **2b.** Schematic presentation of results for the microtiterpllate detection format

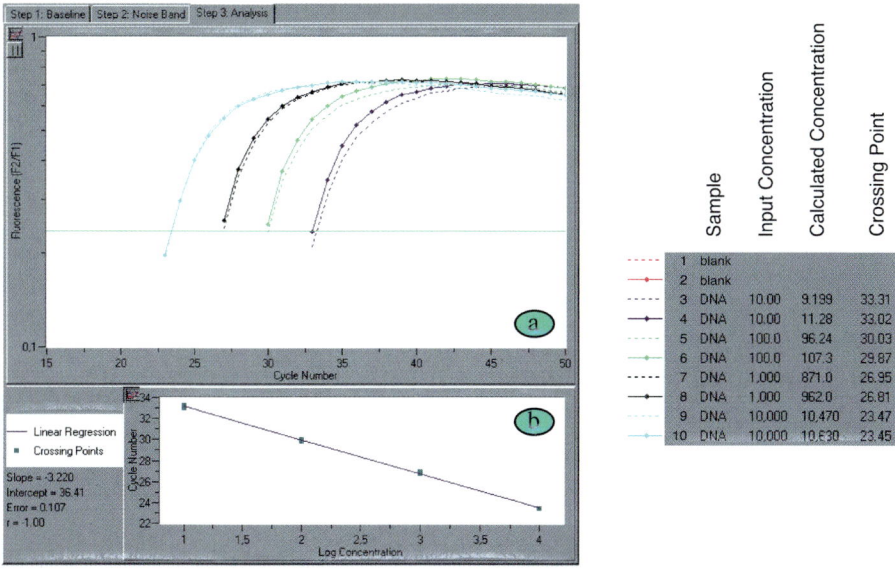

Fig. 3. Real-time quantification PCR analysis. **3a.** Analysis of the amplification plots for the standard concentrations (in duplicate) of T. gondii. **3b.** Standard curve; plot of the crossing point (cycle number) against the input target quantity

Our results showed that PCR amplification of the B1 gene of *T. gondii* on the LightCycler is as sensitive as the conventional PCR used in our laboratory. Moreover, the LightCycler has additional advantages over the conventional PCR, such as: (1) contamination prevention (the closed-tube system decreases the carry-over of PCR products), (2) speed (results can be obtained in less than 1 h, compared with 5 h with the conventional PCR), and (3) quantitative analysis which may have some prognostic implications and may be useful for monitoring drug therapy.

References

1. Roizen N, Swisher CN, Stein MA et al (1995) Neurologic and developmental outcome in treated congenital toxoplasmosis. Pediatrics 95 (1): 11–20
2. Fircker-Hidalgo H, Pelloux H, Muet F, Racinet C, Bost M, Goullier-Fleuret A, Ambroise-Thomas P (1997) Prenatal diagnosis of congenital toxoplasmosis: comparative value of fetal blood and amniotic fluid using serological techniques and cultures. Prenat Diagn 17(9): 831–835
3. Bretagne S, Costa JM, Vidaud M, Tran Van Nhieu J, Fleury-Feith J (1995) Detection of Toxoplasma gondii by competitive DNA amplification of bronchoalveolar lavage samples. J Clin Microbiol 33 (6): 1662–1664
4. Hohlfeld P, Daffos F, Costa JM, Thulliez P, Forestier F, Vidaud M (1993) Prenatal diagnosis of congenital toxoplasmosis with a polymerase-chain-reaction test on amniotic fluids. N Engl J Med 331 (11): 695–699
5. Burg JL, Grover CM, Pouletty P, Boothroyd JC (1989) Direct and sensitive detection of a pathogenic protozoan, Toxoplasma gondii, by polymerase chain reaction. J Clin Microbiol 27 (8): 1787–1792

Development of Quantitative PCR Tests for the Detection of the Orthopox Virus Adsorption Protein Gene (ORF D8L) on the LightCycler

Claus-Peter Czerny*, Michaela Alex, Jana Pricelius, Christiane Zeller-Lue

Introduction

The genus *Orthopoxvirus* (OPV) includes several species of well-known pathogens [13, 21]. Although the Variola and Alastrim viruses were eradicated by a worldwide compulsory vaccination with vaccinia virus, monkeypox and, especially in Europe, cowpox virus strains may cause diseases in humans. Human cowpox is a relatively rare zoonosis and cows are currently not known to be involved. In recent years, virus strains have drawn the attention of the medical profession by causing localized and generalized skin infections in humans [5, 6, 12, 14, 17, 24] as well as in cats [7, 10, 22] or zoo-animals like large felides or elephants [3, 4, 18, 25]. Usually, a low infectivity for healthy persons is observed with benign skin lesions. However, in immunocompromised and non-vaccinated individuals the infection can lead to severe diseases which may end lethally [8, 11, 12].

A permanent natural cowpox virus reservoir has not yet been identified, but based on epidemiological surveys, it is assumed that these virus strains are maintained in wild-life rodents [2, 16, 18]. The virus spreads between the rodents which are considered to be intermediate hosts, whereas, cats, cows, elephants, and human beings are final hosts.

Because of the wide host range but unclear ecology of OPV, it is important to investigate their natural distribution, reservoirs, and infective routes. This makes very sensitive and quick diagnostic tools necessary. Several PCR methods for OPV detection and differentiation have recently been developed [9, 19, 20]. In this connection we evaluate a new quantitative LightCycler PCR for rapid detection of OPV strains, based on the amplification of the vaccinia virus open reading frame D8L-analogues [15], the gene for the 32 kDa adsorption protein.

Materials

Ultracentrifuge Optima L60 (Beckman, Munich) **Equipment**
ALFexpress Automatic DNA Sequencer (Amersham Pharmacia, Freiburg)

---

* Claus-Peter Czerny (✉) (e-mail: claus.czerny@tgd.bayern.de)
  Senator-Gerauer-Str. 23, 85586 Poing, Germany

LightCycler Instrument (Roche Diagnostics, Mannheim)
DNA Star Lasergene, DNA and Primer analysis software (gatc, Konstanz)

Reagents
TaqStart Antibody (Clontech, Heidelberg)
Eagle's MEM containing 5% FCS
Proteinase K, 14 mg/ml (Roche Diagnostics, Mannheim)
Sodium dodecylsulfate (SDS), 20% (Sigma, Munich)
Phenol-Chloroform-Isoamylalcohol 25:24:1 (Roth, Karlsruhe)
Amplification Primers (MWG Biotec, Ebersberg)
Hybridization Probes (TIB MOLBIOL, Berlin)
Sequencing Primers (MWG Biotec, Ebersberg)
LightCycler-DNA Master SYBR Green I (Roche Diagnostics, Mannheim)
LightCycler DNA Master Hybridization Probes (Roche Diagnostics, Mannheim)

Procedure

Sample Preparation
To isolate orthopox virus (OPV) DNA from tissue cultures infected with several virus strains, 500 µl aliquots were exposed to three alternating heating (100°C 5 min) and freezing steps (–70°C) before 5 µl proteinase K, 22.5 µl TEN buffer and 22.5 µl SDS were added. After an incubation of 2 h at 56°C, DNA was extracted from proteins and lipids with the same volume of phenol-chloroform, set on ice for 30 min and centrifugated at ca. $9,000 \times g$ and room temperature for 15 min. The supernatants were carefully collected and mixed with 1/10 volume of a 5 M potassium-acetate solution and the double volume of absolute ethanol. After precipitation over 30 min at –70°C, DNA was pelleted at $17,600 \times g$ and 4°C for 30 min, washed three times with 70% ice-cold ethanol, dried and resuspended in PCR water. This semipurified DNA was stored at –20°C until use.

Additionally, purified DNA was prepared according to the same procedure from cowpox virus KR2 Brighton ultra-centrifuged on sucrose gradients [9].

Primer Design
Primer design was performed under the prerequisite to amplify the whole D8L gene (GeneBank accession number: M35027) in various orthopox virus species. From published sequence data [15, 23] and the data of our group it was known that N- and C-termini of D8L-analogues of several OPV strains were conserved. Primers were designed by the DNA Star computer program. Also, 3′-Pentamer Stability ΔG was set to –13.5 kCal/mol. Six bp of dimer or hairpin duplexing was accepted. The average cut off for mispriming (-ΔG/primer length) was set to 0.76. Hybridization probes were designed by TIB MOLBIOL (Berlin) with a moderate stringency according to published guidelines [1]. For sequencing, the PCR primers were conjugated to Cy5 at the 5′-terminus. Oligonucleotides and characteristics are shown in Table 1.

Development of Quantitative PCR Tests for the Detection of the Orthopox Virus Adsorption Protein Gene (ORF D8L)

Table 1. Oligonucleotide primers

Orthopox Virus D8L Gene (Genebank Accession #NC_002170)				
	Position	Length	GC (%)	T_m (°C)
Primers				
ATGCCGCAACAACTATCTCCT	103544R	21	47.6	67.2
CTAGTTTTGTTTTTCTCGCGAA	102630	22	36.4	62.7
Product	102630–103544	915		
Probes				
TCAATTGGATTCCATTAGATCCGCCAA-F	103137R	27	40.7	66.7
LCRed640-ACGTCTGCACCGTTTGA-TTCAGTATTTTATC-P	103109R	31	38.7	67.7

LightCycler PCR

The following master mix was used with SYBR Green I detection for each 20 µl reaction:

	Volume [µl]	[Final]
LightCycler-DNA Master SYBR Green I	2	1×
MgCl$_2$ (25 mM)	2.4	3 mM
Primers (50 µM each)	0.5+0.5	1.25 µM each
H$_2$O (PCR grade)	9.6	
Total volume	15	

For hot start conditions, 1.6 µl of 0.7 µM TaqStart antibody was added to the DNA Master SYBR Green I solution and incubated for 5 min at room temperature, replacing an equal volume of water.

The following master mix was used with hybridization probe detection for each 20 µl reaction:
- Denaturation at 95°C for 1–2 min
- Amplification

	Volume [µl]	[Final]
LightCycler-DNA Master Hybridization Probes	2	1×
MgCl$_2$ (25 mM)	2.4	3 mM
Primers (50 µM each)	0.5+0.5	1.25 µM each
Probes (3 µM each)	1+1	0.15 µM each
H$_2$O (PCR grade)	7.6	
Total volume	15	

For hot start conditions, 3.2 µl of 0.7 µM TaqStart antibody was added to the DNA Master SYBR Green I solution and incubated for 5 min at room temperature, replacing an equal volume of water.

- Melting Curve Analysis

Parameter	Value		
Cycles	1		
Type	Melting Curve		
	Segment 1	Segment 2	Segment 3
Target temperature [°C]	95	65	95
Incubation time [s]	0	10	0
Temperature transition rate [°C/s]	20	20	0.2
Acquisition mode	None	None	Continuous
Gain		FL1=5	

Cycle Sequencing   Specificity of the amplified PCR products was confirmed by cycle sequencing on an ALFexpress automatic sequencer (Amersham Pharmacia, Freiburg). Purified template DNA, which was also used for LC-PCR, was sequenced with the corresponding Cy5-labelled primers. Sample volumes of 20 µl containing 1.5 µg DNA were prepared with the "Thermo sequenase fluorescent labelled primer cycle sequencing kit with 7-deaza-dGTP" (Amersham Pharmacia, Freiburg), according to the manufacturer's description.

Results

LC-PCR   Titration experiments were performed with purified DNA of cowpox virus KR2 Brighton. A concentration of 1 ng/µl was titrated in $\log_{10}$ dilutions. The D8L gene analogue was amplified with/without hot start conditions and detected by the SYBR Green I or hybridization probes. In the case of the SYBR Green I detection, Fig. 1A,B shows similar amplification curves with approximately the same slope. However, melting curve analysis of the samples without TaqStart (Fig. 1C,1D) shows a significant amount of primer dimer in samples starting with less than 10^{-3} ng/µl DNA, resulting in false positives unless melting curve analysis is performed. These primer dimers dramatically reduced the sensitivity of the PCR for targets with a low copy number and prevented specific amplification below 10^{-3} ng/µl DNA. With TaqStart, the formation of primer dimers could be inhibited down to a DNA concentration of 10^{-6} ng/µl (Fig. 1D).

When the amplified D8L gene was detected by specific probes the use of the TaqStart antibody resulted in an increase of the sensitivity for targets with a low copy number by a factor of 100. The detection limit was 10^{-3} ng/µl without and 10^{-5} ng/µl with hot start technique (Fig. 1E,F).

LC-PCR with Semipurified DNA from Cell Cultures of Various Orthopox Virus Strains   The D8L genes of four OPV reference strains listed in Table 2 were amplified from semipurified DNA and detected by SYBR Green I. A significant DNA increase could be observed from cycle 20 for the virus strains tested. The plateaus were reached at cycle 40 (Fig. 2A,B). A small amount of primer dimer accumulated in the negative control (PCR grade water), as shown in the melting curve analysis

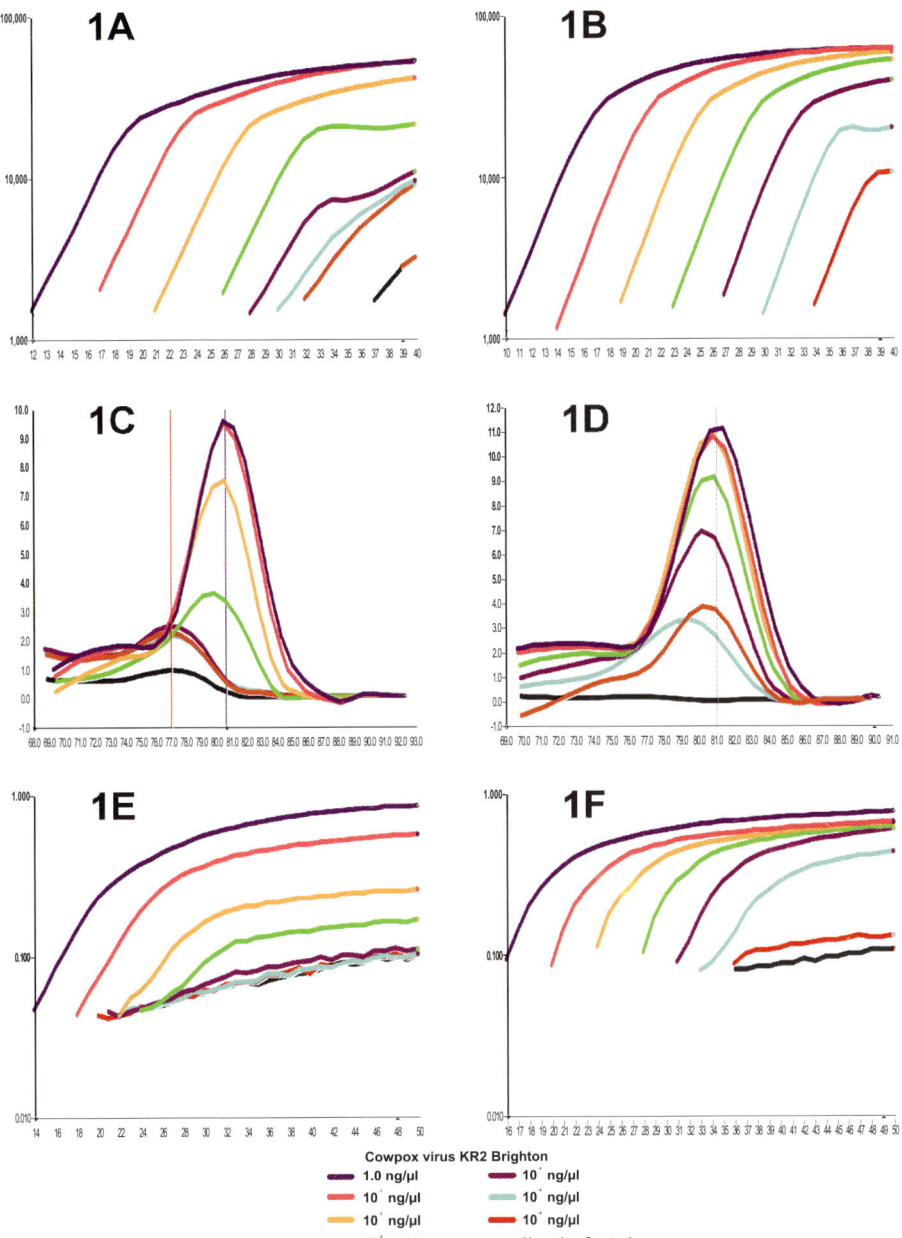

Fig. 1. Quantitative LC-PCR of titrated DNA samples. Purified DNA (1 ng/µl) of cowpox virus KR2 Brighton was serially diluted in log$_{10}$-steps by PCR grade water. LC-PCR was performed for 40–50 cycles using the SYBR Green I without (**A,C**) and with (**B,D**) hot start conditions or the hybridization probe format (**E,F**) as described in "Material and Methods"

(Fig. 2C). The formation of these dimers could be inhibited completely by using a "hot start" technique (Fig. 2B,D). A significant DNA increase could be observed from cycle ten in the case of vaccinia virus Elstree and cowpox virus KR2 Brighton, whereas in the case of ectromelia virus and camelpox virus, the curves increased at cycle 22. The plateaus were reached at cycle 40. These variations depend on distinct DNA concentrations in the semipurified samples. Slight differences in the T_ms may be related to sequence variations between the OPV strains used. Specificity of the PCR products was confirmed by direct cycle sequencing. Table 3 shows the degree of nucleotide homology of the D8L-homologues compared to the published sequence of vaccinia virus Kopenhagen [15]. Additionally, PCR products amplified in the SYBR Green I format were recovered from the capillaries by reverse centrifugation into 1.5-ml Eppendorf tubes. Samples were diluted in sucrose/bromphenol blue buffer and loaded on 1.0% agarose gels. 1.0× TAE buffer contained 3 µl/100 ml of a 1% ethidium bromide solution. Gels were photographed on a Geldoc 1000 image analysing system (BioRAD, Munich; data not shown). Table 4 shows the quantative LC-PCR values of titrated DNA samples.

Comments

Sample Preparation

Sample preparation and isolation of high quality DNA are critical aspects of quantitative PCR on the LightCycler. Standard preparation of semipurified poxvirus DNA from tissue cultures gives good results on the LightCycler with the

Table 2. Orthopox virus reference strains used in the following study to develop quantitative PCR tests for the LightCycler

Orthopox Virus Species		Strain	Reference
OPV bovis	(Cowpox virus)	Brighton KR2	WHO-Reference strain
OPV cameli	(Camelpox virus)	CP1	Ramyar und Hassami, 1972
OPV commune	(Vaccinia virus)	Elstree	WHO-Reference strain
OPV muris	(Mousepox virus)	Ectromelia Munich1	Mahnel, 1974

Table 3. Degree of nucleotide homology in percentages of the D8L-genes of various orthopox virus reference strains compared to vaccinia virus Kopenhagen [15]

Orthopox Virus Strains	Degree of nucleotide homology
Vaccinia virus Kopenhagen	100.00
Vaccinia virus Elstree	99.34
Camelpox virus CP1	97.81
Cowpox virus KR2 Brighton	96.17
Ectromelia virus Munich1	94.86

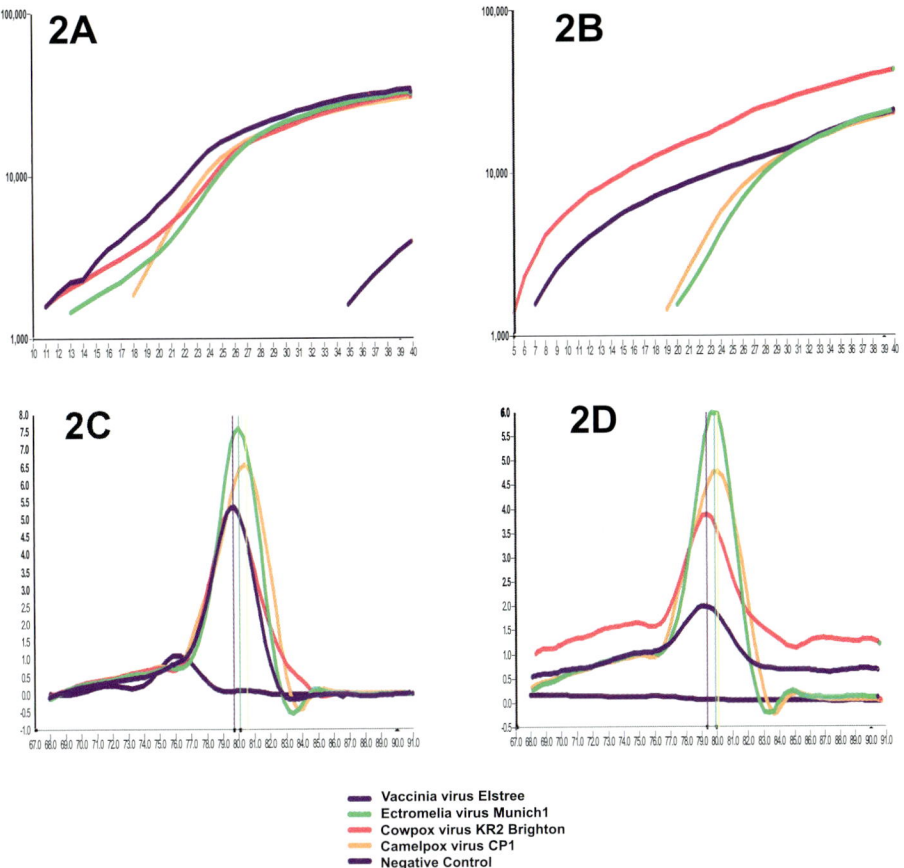

Fig. 2. LightCycler PCR and melting curve analysis of semipurified DNA samples from four OPV reference strains using the SYBR Green I detection system without (**A,C**) and with (**B,D**) hot start conditions

SYBR Green I detection system. Using the hybridization probe detection system with semipurified DNA from different orthopox virus strains led to inconsistent results. It is unusual that the various strains gave good results with SYBR Green I, but not with hybridization probes. The reason could be species-specific differences of up to four nucleotides underneath the hybridization probes. However, with the application of higher purified DNA we achieved satisfactory results. Recently, new kits for the preparation of highly pure viral nucleic acids from tissue culture supernatants have been developed, but in the case of OPV, about 95% of the derivative virus are cell associated.

Table 4. Quantative LC-PCR of titrated DNA samples. The percentages of expected values are calculated by the mean of two amplifications. Genomic orthopox virus DNA was serially titrated in log10-steps. LC-PCR was performed for 40–50 cycles using the SYBR Green I and Hybridization Probe format described in "Material and Methods". The expected values were calculated as concentration of the measured amount of message in the undiluted sample

DNA Concentration	SYBR Green I	Hot start	Hybridization Probes	Hot start
1 ng/µl	110.6	101.2	86.5	89.6
10^{-1} ng/µl	125.5	123.0	97.5	87.8
10^{-2} ng/µl	104.1	104.5	125.6	118.2
10^{-3} ng/µl	76.8	88.4	93.1	99.5
10^{-4} ng/µl		91.2		99.3
10^{-5} ng/µl		99.3		101.1
10^{-6} ng/µl		106.5		

Primer Design

Good primer design is critical for the successful development of reliable quantitative PCR tests. We used a computer aided search for primer sequences with highly stringent binding properties. The primers should amplify the D8L-gene in total, because it is also a good candidate for a species-specific differentiation. N- and C-termini offer conserved sequences, whereas the middle of the gene posseses species-specific variations, even though the size of our amplicon was 915 bp and larger than is usually recommended. However, we could clearly demonstrate that the LightCycler OPV-PCR was highly sensitive. A detection limit of 1 fg DNA/µl with SYBR Green I and hot start conditions corresponds to a detection of 6 DNA molecules.

References

1. Caplin BE, Rasmussen RP, Bernard PS, Witter CT (1999) LightCycler Hybridization Probes. Biochemica 1:5–8
2. Baxby, D (1977) Poxvirus hosts and reservoirs. Arch Virol 55:196–196
3. Baxby D, Shackleton WB, Wheeler J, Turner A (1979) Comparison of cowpox-like viruses isolated from european zoos. Arch Virol 61:337–340
4. Baxby D, Ashton DG, Jones DM, Thomsett LR (1982) An outbreak of cowpox in captive cheetahs: virological and epidemiological studies. J Hyg Camb 89:365–372
5. Baxby D, Bennett M, Getty B (1994) Human cowpox: A review based on 54 cases, 1969–93. Br J Dermatol 131:598–607
6. Baxby D, Bennett M (1997) Cowpox: a re-evaluation of the risk of human cowpox based on new epidemiological information. Arch Virol Suppl 13:1–12
7. Bennett M, Gaskell RM, Gaskell CJ, Baxby D, Kelly DF (1989) Studies on poxvirus infection in cats. Arch Virol 104:19–33
8. Czerny C-P, Eis-Hübinger AM, Mayr A, Schneweis KE, Pfeiff B (1991) Animal poxviruses transmitted from cat to man: current event with lethal end. J Vet Med B 38:421–431
9. Czerny C-P, Johann S, Hölzle L, Meyer H (1994) Epitope detection in the envelope of intracellular mature orthopox viruses and identification of encoding genes. Virology 200: 764–777

10. Czerny C-P, Wagner K, Gessler K, Mayr A, Kaaden O-R (1996) A monoclonal blocking-ELISA for detection of orthopox virus antibodies in feline sera. Vet Microbiol 52:185–200
11. Czerny C-P, Zeller-Lue C, Eis-Hübinger AM, Kaaden O-R, Meyer H (1997) Characterization of a cowpox-like orthopox virus which had caused a lethal infection in man. Arch Virol Suppl 13:13–24
12. Eis-Hübinger AM, Gerritzen A, Schneweis KE, Pfeiff B, Pullmann H, Mayr A, Czerny C-P (1990) Fatal cowpox-like virus infection transmitted by cat. Lancet 336:880
13. Fenner F, Wittek R, Dumbell KR(eds). (1989) The Orthopoxviruses. Academic Press, San Diego, California, pp 1–432
14. Egberink HF, Willemse A, Horzinek MC (1986) Isolation and identification of a poxvirus from a domestic cat and a human contact case. J Vet Med B 33:237–240
15. Goebel SJ, Johnson GP, Perkus ME, Davis SW, Winslow JP, Paoletti E (1990) The complete DNA sequence of vaccinia virus. Virology 179:247–266
16. Kaplan C, Healing TD, Evans N, Healing L, Prior A (1980) Evidence of infection by viruses in small British field rodents. J Hyg 84:285–294
17. Klingebiel T, Vallbracht A, Döller G, Stierhof YD, Gerth HJ, Glashauser E, Herzau V (1988) A severe human cowpox infection in South Germany. Pediatr Infect Dis J 7:883–885
18. Marennikova SS, Shelukhina EM, Efremova, EV (1984) New outlook on the biology of cowpoxvirus. Acta Virol 28:437–444
19. Meyer H, Pfeffer M, Rziha HJ (1994) Sequence alterations within and downstream of the A-type inclusion protein genes allow differentiation of orthopox virus species by polymerase chain reaction. J Gen Virol 75:1975–1981
20. Meyer H, Ropp S, Esposito JJ (1997) Gene for A-type inclusion body protein is useful for a polymerase chain reaction assay to differentiate orthopox viruses. J Virol Methods 64:217–221
21. Murphy FA, Fauquet CM, Bishop DHL, Ghabraial SA, Jarvis AW, Martelli GP, Mayo MA, Summers MD (eds) (1995) In: Virus Taxonomy. Sixth Report of the International Committee on Taxonomy of Viruses. Springer Verlag, Heidelberg Berlin New York, pp 79–91
22. Naidoo J, Baxby D, Bennett M, Gaskell RM, Gaskell CJ (1992) Characterization of orthopoxviruses isolated from feline infections in Britain. Arch Virol 125:261–272
23. Niles EG, Condit RC, Caro P, Davidson K, Matusick L, Seto J (1986) Nucleotide sequence and genetic map of the 16-kb vaccinia virus Hind III D fragment. Virology 153, 96–112
24. Stolz W, Götz A, Thomas P, Ruzicka T, Süss R, Landthaler M, Mahnel M, Czerny C-P (1996) Characteristic but unfamiliar – The cowpox infection transmitted by a domestic cat. Dermatology 193:140–143
25. Zwart P, Gispen R, Peters C (1971) Cowpox in Okapi "Okapi johnstoni" at Rotterdam Zoo. Br Vet J 127:20–24

Plant Gene Products and Miscellaneous   VI

Quantification of Genetically Modified Soybeans
in Food with the LightCycler System 383
Klaus Pietsch, Hans-Ulrich Waiblinger

**Real-Time PCR Monitoring of Estuarine Water Samples for *Pfiesteria piscicida*:
A Dinoflagellate Associated with Fish Kills and Human Illness** 391
Holly Bowers, Torstein Tengs, Mark Herrmann, David Oldach

Quantification of Retrotransposon XIR-2.5 Copy Number
in Genomes of *Poeciliidae* Species 399
Meinhard Hahn, Christiane Thömmes, Jochen Wilhelm,
Jamilah Michel

Quantification of Genetically Modified Soybeans in Food with the LightCycler System

Klaus Pietsch*, Hans-Ulrich Waiblinger

Introduction

Worldwide, the use of genetically modified plants is increasingly important. The European Community (EC) has already approved herbicide tolerant soybeans (Roundup Ready Soybeans, RRS, Monsanto) and insect tolerant corn (Bt-corn, Novartis), and further approvals of genetically modified plants for food production (tomatoes, radicchio, rapeseed, soybeans and corn) are pending.

In 1998, the EC adopted labeling regulations for foods and food ingredients produced from RRS and Bt-corn. Labeling is required if DNA or proteins resulting from genetic modification are present.

According to the preface of the regulation (EC No. 1139/98) labeling due to minor contamination would not be necessary if tolerable limits for genetically modified DNA and protein could be established. For this purpose, methods that quantify genetically modified moleculs in food would be useful. Present methods allow the detection of genetically modified molecules in food [1, 2], but they do not allow any quantification.

Recently, a competitive PCR method for the detection and quantification of Roundup Ready Soybeans has been developed [3, 4]. However, this semiquantitative PCR method using internal standards and competitors is time-consuming and difficult to evaluate. The LightCycler system allows quantification of target DNA by measuring fluorescence in the PCR log-linear phase.

In the present study we describe a method for the real time quantification of Roundup Ready Soybeans in food with the new LightCycler system.

Materials

GeneQuant II RNA/DNA Calculator (Pharmacia, Freiburg)
LightCycler Instrument

Equipment

Wizard DNA Extraction Kit (Promega, Heidelberg)
Amplification Primer (BIG Biotech, Freiburg)
Certified Standard GMO Soybean (Fluka, Deisenhofen)

Reagents

---

* Klaus Pietsch (✉) (e-mail: pietsch@clua.cvuafr.bwl.de)
  Chemische und Veterinäruntersuchungsamt Freiburg, Bissierstr. 5, 79114 Freiburg, Germany

TaqMan Probes (TIB MOLBIOL, Berlin)
LightCycler-DNA Master Hybridization Probes (Roche Diagnostics, Mannheim)

Procedure

Sample Preparation

In accordance with the Swiss Food Manual [5], 300 mg of homogenized GMO Standard Material containing 5% Roundup Ready Soybeans and a food sample (soybean diet) were incubated with 860 µl TNE buffer [10 mM Tris (pH 8.0), 150 mM NaCl, 2 mM EDTA and 1% (w/v) sodium dodecyl sulphate], 100 µl 5 M guanidine hydrochloride (Sigma, Deisenhofen) and 40 µl (20 mg/ml) proteinase K (Merck, Darmstadt) on a thermomixer at 58°C for at least 3 h. After centrifugation, 500 µl of the supernatant were transferred to a new tube and incubated with 5 µl RNase (10 mg/ml) (Fluka) at 60°C for 10 min. The extracted DNA was purified using the Wizard Protocol (Promega). The DNA was finally eluted in 50 µl Tris buffer (10 mM, pH 9.0). DNA concentrations were determined spectrophotometrically, using the GeneQuant II RNA/DNA Calculator (Pharmacia).

Oligonucleotides

The primers and probes that were used to perform real time PCR are listed in Table 1. The oligonucleotides were stored at –20°C.

Table 1. Oligonucleotides[6]

	Position	Length	GC (%)	T_m (°C)
Total Soybean DNA (Soybean lectin – Genebank Accession #K00821)				
Primers				
GACGCTATTGTGACCTCCTC	1131	20	55	62.4
TGTCAGGGGCATAGAAGGTG	1311R	20	55	63.9
Product	1131-1311	181		
TaqMan Probe				
FAM-CAACTCAATAAGGTTGACGAAAACGGC-TAMRA	1161	27	44.4	67.1
Roundup Ready DNA (Genebank Accession #X04879 and #M21084)				
Primers				
TGATGTGATATCTCCACTGACG (X04879)	200	22	45.5	61.8
TGTATCCCTTGAGCCATGTTGT (M21084)	62R	22	45.5	64.2
Product		172		
TaqMan Probe				
FAM-CCCACTATCCTTCGCAAGACCCT-TAMRA		23	56.5	68.1

LightCycler PCR

Each PCR reaction (20 µl total volume) contained the following master mix

	Volume [µl]	Final
LightCycler-DNA Master Hybridization Probes	2	1x
MgCl$_2$ (25 mM)	2.4	4 mM
Primers (5µM each)	1 + 1	0.25 µM
TaqMan Probe (1.5µM)	2	0.15 µM
H$_2$O (PCR grade)	6.6	

A total of 15 µl of master mix and 5 µl of sample DNA were added in each capillary. Sealed capillaries were centrifuged in a microcentrifuge and placed into the LightCycler rotor. The control reaction and estimation of total soybean DNA was performed using the primers targeting the soybean lectin gene (Table 1). For detection of Roundup Ready Soybeans, a primer pair specific for the RRS gene (transgenic construct between the cauliflower mosaic virus 35 S promotor and the Petunia hybrida 5-enolpyruvylshikimate-3-phosphate synthase singal peptide) was used (Table 1).

The following LightCycler protocol was used for total soybean DNA estimation (conditions for the RRS TaqMan Probe are in parentheses):
- Denaturation for 1 min at 95°C
- Amplification

Parameter	Value		
Cycles	40 (50)		
Type	Quantification		
	Segment 1	Segment 2	Segment 3
Target temperature (°C)	95	60 (58)	72
Incubation time (s)	5	15	8
Temperature transition rate (°C/s)	20	20	20
Acquisition mode	None	Single	None
Gains	F1=3	F2=10	F3=10

- Cooling for 30 s at 40°C

Results

The concentration of purified soybean DNA was measured photometrically and diluted from 80,000 to 80 copies per reaction. Under optimum conditions, a single copy gene (lectin) can be detected in 2.7 pg soybean genomic DNA (approx. 1 soybean genome equivalent) [7]. In separate assays, a lectin gene sequence and a RRS sequence of the samples were amplified. The formation of the amplicons was measured fluorimetrically each cycle after hydrolysis of a specific FAM/TAMRA labeled probe (TaqMan Probe). The observed fluorescence is proportional to the amount of hydrolyzed probe generated during the ongoing PCR process.

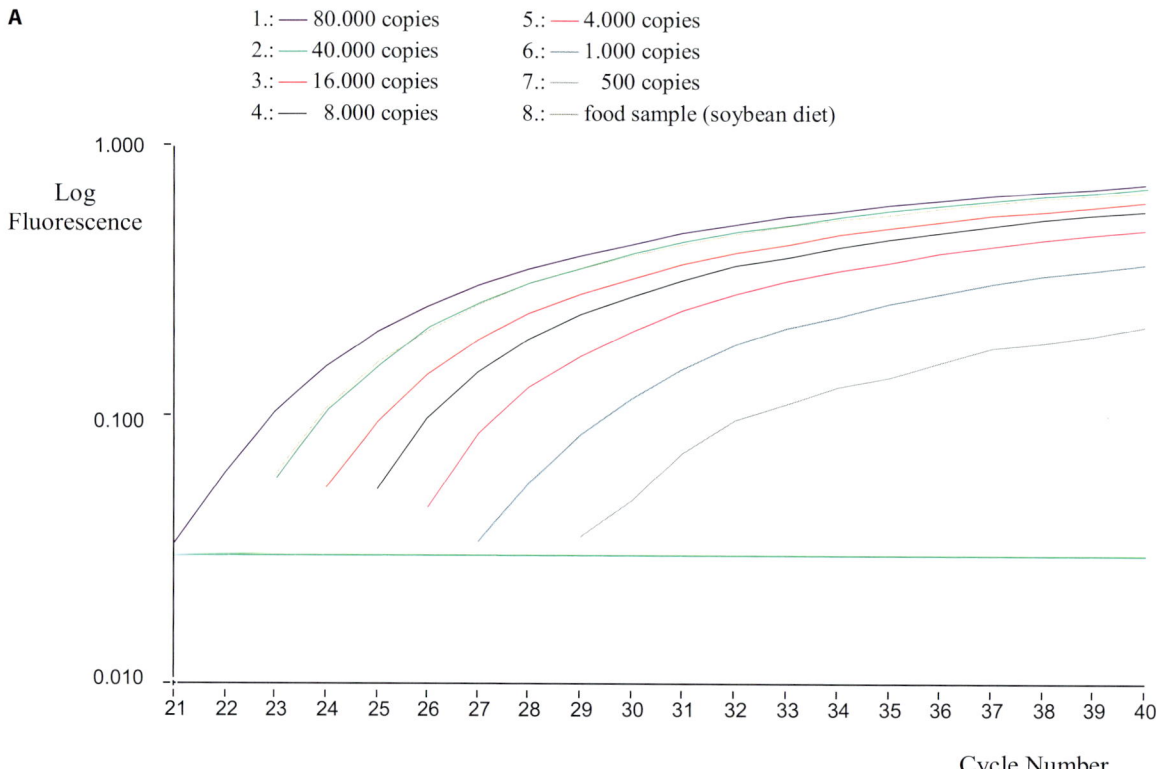
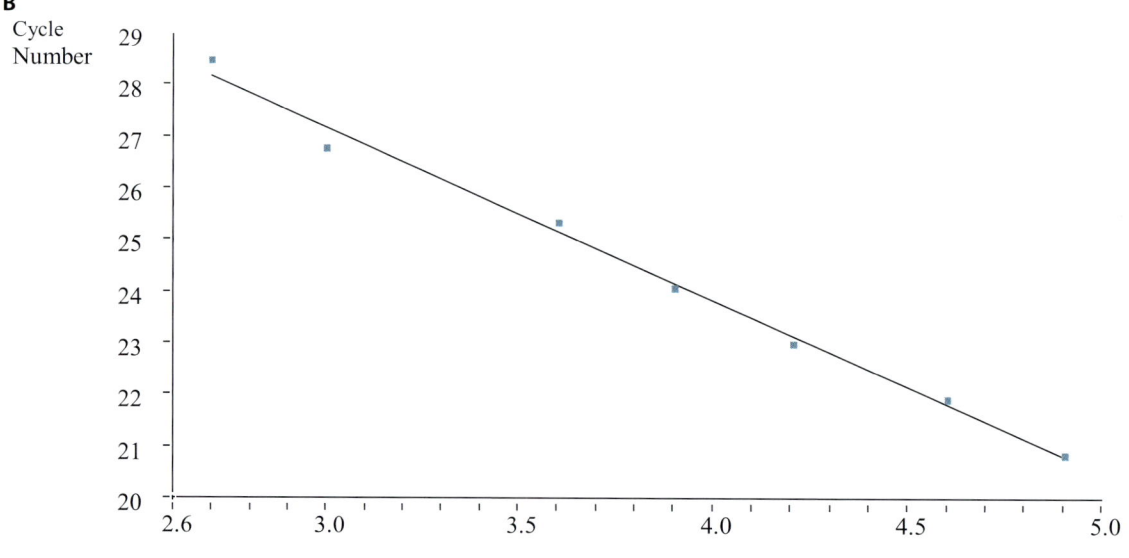

Figure 1 shows the amplification of the lectin gene from different starting amounts of soybean DNA in the GMO standard and the food sample DNA. After PCR is completed, the noise band is set and the "fit point method" used to calculate crossing points (cycle numbers) that are inversely proportional to the log of the initial template concentration. Figure 1A shows the logarithmic plot of all data above the noise band during the amplification of the lectin gene. In Fig. 1B, the crossing points of the GMO standards are plotted against the logarithm of the DNA concentration. The copy number of the lectin gene in the food sample was obtained by interpolation form the standard curve.

In a second PCR, the RRS gene was amplified from diluted standard DNA (containing 5% RRS) and the food sample DNA.

Figure 2A shows the logarithmic plot of all data above the noise band during the amplification of the RRS gene. The control (water) showed, as expected, no increase of fluorescence. In Fig. 2B, the crossing points of the GMO standards are plotted against the logarithm of the DNA concentration. The copy number of the RRS gene in the food sample was interplated from the standard DNA dilutions.

Table 2 shows the photometrically determined and the LightCycler calculated copy numbers of the lectin gene and the RRS gene in the GMO standards. The standards ranged from 80,000 to 500 copies for the lectin gene and from 4,000 to 200 for the RRS gene.

For the determination of the amount of RRS in the food sample, the copy number of the RRS gene was divided by the copy number of the lectin gene. The percentage of RRS to total soybean DNA in the food sample was 1.3% (Table 2).

Comments

The present study describes a LightCycler system for the quantification of the amount of Roundup Ready Soybeans (RRS) in food samples, in order to control maximum limits for genetically modified organisms (GMO) in food. Both lectin and RRS amplification yielded highly reproducible results with a sensitivity of approximately 0.1% RRS DNA content.

The LightCycler system can also be used for the determination of RRS DNA in complex food (i.e., biscuits). We conclude that the LightCycler system will allow convenient PCR analysis for a wide range of food testing. This includes not only quantification of GMO but also the differentiation of animal and plant species.

Fig. 1 A,B. Quantification of the lectin gene in Standard GMO soybean and food sample DNA. (**A**) A logarithmic plot of the fluorescence data above the noise band during the amplification of the lectin gene in soybean standard and food sample DNA. The dilution of the external standard DNA ranged from 80,000 to 500 copies of the lectin gene. The fluorescence signals of the food sample and the 40,000 copy standard are not distinguishable. (**B**) The crossing points (cycle numbers), plotted against the log concentration (copy number) of the GMO standard DNA. The calculated values and the copy number of the lectin gene in the food sample are listed in Table 2

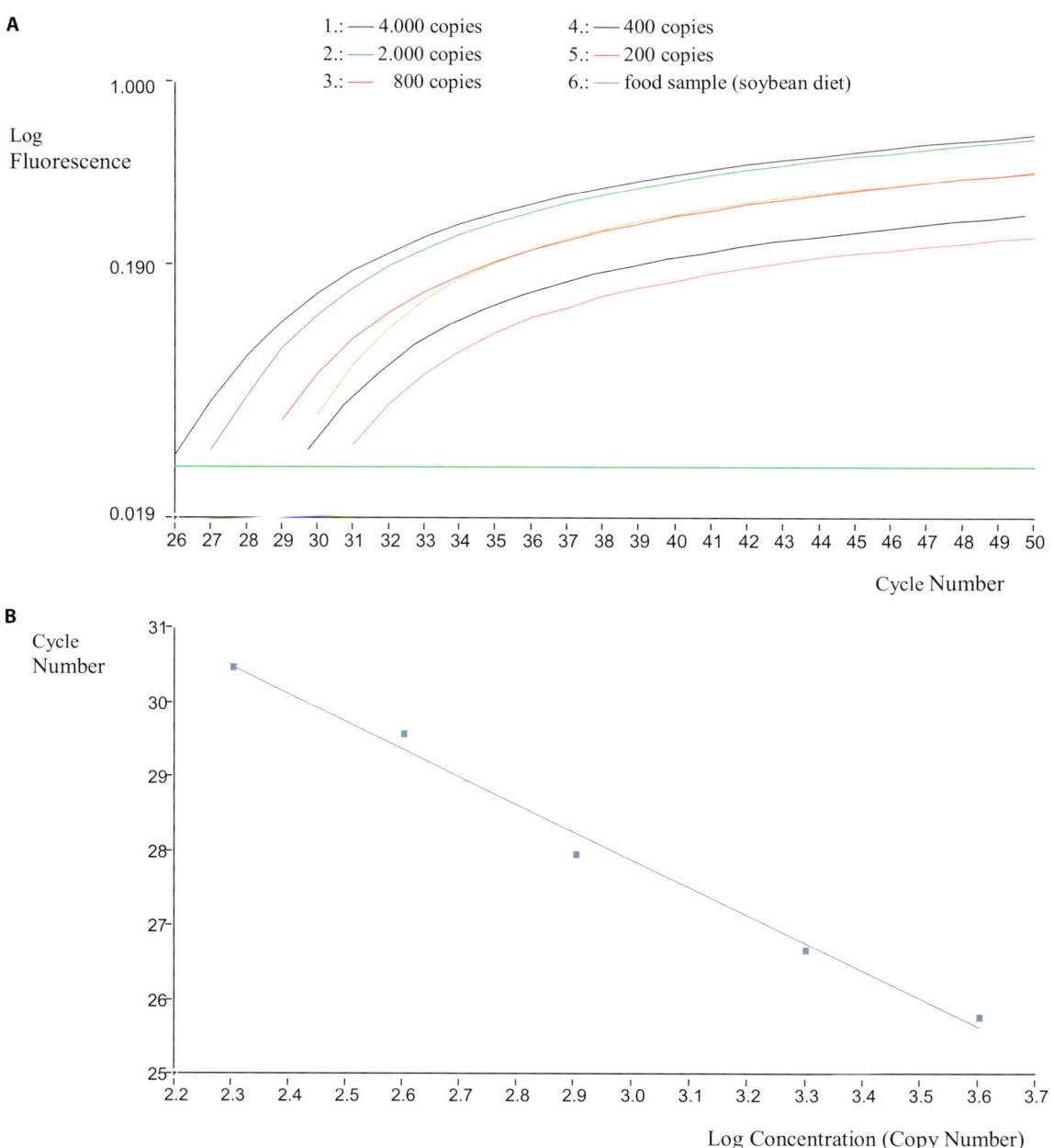

Fig. 2 A,B. Quantification of RRS gene in Standard GMO soybean and food sample DNA. (**A**) A logarithmic plot of the fluorescence data above the noise band during the amplification of the RRS gene in soybean standard and food sample DNA. The dilution of the external standard DNA ranged from 4,000 to 200 copies of the RRS gene. (**B**) The crossing points (cycle numbers), plotted against the log concentration (copy number) of the GMO standard DNA. The calculated values and the copy number of the RRS gene in the food sample are listed in Table 2

Table 2. Determination of the amount of Roundup Ready Soybeans in food

Sample	Lectin-PCR			Roundup Ready-PCR			Roundup Ready content (%)
	Copy number (photometric)	Copy number (calculated)	Crossing point	Copy number (photometric)	Copy number (calculated)	Crossing point	
5% RRS	80,000	80,000	20.8	4,000	4,000	25.8	4.6
5% RRS, 1:2	40,000	37,000	21.9	2,000	2,000	26.7	5.6
5% RRS, 1:5	16,000	18,000	23.0	800	900	28.0	5.3
5% RRS, 1:10	8,000	8,000	24.1	400	400	29.6	4.1
5% RRS, 1:20	4,000	4,000	25.4	200	200	30.5	5.7
5% RRS, 1:80	1,000	1,000	26.8	50	Not determined		
5% RRS, 1:160	500	406	28.5	25	Not determined		
Food sample "soybean diet"		40,000	21.8		500	28.9	1.3

References

1. Pietsch K, Waiblinger HU, Brodmann P, Wurz A (1997) Screening method for the detection of genetically modified plants. Dtsch Lebensm Rundsch 93: 35–38
2. Waiblinger HU, Wurz A, Freyer R, Pietsch K (1999) Specific detection of genetically modified rapeseed in honey. Dtsch Lebensm Rundsch 95: 44–48
3. Studer E, Rhyner C, Lüthy J, Hübner P (1998) Quantitative competitive PCR for the detection of genetically modified soybean and maize. Z Lebensm Unters Forsch A 207, 207–213
4. Pietsch K, Bluth A, Wurz A, Waiblinger HU (1999) Komptetitive PCR zur Quantifizierung konventioneller und transgener Lebensmittelbestandteile. Dtsch Lebensm Rundsch 95: 57–59
5. Swiss Food Manual (Schweizerisches Lebensmittelbuch) (1998) Molekularbiologische Methoden. Eidgenössische Drucksachen- und Materialzentrale, Bern
6. Bluth A (1998) Entwicklung von zwei PCR-gestützten Methoden zur Quantifizierung von DNA aus gentechnisch veränderten Organismen in Lebensmitteln. Thesis, University of Freiburg
7. Graham MJ, Nickell CD, Rayburn AL (1994) Relationship between genome size and maturity group in soybean. Theor Appl Genet 88:429–432

Real-Time PCR Monitoring of Estuarine Water Samples for *Pfiesteria piscicida*: A Dinoflagellate Associated with Fish Kills and Human Illness

HOLLY BOWERS, TORSTEIN TENGS, MARK HERRMANN, DAVID OLDACH*

Introduction

Pfiesteria piscicida, a heterotrophic estuarine dinoflagellate, has been associated with fish kills in North Carolina and Maryland [2, 3]. The rivers most affected in Maryland are tidal tributaries of the Chesapeake Bay, where many residents make their living on the water, either as commercial fishermen or in the recreation industry. During the summer of 1997, Maryland experienced a series of fish kills, mostly affecting Atlantic menhaden (*Brevoortia tyrannus*), with dramatic impact on the seafood and tourism industries of the State [5]. In addition, numerous persons that came into contact with water containing *Pfiesteria piscicida* developed skin rashes, flu-like symptoms, and cognitive problems [4, 7]. Due to the potential human health risks, extensive loss of fish, and economic impacts related to fish kills, the Maryland Department of Natural Resources (DNR) developed a comprehensive water quality monitoring program, initiated in 1998 and fully deployed in 1999 and 2000, to determine ecological parameters associated with *Pfiesteria piscicida* blooms. As part of this effort, water samples are collected monthly from stations throughout tributaries of the Chesapeake Bay and analyzed by our laboratory for the presence of the organism.

In analyzing environmental water samples for *Pfiesteria piscicida*, our laboratory routinely uses a Taqman probe assay optimized for use on the LightCycler. The specificity of this assay system has been demonstrated against a panel of estuarine dinoflagellates, including *Pfiesteria piscicida* and a wide range of other taxa [1]. Use of the LightCycler, in comparison to non-real-time PCR with the Perkin Elmer 9600, has not only increased the specificity of our reactions through elimination of non-specific amplification, but has also decreased the time lapse between processing of water samples and results.

In addition to the Taqman assay, we routinely use a SYBR Green I protocol for detection of dinoflagellate DNA in processed water samples. This assay, performed on samples negative with the *Pfiesteria piscicida* Taqman probe, is used to confirm the integrity of our DNA extraction method. (Dinoflagellates are essentially ubiquitous in estuarine waters, making it possible to use primers designed to act as 'universal dinoflagellate amplification primers' for

---

* David Oldach (✉) (e-mail: oldach@umbi.umd.edu)
   Institute of Human Virology, 725 West Lombard Street, Baltimore, MD 21201, USA

process controls, much as one might target actin genes as controls in mammalian systems).

Materials

Equipment
250 ml polyethylene bottles (VWR, Bridgeport NJ)
Filtering apparatus; 15 ml glass funnel and fritted glass support (VWR, Bridgeport NJ)
5 µM hydrophilic, low protein binding Durapore filters; 25 mm diameter (Millipore, Bedford MA)
Microcentrifuge
Vacuum manifold (Qiagen, Valencia CA)
Primer Express software
LightCycler

Reagents
Acid Lugol Solution (Sigma, St. Louis MO)
DNeasy Plant Kit (Qiagen, Valencia CA)
Amplification primers (Operon, Alameda CA)
Taqman probe (Operon, Alameda CA)
Taq polymerase and PCR Buffer (Life Technologies, Rockville MD)
50 mM $MgCl_2$ (Life Technologies, Rockville MD)
10× BSA (Idaho Technology, Salt Lake City, UT)
10× SYBR Green I (Idaho Technology, Salt Lake City, UT)
10× PCR buffer with 30 mM $MgCl_2$ (Idaho Technology, Salt Lake City, UT)
10× dNTPs (Idaho Technology, Salt Lake City, UT)
TaqStart antibody (Clontech, Palo Alto CA)
Enzyme diluent (Idaho Technology, Salt Lake City, UT)

Procedure

Sample Preparation
Surface water samples are collected in 250-ml bottles and either maintained unfixed or preserved with 1% acid-Lugol's iodine solution. Approximately 40 ml (depending on turbidity) of the sample is vacuum filtered through a 5-µM filter. The protocol supplied with the Qiagen DNeasy Plant Kit is followed, with minor modification, for extraction of DNA from algal species retained on the filter. Briefly, after addition of lysis buffer and RNase A, the filter is incubated for 10 min at 65°C. Precipitation buffer is added and the mixture is incubated on ice for 5 min. The lysate and filter are transferred to a Qiashredder Spin Column and centrifuged for 2 min at 10,000 rpm. Binding buffer and ethanol are added to the lysate, mixed, and vacuumed through a DNeasy Spin column attached to a manifold. The column is washed twice with wash buffer and spun for 1 min at 10,000 rpm to remove residual wash buffer. DNA is eluted with 100 µl or 200 µl of elution buffer and stored at −20°C.

Primer Design

Dinoflagellate Assay with SYBR Green I

A PCR primer was designed to anneal to a unique 18 S ribosomal DNA sequence region, conserved among dinoflagellate species but unlikely to anneal efficiently to a wide range of other marine and estuarine eukaryotes. When used in conjunction with a 'universal' 3′ primer targeted to the small subunit ribosomal (18 S) gene sequence (a modified version of primer B developed by Medlin et al. 1988), a product of approximately 141 bp is amplified from all dinoflagellate species tested to date. This primer pair has a high specificity for dinoflagellates, and when it was tested against a panel of other protists, amplification was not evident (data not shown; the apicomplexan organism *Cryptosporidium* is an exception; others may exist).

Amplicons generated with the dinoflagellate-specific assay are sequenced as needed, and primers targeted to the derived sequences have been utilized in conjunction with a 5′ 'universal' 18 S ribosomal DNA sequence primer, to produce near full length 18 S amplicons. This strategy is made necessary by the complex organism pool present in all environmental samples and many laboratory dinoflagellate cultures.

Pfiesteria piscicida Assay with Taqman Probe

Amplicons generated with the dinoflagellate-specific assay were sequenced and the sequence data was used in conjunction with the heteroduplex mobility assay to derive the 18 S sequence of the *Pfiesteria piscicida* genome [8]. A primer pair and Taqman probe (Primer Express software) with specificity for this organism were designed and tested against a variety of other dinoflagellate cultures [1] (Table 1).

Table 1. Primer and probe sequences used in SYBR Green I and Taqman assays

Pfiesteria piscicida 18S-DNA (Genebank # AF077055)				
	Region	Length	GC (%)	T_m (°C)
Dinoflagellate-specific primers:				
CGATTGAGTGATCCGGTGAATAA	18 S rDNA	23	43	63.1
TGATCCTTCTGCAGGTTCACCTAC*	18 S rDNA	24	50	66.5
Product		141		
Pfiesteria piscicida-specific primers:				
CAGTTAGATTGTCTTTGGTGGTCAA	107	25	40	63.0
AGCTGATAGGTCAGAAAGTGAT-ATGGTA	320R	28	39	63
Product	107–320	214		
Hydrolysis Probe				
FAM-CATGCACCAAAGCCCGACTTC-TCG-TAMRA	170	25	56	70.3

*modified version of primer B developed by Medlin et al. (1988)

Master Mixes The following master mixes were used:
- Master Mix for SYBR Green I Assay

	Volume [μl]	[Final]
Dinoflagellate-specific primers (10 μM each)	2	1 μM
TaqStart Antibody mixture*	1	0.04 U/μl
10× SYBR Green I	1	1×
10× PCR buffer with 30 mM MgCl$_2$	1	3 mM
dNTPs (2 mM each)	1	0.2 mM each dNTP
H$_2$O (PCR grade)	1	–
DNA template (~10 ng/μl)	3	3 ng/μl

*Taq polymerase (5 U/μl) is mixed in equal proportions with TaqStart antibody and allowed to incubate at room temperature for 5 min, then 4 μl is added to 21 μl enzyme diluent for final TaqStart antibody mixture.

- Master Mix for Taqman Probe Assay

	Volume [μl]	[Final]
Pfiesteria piscicida-specific primers (2 μM each)	1+1	0.2 μM
Pfiesteria piscicida-specific Taqman probe (3 μM)	0.5	0.15 μM
Taq polymerase (5 U/μl)	0.2	0.1 U/μl
MgCl$_2$ (50 mM)	0.8	4 mM
dNTPs (2 mM each)	1	0.2 mM each dNTP
BSA (2.5 mg/ml)	1	0.25 mg/ml
PCR Buffer	1	1×
PCR Grade Water	0.5	–
DNA template (~10 ng/μl)	3	3 ng/μl

Master mix containing genomic DNA was added to a capillary tube and centrifuged for 2 s at 10,000 rpm. Capillaries were then sealed and placed into the LightCycler rotor.

The following SYBR Green I protocol was used for detection of dinoflagellate DNA in environmental water samples:
- Cycling Program for SYBR Green I Assay

Parameter	Value		
Cycles	45		
Type	Quantification		
	Segment 1	Segment 2	Segment 3
Target temperature [°C]	94°C	51°C	72°C
Incubation time [s]	0	0	10
Temperature transition rate [°C/s]	20	20	1
Acquisition mode	None	None	Single
Gain		F1=16	

- Melting Curve Analysis for SYBR Green I Assay

Parameter	Value		
Cycles	40		
Type	Quantification		
	Segment 1	Segment 2	Segment 3
Target temperature [°C]	97°C	45°C	97°C
Incubation time [s]	20	20	0
Temperature transition rate [°C/s]	20	20	0.2
Acquisition mode	None	None	Continuous
Gain		F1=16	

The following Taqman protocol was used for detection of *Pfiesteria piscicida* in cultures and environmental water samples:

- Cycling Program for Taqman Assay

Parameter	Value	
Cycles	50	
Type	Quantification	
	Segment 1	Segment 2
Target temperature [°C]	94°C	60°C
Incubation time [s]	0	20
Temperature transition rate [°C/s]	20	20
Acquisition mode	None	Single
Gain	F1=2	

Results

Detection of dinoflagellate DNA using the SYBR Green I protocol, incorporating the dinoflagellate-specific primers has routinely been used to ensure the integrity of our DNA extraction method. We have been able to successfully amplify dinoflagellate DNA from approximately 99% of estuarine water samples assayed to date (>1000 samples collected throughout 1998–2000 in Maryland waters). Figure 1 represents a typical SYBR Green I melting curve, using DNA derived from cultures and environmental water samples.

In order to increase specificity in detecting the *Pfiesteria piscicida* organism in estuarine water samples, we developed a Taqman probe. The sensitivity was tested against a panel of *Pfiesteria piscicida* samples representative of various environmental conditions, including pure cultures identified by scanning electron microscopy, and environmental samples either unmodified or spiked with a known number of *Pfiesteria piscicida* organisms (Fig. 2). Figure 3 depicts two samples positive with the probe, found during routine screening of environmental samples collected by the Maryland DNR. To date, more than 1000 environ-

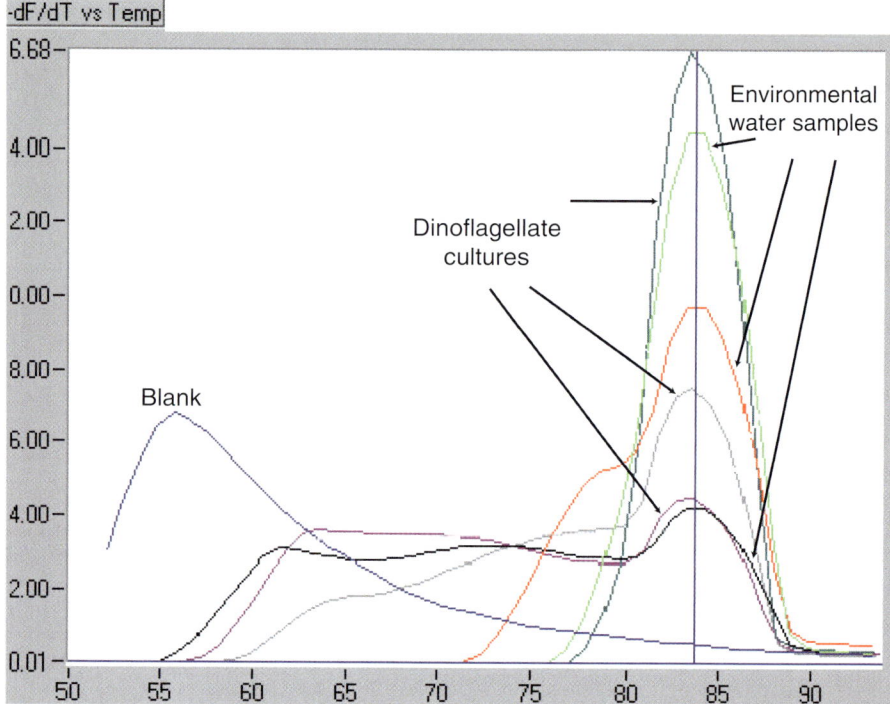

Fig. 1. SYBR Green I LightCycler melting protocol, using dinoflagellate-specific primers and DNA derived from cultures and environmental estuarine water samples. The blank corresponds to no DNA template

mental samples collected around the State have been analyzed with the Taqman protocol, which has revealed several affected river systems. In Fig. 4, a serial dilution of plasmid with the full-length 18 S DNA sequence for *Pfiesteria piscicida* inserted was used as the template for the Taqman assay. Plasmid DNA is used to create standard curves for determining concentrations of *Pfiesteria* organisms in archived DNA from the routine monitoring samples.

Comments

The association of *Pfiesteria piscicida* with fish and human health problems in Maryland generated the need for a monitoring program by State officials. Most tests, including fish kill bioassays and identification by scanning electron microscopy, are costly and labor intensive. The implementation of a molecular detection assay using the LightCycler has proven invaluable in providing results in a timely manner from routine environmental monitoring and rapid response samples collected in areas of dead or lesioned fish by the Maryland DNR. In addi-

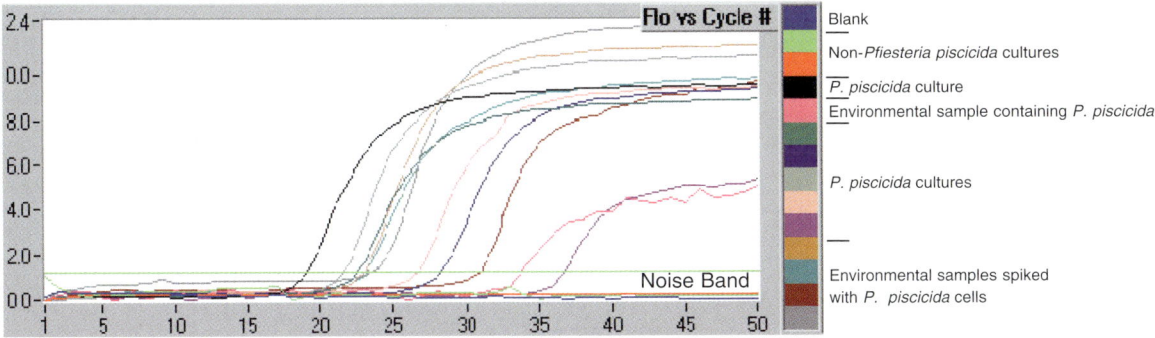

Fig. 2. LightCycler amplification protocol using *Pfiesteria piscicida*-specific primers and Taqman probe on DNA derived from cultures and environmental estuarine water samples spiked with *Pfiesteria piscicida*. Samples below the noise band represent the no DNA template and non-*Pfiesteria piscicida* culture controls

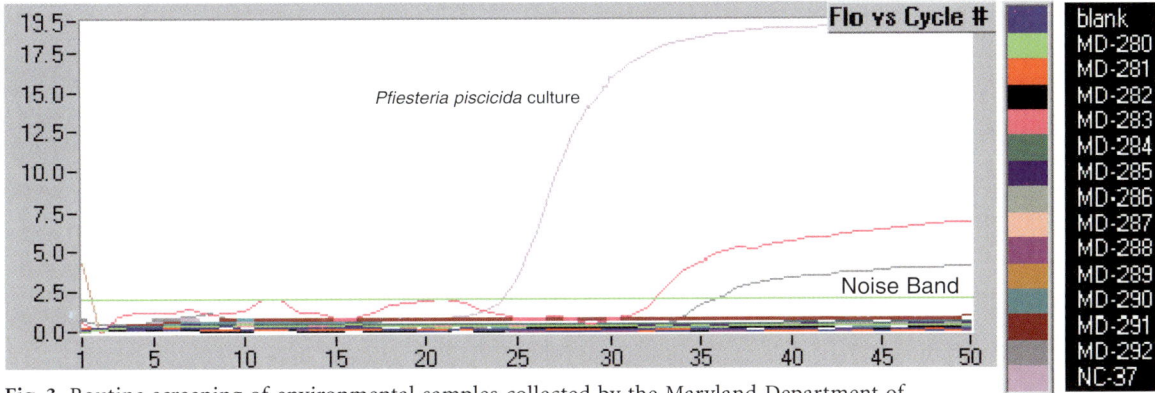

Fig. 3. Routine screening of environmental samples collected by the Maryland Department of Natural Resources from various tributaries of the Chesapeake Bay. *Pfiesteria piscicida* was detected with the Taqman probe in two samples from the Transquaking River

tion to reducing the amount of time needed to produce results by approximately one fourth, we have increased the confidence and specificity of our results through development of a Taqman probe assay for *Pfiesteria piscicida*. With this high level of specificity we are able to report the detection of *Pfiesteria piscicida* in environmental samples collected through the Maryland DNR's water quality monitoring program, as well as through cultures obtained from collaborating laboratories. This capability will enhance ongoing efforts to characterize the organism, its potential toxin production, and factors influencing *Pfiesteria* blooms and their outcomes.

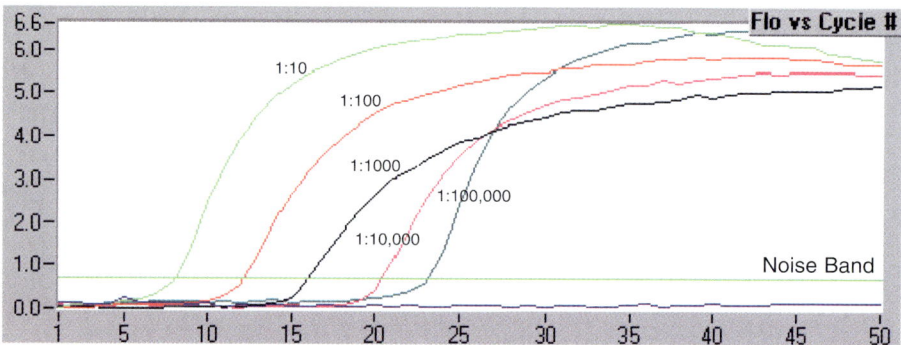

Fig. 4. Taqman assay performed on serial dilution of plasmid containing the 18 S ribosomal DNA sequence of the *Pfiesteria piscicida* genome. A standard curve, used for quantitative analysis of samples containing *P. piscicida,* may be developed, utilizing the points at which each dilution intersects the noise band

References

1. Bowers HA, Tengs T, Glasgow HB Jr, Burkholder JM, Rublee PA, Oldach DW (2000) Development of real-time PCR assays for rapid detection of *Pfiesteria piscicida* and related dinoflagellates. Appl Environ Microbiol (in press)
2. Burkholder JM, Noga EJ, Hobbs CW, Glasgow HB Jr, Smith SA (1992) New "phantom" dinoflagellate is the causative agent of major fish estuarine fish kills. Nature 358: 407–410
3. Burkholder JM, Glasgow HB Jr (1997) *Pfiesteria piscicida* and other toxic *Pfiesteria*-like dinoflagellates: behavior, impacts, and environmental controls. Limnol Oceanogr 42: 1052–1074
4. Grattan LM, Oldach D, Perl TM, Lowitt MH, Matuszak DL, Dickson C, Parrott C, Shoemaker RC, Kauffman CL, Wasserman MP, Hebel JR, Charache P, Morris JG Jr (1998) Learning and memory difficulties after environmental exposure to waterways containing toxin-producing *Pfiesteria* or *Pfiesteria*-like dinoflagellates. Lancet 352: 532–539
5. Lipton DW (1999) *Pfiesteria's* economic impact on seafood industry sales and recreational fishing. Proceedings of the Conference, Economics of Policy Options for Nutrient Management and *Pfiesteria*. Gardner BL, Koch L (eds) Center for Agricultural and Natural Resource Policy, University of Maryland, College Park. 35–38
6. Medlin L, Elwood HJ, Stickel S, Sogin ML (1988) The characterization of enzymatically amplified eukaryotic 16S-like rRNA-coding regions. Gene 71: 491–499
7. Oldach DW, Grattan LM, Morris JG Jr (1999) *Pfiesteria piscicida* and human health, p 135–151. In: Scheld WM, Craig WA, Hughes JM (eds) Emerging Infections 3. ASM Press, Washington, DC
8. Oldach DW, Delwiche CF, Jakobsen KS, Tengs T, Brown EG, Kempton JW, Schaefer EF, Bowers HA, Glasgow HB Jr, Burkholder JM, Steidinger KA, Rublee PA (2000) Heteroduplex mobility assay-guided sequence discovery: elucidation of the small subunit (18 S) rDNA sequences of *Pfiesteria piscicida* and related dinoflagellates from complex algal culture and environmental sample DNA pools. Proc Natl Acad Sci USA 97: 4303–4308

Quantification of Retrotransposon XIR-2.5 Copy Number in Genomes of *Poeciliidae* Species

MEINHARD HAHN*, CHRISTIANE THÖMMES, JOCHEN WILHELM, JAMILAH MICHEL

Introduction

Recently quantitative PCR techniques are standard methods used for the precise quantification of low copy DNA target sequences [1], e.g. single or few DNA target molecules (viruses, bacteria) in biological samples or single copy genes in large genomes. In contrast, the precise quantification of high copy number elements in DNA genomes, e.g. transposons or retrotransposons, by quantitative PCR is not well represented in literature. For these targets so far conventional quantification techniques were used which are either time consuming and need large amounts of high molecular genomic DNA (Southern blots) [2] or expensive technical equipment (fluorescence *in situ* hybridization) [3], result in data of limited precision (c_0t-analysis) [4] or have only a low dynamic range (*in situ* hybridization of radioactively labeled probes to interphase or metaphase chromosomes) [5].

Here we present a powerful new approach for retrotransposon quantification using quantitative rapid cycle real-time PCR on the LightCycler combined with the SYBR-Green detection format which avoids these limitations. We analyzed the retrotransposon XIR-2.5 [2, 5], a DNA element of 2.5 kb, which was first isolated and described in the genomes of fishes of the genus *Xiphophorus*, e.g. the well-known platyfish and sword-tails, which are viviovoparous small fishes of the family *Poeciliidae*. These fishes represent the oldest animal models used for the exploration of the hereditary and somatic causes of cancer [6], especially, the genetic origins of melanoma [5, 7]. Recent work points out that XIR-2.5 is the prototype of a new genetic class of oncodeterminants, called paragenetic suppressors of suppressor genes [2, 5, 8]. In this context, the copy number of XIR-2.5 in the genomes of different *Poeciliidae* species and subspecies as well as related genera is of interest. Therefore, we used the LightCycler real-time PCR technique to determine the XIR-2.5 copy number within different species, subspecies and hybrids of *Xiphophorus*, and also species of additional genera of *Poeciliidae* living at Mexico, Central America and northern parts of South America.

---

* Meinhard Hahn (✉) (e-mail: Meinhard.U.Hahn@chemie.bio.uni-giessen.de)
  Institute of Biochemistry, FB 08, Justus-Liebig-University Giessen, Heinrich-Buff-Ring 58, 35392 Giessen, Germany

Materials

Equipment   LightCycler instrument and software 3 (Roche Diagnostics, Mannheim, Germany)
LightCycler capillaries, centrifuge adapters and cooling blocks (Roche Diagnostics)
OLIGO primer analysis software 5.0 (National Biosciences Inc., Plymouth, USA)

Fishes   Fishes of the family *Poeciliidae* were received by a local breeder of aquarium fishes (*Girardinus falcatus*; *G. metallicus*; *Heterandria formosa*; *Lebistes reticulatus*, guppy; *Priapella compressa*; *Xiphophorus guentheri*; *X. variatus*) and by Prof. Dr. Fritz Anders and Dr. Annerose Anders, Institute of Genetics, Justus-Liebig-University Giessen (*X. helleri catemaco*; *X. helleri*, swordtail; *X. maculatus*, platyfish; *X. variatus*; hybrids of *X. helleri* × *X. maculatus*, backcrossed against *X. helleri* for more than 50 generations). The DNA of a minimum of two individuals for each species or strain or fish line was analyzed by quantitative LightCycler PCR to investigate if there are significant variations of the genomic XIR 2.5 concentration within different individuals of the tested species or strains.

Plasmid   pXIR-2.5, linearized with *Eco*RV. The plasmid pXIR-2.5, used as external DNA standard, is a derivative of the phagemid pBluescript Sk(+) with a length of 5786 bp containing the 2.5 kbp spanning xiphophorine retrotransposon XIR-2.5 [8].

Reagents   Deoxynucleotides (dATP, dCTP, dGTP, dTTP), PCR-grade (Roche Diagnostics)
Taq DNA polymerase 10 × reaction buffer (100 mM Tris-HCl, 15 mM $MgCl_2$, 500 mM KCl, pH 8.3 at 20°C) (Roche Diagnostics)
Bovine serum albumin (BSA), 20 mg/ml, special quality for molecular biology (Roche Diagnostics)
$MgCl_2$, 25 mM (Roche Diagnostics)
Taq DNA polymerase, recombinant, 5 U/µl (Roche Diagnostics)
Oligodeoxynucleotides / PCR primers, HPSF-grade (MWG-Biotech AG, Ebersberg, Germany)
SYBR Green I, 10,000 × conc. in DMSO (Roche Diagnostics)
Pure water (Merck Eurolab GmbH, Darmstadt, Germany)
TPE buffer (80 mM Tris-phosphate, pH 8.0, 2 mM EDTA)
Dilution buffer (10 mM Tris-HCl, pH 8.5)
RNase A, DNase-free, 7,000 U/ml (QIAGEN GmbH, Hilden, Germany)
QIAamp DNA Mini Kit (QIAGEN)
QIAGEN Plasmid Maxi Kit (QIAGEN)
QIAquick PCR Purification Kit (QIAGEN)

Procedure

Genomic DNA Preparation   Fishes were anaesthetized in ice water and decapitated with surgical scissors. Then the brains were dissected, cut into pieces and their genomic DNA isolated using the QIAamp DNA Mini Kit according to the manufacturer's protocol, including an RNase A incubation step for complete degradation and removal of

RNA. RNA contaminations interfere on the UV-spectroscopic DNA quantification which would result in an underestimation of the copy number per genome. The genomic DNA was eluted in kit buffer AE and UV-absorbance of DNA solutions was registered in the range of 240–320 nm. The DNA concentrations were calculated using the dsDNA-specific relation 1 $A^{260 \text{ nm}} \triangleq$ 50 µg/ml. DNA stock solutions of 10 ng/µl were prepared by diluting the DNA samples in dilution buffer.

Preparation of Plasmid DNA Standard

RNA-free plasmid DNA pXIR-2.5 was isolated from a 500 ml over night culture of the transformed *E. coli* strain HB 101 according to the manufacturer's protocol of QIAGEN Plasmid Maxi Kit. 50 µg plasmid DNA was linearized completely by digestion with a restriction enzyme which introduces only one cut per plasmid (e.g. *Eco*R V). When cloned sequences are used (e.g. plasmid DNA) for external standardization, it is important to use completely linearized DNA because circular and supercoiled plasmid DNA is amplified with a lower efficiency than linear fragments [9]. The linearized plasmid DNA was purified using the QIAquick PCR Purification Kit and eluted from the adsorption columns in the dilution buffer. The molar concentration of XIR-2.5 copies was quantified by UV-spectroscopy (see above). A 10 nM stock solution of this DNA was used to make a dilution series in the range of 100 pM – 1 aM in dilution buffer. This standard DNA master dilution series was aliquoted and stored at –20°C.

LightCycler-PCR

SYBR Green I reaction master mix for each 10 µl reaction:

	Volume [µl]	[Final]
10 × *Taq* reaction buffer (containing 15 mM MgCl$_2$)	1.0	1 ×
MgCl$_2$ (25 mM)	1.0	4 mM
BSA (5 mg/ml) [10]	1.0	0.5 mg/ml
dNTP mix (2 mM each)	1.0	0.2 mM each
Primer mix (5 µM each)	1.0	0.5 µM each
SYBR Green I (1 : 1000 diluted in H$_2$O)	0.33	1 : 30,000
Taq DNA polymerase (5 U/µl)	0.1	0.5 U/10µl
H$_2$O	2.57	
Total volume	8	

The diluted SYBR Green I solution (1 : 1000 in water) should not be exposed to repeated cycles of freezing / thawing. While preparing the reaction mix all steps have to be carried out in the cold (4°C) and by using the pre-chilled centrifuge adapters in an aluminum cooling block for minimization of primer dimer formation. Primer dimer formation can also be avoided by suitable hot start PCR techniques e.g. addition of *Taq* Start Antibody (for details see recommendations of the manufacturer Clontech).

For a series of *n* PCRs, depending on the number of reactions, the (*n*+1) or (*n*+2) amount of SYBR Green I reaction master mix was prepared, first of all by thoroughly mixing all components without *Taq* DNA polymerase. The enzyme was added to the master mix immediately before 8 μl of master mix and 2 μl of DNA template (10 ng/μl) were pipetted into each LightCycler capillary. Then the PCR was carried out. For analyzing unknown samples in fivefold parallel reactions, the corresponding amounts of master mix and sample DNA were mixed and dispensed into five capillaries. Capillaries were sealed, centrifuged in a microcentrifuge at low speed (3,000 rpm for 10 s) using the centrifuge adapters and transferred into the LightCycler sample carousel. The following protocol was used for analysis of amplification, detection of SYBR Green I fluorescence and melting curve analysis:

- Denaturation at 95°C for 30 s
- Amplification

Parameter	Value			
Cycles	50			
Type	Quantification			
	Segment 1	Segment 2	Segment 3	Segment 4
Target temperature (°C)	95	55	72	82
Incubation time [s]	0	5	10	0
Temperature transition rate [°C/s]	20	20	20	20
Acquisition mode	None	None	None	Single
Gains	F1=5			

- Melting Curve Analysis

Parameter	Value	
Cycles	1	
Type	Melting Curve	
	Segment 1	Segment 2
Target temperature (°C)	72	95
Incubation time [s]	10	0
Temperature transition rate [°C/s]	20	0.1
Acquisition mode	None	Continuous
Gains	F1=5	

Gel Analysis of PCR Product

When the amplification process was finished, PCR products were aquired from the desealed inverted capillaries by centrifugation into reaction tubes of 0.5 ml. The samples were mixed with 2.5 μl of 5 × gel loading buffer and electrophoretically separated in a 15% polyacrylamide gel using the TPE buffer system. The gels were stained with ethidium bromide and UV-induced fluorescence of the gels was documented by a video documentation system.

Analysis of LightCycler Data

The quantitative analysis of the raw data was done by LightCycler Software 3. The background readings were subtracted from the raw data of channel 1 and were analyzed using the fit points method. The threshold value was adjusted to minimize the error of the calibration curve.

The molar XIR-2.5 concentration of unknown DNA samples was determined by LightCycler Software. In the next step the genome copy number of the retrotransposon was calculated by dividing the number of XIR-2.5 copies per reaction volume by the number of nuclear DNA equivalents (1.2 pg per diploid genome, see comments) of the target DNA.

Calculation of XIR-2.5 Genome Copy Number

Results

The assay we employed makes it possible to measure a large dynamic range of XIR-2.5 concentrations (> 4 orders of magnitude) (Fig. 1a, 1b) with a low variance of copy numbers (usually < 5%) (Fig. 2), and is highly reproducible, easily practicable, very rapid (< 20 min for amplification) and allows a high sample throughput (analysis of > 20 animals in five-fold replication per day).

When plasmid DNA or genomic DNA was used as template, one single specific PCR product was amplified as shown by analysis of its T_m-value (87°C) (Fig. 3) and as visualized by gel electrophoresis (Fig. 4). Even in the case of *Xiphophorus* DNA containing several thousands of XIR-2.5 copies per genome only one discrete fragment is amplified (Fig. 4) which might implicate an evolutionary conservation of the transposon copies or a selection for functional copies.

Due to the high phenotypic variability in the family of *Poeciliidae*, it has been a problem to systematically group them. The genomic XIR-2.5 content could possibly serve as a tool to solve this systematic problem to some extent [11]. Large differences in genomic XIR-2.5 copy number were demonstrated for different species and genera of *Poeciliidae* (Fig. 5). All tested *Xiphophorus* species, closely related to each other, contain several hundreds or thousands of retrotransposon copies. *X. maculatus*, the classical species of the tumor model system, has the highest retrotransposon load, whereas fishes of genera *Girardinus*, *Heterandria* or *Lebistes* posses only low amounts of this genetic element. Remarkably, *P. compressa* shows similar XIR-2.5 copy numbers as *Xiphophorus* species which may imply that this species of genus *Priapella* is more closely related to *Xiphophorus* and should be regrouped into this systematic group. Therefore, these results should be considered carefully in view of evolutionary aspects of this fish family.

Table 1. Oligonucleotides

XIR-2.5 (numbering according to [8])					
	Position	ORF	Length	GC (%)	T_m (°C)
TGC TCG CTG CTG GCT GAA CG	1517	2	20	65	72.0
TCG ATG CGT CGA GGA GAT AC	1739 R	3	20	55	65.0
Product	1517–1739	2–3	223	50	

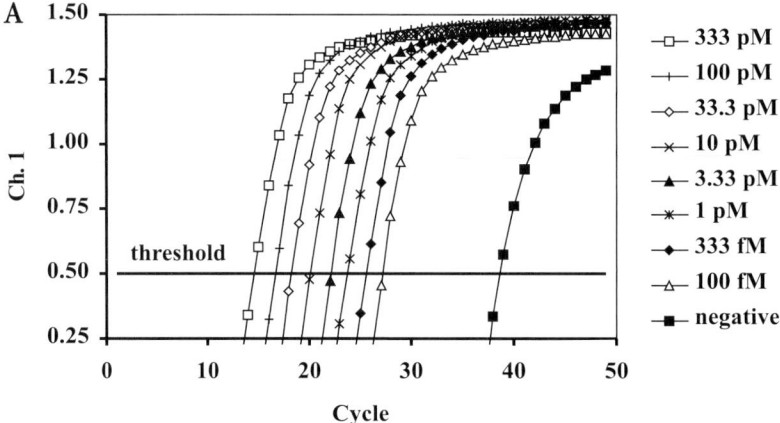

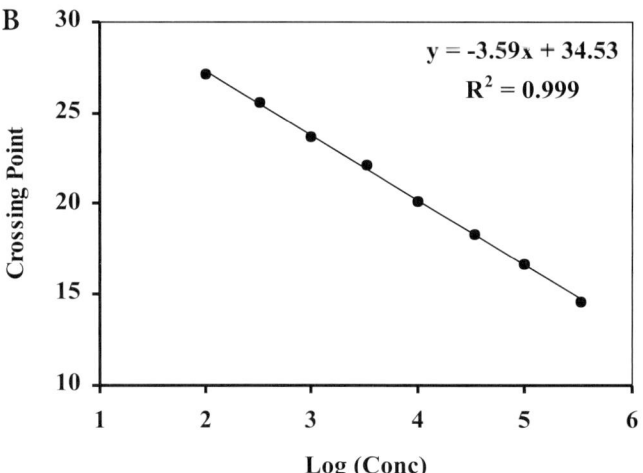

Fig. 1. Detection of a dilution series of the pXIR-2.5 standard DNA. (**A**) Logarithm of fluorescence signal intensity (channel 1) during the amplification of the 223 bp fragment. Although the without-template negative control gave a primer dimer specific signal, reactions in the analyzed concentration range (down to femto-molar concentrations) resulted in no primer dimer formation (data not shown). (**B**) Calibration curve of logarithmic concentrations of pXIR-2.5 standard DNA versus crossing point

These experiments were carried out to clarify whether the "anticipation" of melanoma formation observed in *Xiphophorus* hybrid fishes correlates with the amount of genomic XIR-2.5 copies. When hybrid fishes of *X. maculatus* and *X. helleri* are exposed to genotoxic X-rays or UV-B irradiation their descendents (backcrosses against *X. helleri*) suffer of melanoma formation. The incidence as well as the severity of the disease increases dramatically from generation to generation

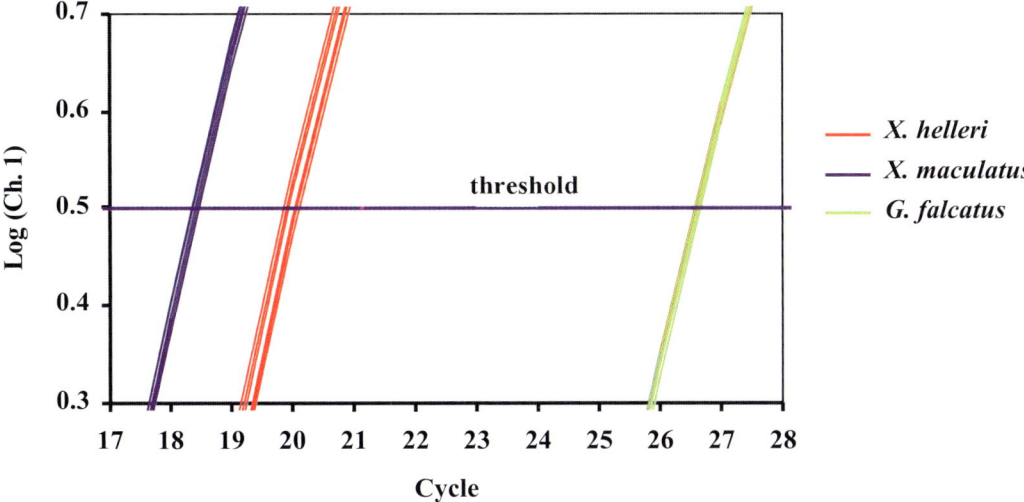

Fig. 2. Quantitative PCR of genomic fish DNAs. Logarithmic plot of the amplification specific fluorescence curves are shown for genomic DNA samples of three *Poeciliidae* species, each in a five-fold approach. It shows the critical curve segments around the threshold value to illustrate the high precision and reproducibility of quantifications

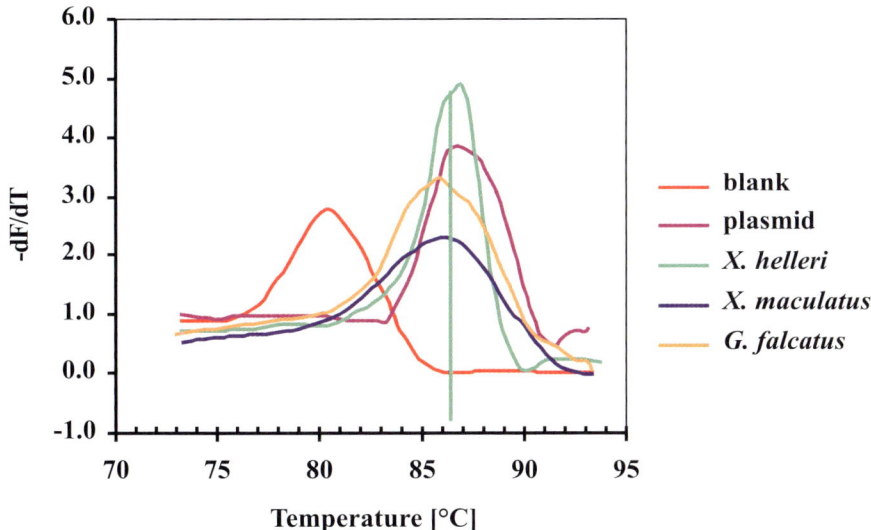

Fig. 3. Melting curve analysis of species specific PCR-products. The exemplary set of melting curves shows data for PCR-products of XIR-2.5 standard DNA template (plasmid) but also for *X. helleri* and *X. maculatus*, prototypes of XIR-2.5 hosts with high genomic copy number, while *G. falcatus* represents a species of a different genus with very low genomic copy number. The blank sample shows the melting behaviour of primer dimers

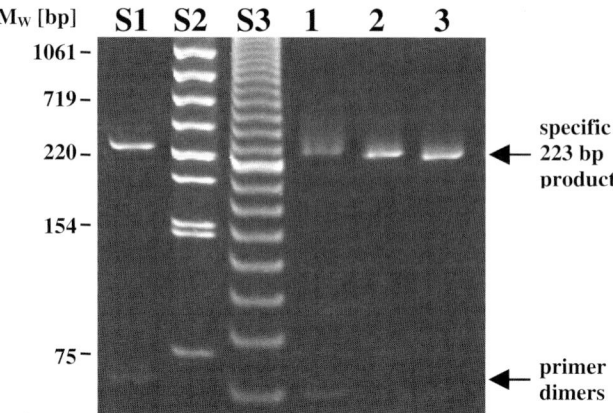

Fig. 4. Gel electrophoretic analysis of PCR products amplified in the LightCycler. The genomic DNA samples of (1) *G. falcatus*, (2) *X. helleri* and (3) *X. maculatus* were amplified on the Light-Cycler (data shown in Fig. **2** and **3**). The XIR-2.5 specific PCR products were separated in a 15% TPE-PA-gel and stained with ethidium bromide. S1: PCR product of the 1 pM plasmid standard, S2: DNA fragment length standard, S3: 20 bp DNA fragment ladder

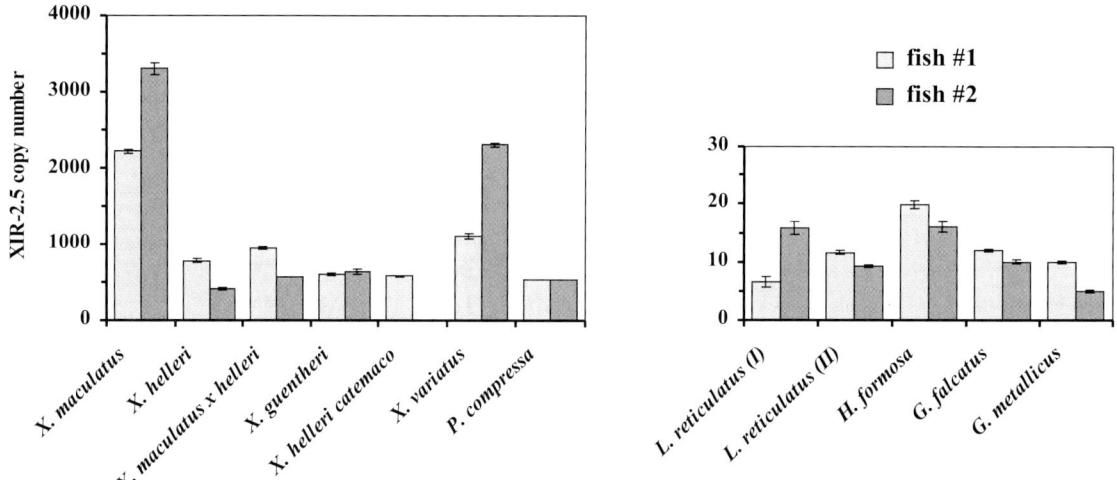

Fig. 5. Genome copy number of retrotransposon XIR-2.5. The diagrams show the number of XIR-2.5 copies in the diploid genomes of different genera and species of fish family *Poeciliidae*. For each species two individuals (exception: *X. h. catemaco*) were analyzed in five-fold estimation, mean value and standard deviation are shown. Pay attention to the different scales of the two diagrams. *L. reticulatus* (I): fishes of a Venezuelan wild population; *L. reticulatus* (II): fishes of Rio Tefe wild population

without any further genotoxic treatment [2, 5, 8], even after more than 50 generations of backcrossing, an effect which is called anticipation. Our data show that this phenomenon of anticipation is not correlated with an increase of genomic XIR-2.5 copy number. Therefore, this phenomenon of anticipation is caused due to a different activity of XIR-2.5 [8].

Comments

- For the first time the genomic copy number of a vertebrate retrotransposon was quantified precisely and sensitively by rapid-cycle real-time PCR. This approach, specific for the retrotransposon XIR-2.5, makes use of the LightCycler capillary PCR as well as the principle of externally standardized quantitative PCR. For this purpose, the linearized plasmid pXIR-2.5, containing the retrotransposon XIR-2.5, was used as the external standard DNA.
- During the amplification process the amount of amplicons was followed by real-time detection of dsDNA-specific SYBR Green fluorescence once in each cycle at 82°C, a temperature well above the melting point of any primer dimers (T_m 79°C), which were observed in some cases when DNA samples with very low genomic XIR-2.5 copy number were amplified. Thereby, we could exclude the interfering fluorescence signal of primer dimers.
- The retrotransposon copy number per diploid fish genome was calculated assuming a value of 1.2 pg DNA per diploid genome for all tested *Poeciliidae* species. This value is already reported in literature for different *Xiphophorus* species [12].
- The XIR 2.5 concentration of each DNA-sample was analyzed and quantified in five PCR amplification experiments. The mean value of these quantifications is used as a precise and reliable result for the genomic XIR 2.5 concentration.
- All amplifications were carried out in 10 µl reaction volumes. The results obtained were precise and reproducible, with a low coefficient of variation of copy numbers (typically about ± 5 %). The smaller reaction volumes reduce the consumption of reagents and, therefore, the costs of the quantification experiments.

Applications

This is a new approach for quantification of genomic high-copy number elements (e.g. transposons, retrotransposons, inserted retroviral DNA-genomes, LINE, SINE) and repetitive genome segments, with high precision and a large dynamic range of concentration.

References

1. Kochanowski B, Reischl U (eds.) (1999) *Quantitative PCR Protocols*, Humana Press, Totowa, New Jersey
2. Roushdy J, Michel J, Petry H, Anders A, Anders F (1999) Paragenetic suppressors of suppressor genes - a new class of oncodeterminants. *J Cancer Res Clin Oncol* 125: 123–133
3. Dorin JR, Emslie E, Hanratty D, Farrall M, Gosden J, Porteous DJ (1992) Gene targeting for somatic cell manipulation: rapid analysis of reduced chromosome hybrids by *Alu*-PCR fingerprinting and chromosome painting. *Hum Mol Genet* 1: 53–59
4. Britten RJ, Kohne DE (1968) Repeated sequences in DNA. Hundreds of thousands of copies of DNA sequences have been incorporated into the genomes of higher organisms. *Science* 161: 529–540
5. Anders A, Petry H, Fleming C, Petry K, Brix P, Lüke W, Gröger H, Schneider E, Kiefer J, Anders F (1994) Increasing melanoma incidence: putatively explainable by retrotransposons. Experimental contributions of the xiphophorine Gordon-Kosswig melanoma system. *Pigment Cell Res* 7: 433–450
6. Kosswig C (1928) Über Kreuzungen zwischen den Teleostiern *Xiphophorus helleri* und *Platypoecilus maculatus*. *Z Indukt Abstammungs Vererbungsl* 47: 150–158
7. Nairn RS, Morizot DC, Kazianis S, Woodhead AD, Setlow RB (1996) Nonmammalian models for sunlight carcinogenesis: genetic analysis of melanoma formation in *Xiphophorus* hybrid fish. *Photochem Photobiol* 64: 440–448
8. Michel J (1999) *Retrotransposons in der Antizipation von Neoplasien: Untersuchungen am* Xiphophorus-*Modell der Melanombildung*. Inauguraldissertation, Justus-Liebig-Universität Gießen
9. Piatak M Jr, Luk K-C, Williams B, Lifson JD (1993): Quantitative competitive polymerase chain reaction for accurate quantitation of HIV DNA and RNA species. *BioTechniques* 14: 70–81
10. Wittwer CT, Herrmann MG, Moss AA, Rasmussen RP (1997): Continuous fluorescence monitoring of rapid cycle DNA amplification. *BioTechniques* 22: 130–138
11. Morizot DC, Siciliano (1982) Protein polymorphisms, segregation in genetic crosses and genetic distances among fishes of the genus *Xiphophorus* (*Poeciliidae*). *Genetics* 102: 539–556
12. Schwab M (1980) Sensitivität für Induktion von Neoplasmen durch N-Methyl-N-nitrosoharnstoff (MNH) als Folge der Kombination von Artgenomen bei *Xiphophorus*. Habilitationsschrift, Justus-Liebig-Universität Gießen